Mathematics for Industry

Volume 1

For further volumes:
http://www.springer.com/series/13254

Aims & Scope

The meaning of "Mathematics for Industry" (sometimes abbreviated as MI or MfI) is different from that of "Mathematics in Industry" (or of "Industrial Mathematics"). The latter is restrictive: it tends to be identified with the actual mathematics that specifically arises in the daily management and operation of manufacturing. The former, however, denotes a new research field in mathematics that may serve as a foundation for creating future technologies. This concept was born from the integration and reorganization of pure and applied mathematics in the present day into a fluid and versatile form capable of stimulating awareness of the importance of mathematics in industry, as well as responding to the needs of industrial technologies. The history of this integration and reorganization indicates that this basic idea will someday find increasing utility. Mathematics can be a key technology in modern society.

The series aims to promote this trend by (1) providing comprehensive content on applications of mathematics, especially to industry technologies via various types of scientific research, (2) introducing basic, useful, necessary and crucial knowledge for several applications through concrete subjects, and (3) introducing new research results and developments for applications of mathematics in the real world. These points may provide the basis for opening a new mathematics-oriented technological world and even new research fields of mathematics.

Masato Wakayama · Robert S. Anderssen
Jin Cheng · Yasuhide Fukumoto
Robert McKibbin · Konrad Polthier
Tsuyoshi Takagi · Kim-Chuan Toh
Editors

The Impact of Applications on Mathematics

Proceedings of the Forum of Mathematics for Industry 2013

Springer

Editors
Masato Wakayama
Yasuhide Fukumoto
Tsuyoshi Takagi
Institute of Mathematics for Industry
Kyushu University
Fukuoka
Japan

Robert S. Anderssen
CSIRO (Commonwealth Scientific and Industrial Research Organisation)
Canberra, ACT
Australia

Jin Cheng
School of Mathematical Sciences
Fudan University
Shanghai
China

Robert McKibbin
Institute of Natural & Mathematical Sciences
Massey University
Palmerston North
Auckland
New Zealand

Konrad Polthier
Institute of Mathematics
Free University of Berlin
Berlin
Germany

Kim-Chuan Toh
Department of Mathematics
National University of Singapore
Singapore
Singapore

ISSN 2198-350X ISSN 2198-3518 (electronic)
ISBN 978-4-431-54906-2 ISBN 978-4-431-54907-9 (eBook)
DOI 10.1007/978-4-431-54907-9
Springer Tokyo Heidelberg New York Dordrecht London

Library of Congress Control Number: 2014943754

Printed on acid-free paper

Springer is part of Springer Science+Business Media (www.springer.com)

In Memory

It is with deep sadness and regret that we note the sudden and unexpected demise of Geoffrey (Geoff) Norman Mercer on Saturday, April 12, 2014, at the age of 51. For us, the editorial members of this series Mathematics for Industry, and the many others who had the highest regards for Geoff's unselfish and creative contributions to colleagues, to the mathematical and epidemiological communities and to cutting edge research, it is a devastating loss. The importance of his various contributions and insightful ideas, which are already having significant impact on research-education activity of Mathematics-for-Industry, will never be forgotten.

Preface

This book is the post-proceedings of the Forum "Math-for-Industry 2013," the focus of which was "The Impact of Applications on Mathematics." This phrase represents an appropriate framework in which to highlight how real-world problems, over the centuries and today, have influenced and are influencing the development of mathematics and, thereby, how mathematics takes up the consequences of such impacts to advance mathematics for the benefit of mathematics and its applications. It is this process that underlies not only the key role that Mathematics-for-Industry plays in fostering productive and successful interaction between industry personnel and mathematicians, but also the cross-fertilization and collaboration that it stimulates in having mathematics involved with the advancement of science and technology.

The various chapters of this volume illustrate different aspects of how, starting with an inquiry or question from industry, there is a logical sequence of conceptualization and model formulation central to the resulting mathematical decision making that focuses on answering a question for the benefit of industry. The question under examination about the industrial problem focuses and drives how the associated mathematics must be unraveled, since there is no uniqueness in the ways that a specific mathematical model/construct/equation can be utilized.

Even when the industry problem leads to standard mathematics, there is an "impact." To answer the question raised by the application, the appropriate interpretation of the mathematics within the context of the application must be identified. This is directly reflected in the fact that the same basic mathematical equations arise in quite unrelated applications.

The impact on mathematics from or by applications is universal. If one goes back into the history of mathematics, one finds that practical problems (applications) played a crucial role in its early development. This is true even today for much of mathematical research. Even within pure mathematics, which quite often possesses highly abstract components, this fact has validity in that one aspect of mathematical research is the identification of (e.g., simpler, constructive) proofs, which unveil the hidden secrets of mathematics and the significance of classical results.

The nature of the interaction is aptly summarized in the following comment by V. I. Arnold: "My best pure mathematics was in applied mathematics and my best applied mathematics was in pure mathematics."

The chapters of this volume discuss the following aspects: specific examples of how the answering of a question, coming from industry, engendered new mathematical activity; how the same mathematics is central to the solution of quite different applications; how the answering of an industrial question requires deep thought about the essence mathematically of the application from which it came; how mathematics has built on the mathematics initially coming from the needs of applications; and how to foster young researchers in this vision.

From a Mathematics-for-Industry perspective, an equally important role of the Forum is the choice of topics, which engage and mentor all students of mathematics, pure and applied, about the nature of mathematics as real-world and social activities. From this point of view, we have held the Forum "Math-for-Industry" (FMI) annually from 2009, as readers can see in the short history of 5 years shown in the illustration following this Preface. A key goal is to foster and motivate participants' professional involvement with industrial mathematics as a source of interesting real-world problems for which new mathematical perspectives are required and from which new research themes may even materialize, as a source for internship topics for students involved with collaborative projects with industry, and as a source for research subjects for graduates studying for a higher degree (M.Sc. or Ph.D.). In this way, all mathematics students have their eyes opened to the opportunities for all aspects of mathematics to contribute to the solution of real-world problems in terms not only of assisting industry but also of advancing the consequential impact on the development of new understanding about and opportunities for mathematics.

Finally, we would like to thank the Scientific Board of the Forum from industry: Hirokazu Anai (Fujitsu Laboratories, Ltd.), Ken Anjyo (OLM Digital, Inc.), Yasuko Fukuzawa (Yokohama Research Laboratory, Hitachi), Jonathan Hosking (IBM T. J. Watson Research Center), Akira Takada (Asahi Glass Co., Ltd.), and Takeshi Yamada (NTT Communication Science Laboratories). Without their cooperation and support, we would never have experienced the great excitement and success of this Forum. Moreover, we would like to express our deep appreciation for the great help of the conference secretaries, especially Seiko Sasaguri and Tsubura Imabayashi, during the Forum and its organization.

Fukuoka, March 2014

Masato Wakayama
On behalf of the Organizing
Committee of the Forum
Math-for-Industry 2013
and the Editorial Committee
of the Proceedings

Contents

FMI2008	FMI2009	FMI2010	FMI2011	FMI2012	FMI2013
Tokyo	Fukuoka	Fukuoka	Honolulu	Fukuoka	Fukuoka
Sep. 16-17	Nov.9-13	Oct. 21-23	Oct.24-28	Oct. 22-26	Nov. 4-8
The 1st Forum:Consortium Math For Industry	CasimirForce, Casimir Operators and the Riemann Hypothesis	Information Security, Visualization, and Inverse Problems, on the basis of Optimization Techniques	TSUNAMI - Mathematical Modelling-Using Mathematics for Natural Disaster Prediction, Recovery and Provision for the Future -	Information Recovery and Discovery	-The Impact of Applications on Mathematics -

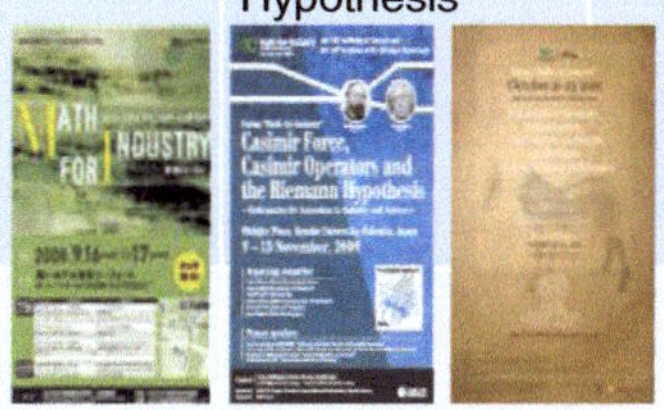

Forum "Math-for-Industry" 2013

-The Impact of Applications on Mathematics-

November 4 to 8, 2013, Fukuoka

	November 04, 2013		November 05, 2013		November 06, 2013		November 07, 2013		November 08, 2013
9:50	Registration Opening Ceremony		Registration		Registration		Registration		Registration
10:00 - 10:50	**Andrea Parmeggiani** Université Montpellier II, France **Exclusion Processes on Networks as Models for Cytoskeletal Transport and Intracellular Traffic**	9:45 - 10:35	**Kaoru Sato** JFE Steel Corporation, Japan **The Continuing Challenge of Steel —How to Win Mathematicians and Influence Scientists in Other Disciplines—**	9:45 - 10:35	**Masakazu Kojima** Chuo University/JST CREST, Japan **Global Optimization via Conic Linear Programming Relaxation**	9:45 - 10:35	**Georges-Henri Cottet** Université de Grenoble and CNRS, France **Mathematics, Hybrid Computing and HPC**	9:45 - 10:35	**John Lewis** Victoria University, New Zealand **Principal Component Analysis and Laplacian Splines: steps toward a uni ed model**
10:55 - 11:25	**Graeme Wake** Massey University, New Zealand **Tumour Cell Biology and Some New Non-local Calculus**	10:40 - 11:20	**José A. Cuminato** Universidade de São Paulo, Brazil **Implicit Methods for Simulating Low Reynolds Number Free Surface Flows: Improvements on MAC-type Method**	10:40 - 11:20	**Hirokazu Anai** FUJITSU LABORATORIES LTD., Japan **Applied Algebraic Geometry in Model Based Design for Manufacturing**	10:40 - 11:20	**Dietmar Hömberg** WIAS/TU, Germany **Modelling, Analysis and Simulation of Multifrequency Induction Hardening**	10:40 - 11:20	**Winston Sweatman** Massey University, New Zealand **Some Australian and New Zealand Mathematics-in-Industry Study Group (MISG) projects with a focus on the Steel Industry**
11:30 - 12:20	**Wil Schilders** Technische Universiteit Eindhoven, The Netherlands **Industrial Mathematics in Europe**	11:25 - 12:15	**Naonori Ueda** NTT Communication Science Labs., Japan **Bayesian Meta-learning and its Application to High-Level Real Nursing Activity Recognition Using Accelerometers**	11:25 - 12:15	**Heinz Bauschke** University of British Columbia, Canada **New Developments in Splitting Methods for Road Design Optimization**	11:25 - 12:15	**Kazuyuki Aihara** University of Tokyo, Japan **Mathematical Approach to Personalized Medicine**	11:25 - 12:15	**Philip Broadbridge** La Trobe Universit, Australia **Applications of Integrable Nonlinear Diffusion Equations in Industrial Modelling**
	LUNCH		LUNCH		LUNCH		LUNCH		LUNCH
13:50 - 14:30	**Osamu Saeki** IMI, Kyushu University **Visualizing Multivariate Data Using Singularity Theory**	13:45 - 14:15	**Ryo Yoshida** The Institute of Statistical Mathematics, Japan **Bayesian Statistics for Designing Systems, Molecules and Others**	13:45 - 14:15	**Go Kato** NTT Communication Science Labs., Japan **Analytical Optimization of Local Quantum Operation and Classical Communication**	13:45 - 14:15	**Sohei Tasaki** Tohoku University, Japan **Phase-separating Elastic System of Mixed Lipid Bilayers**	13:45 - 14:15	**Takaaki Shiratori** Microsoft Research Asia, China **User Interfaces for Creating Character Animation**
14:35 - 15:45	**Monique Chyba** University of Hawai'i at Manoa, U.S.A **Bernard Bonnard** Institut Mathématique de Bourgogne, France **Control and Optimization Techniques in Nuclear Magnetic Resonance**	14:20 - 15:00	**Murray Cameron** IDTC and University of Technology Sydney, Australia **Changing Research Training for Mathematicians for Industry**	14:20 - 15:00	**Tomoyuki Shirai** IMI, Kyushu University **Applications of Determinantal Point Processes**	14:20 - 15:00	**Eric Bonnetier** Université Joseph Fourier, France **Echoscan: Electrical Impedance Tomography by Elastic Perturbation**	14:20 - 15:00	**Evgeny Verbitskiy** Leiden University, The Netherlands **Thermodynamic Ideas in Information Theory**

	COFFEE BREAK		COFFEE BREAK		COFFEE BREAK		COFFEE BREAK		COFFEE BREAK
16:05 - 16:35	Hisayoshi Sato Keisuke Hakuta Hitachi Ltd., Yokohama Research Laboratory, Japan **Cryptographic Technology for Benefiting from Big Data**	15:20 - 16:05	**Young Reseacher Session** Yusuke Imoto **Truncation Error Analysis of Approximation Operators on Particle Methods** Takuya Hayashi **Cryptanalysis of Pairing-based Cryptosystems over Small Characteristic Fields** Shun'ichi Yokoyama **Lit-sphere Extension for Artistic Rendering - After FMI2011 -**	15:20 - 15:50	Shuai Lu Fudan University, China **Parameter Identification in Non-isothermal Nucleation and Growth Processes**	15:20 - 15:50	Shingo Saito Faculty of Arts and Science, Kyushu University **Bayesian Approach to Measuring Parameter and Model Risk in Loss Ratio Estimation**	15:20 - 16:10	Shinichiro Nakamura RIKEN Research Cluster for Innovation, Nakamura Laboratory, Japan **Expectation to Mathematics, from an Industrial Basic Researcher**
16:40 - 17:30	Dooho Choi Electronics and Telecommunications Research Institute, Korea **Secure Crypto Implementation and Mathematics**	16:10 - 18:10	Poster Session	15:55 - 18:00	Poster Session Voting	15:55 - 16:45	Frank de Hoog Commonwealth Scientific and Industrial Research Organisation, Australia **Simple Mathematical Models for Complex Industrial Processes**	16:15	**Closing** **Chair: Masato Wakayama**
17:45	Welcome Party (Snack & Beverages)					16:50 - 17:30	John Hearne RMIT University, Australia **Managing the PhD Experience –More than Just Research**		
				18:30	Banquet at 3F Boardroom Hilton Fukuoka Sea Hawk	17:35 - 17:55	Poster Session Award Ceremony		

Modelling Collective Cytoskeletal Transport and Intracellular Traffic

Andrea Parmeggiani, Izaak Neri and Norbert Kern

Abstract Biological cells require active fluxes of matter to maintain their internal organization and perform multiple tasks to live. In particular they rely on cytoskeletal transport driven by motor proteins, ATP-fueled molecular engines, for delivering vesicles and biochemically active cargoes inside the cytoplasm. Experimental progress allows nowadays quantitative studies describing intracellular transport phenomena down to the nanometric scale of single molecules. Theoretical approaches face the challenge of modelling the multiscale, out-of-equilibrium and non-linear properties of cytoskeletal transport: from the mechanochemical complexity of a single molecular motor up to the collective transport on cellular scales. We will present some of our recent progress in building a generic modelling scheme for cytoskeletal transport based on lattice gas models called "exclusion processes". Interesting new properties arise from the emergence of density inhomogeneities of particles along the network of one dimensional lattices. Moreover, understanding these processes on networks can provide important hints for other fundamental and applied problems such as vehicular, pedestrian and data traffic, or ultimately for technological and biomedical applications.

Keywords Molecular motors · Transport · Traffic phenomena · Networks · Cytoskeleton · Biological physics · Statistical mechanics · Stochastic process · Nonlinear phenomena

---

A. Parmeggiani (✉) · I. Neri · N. Kern
L2C, UMR 5221 CNRS, Université Montpellier 2, Pl. E. Bataillon,
34095 Montpellier Cedex 5, France
e-mail: andrea.parmeggiani@univ-montp2.fr

I. Neri
e-mail: izaakneri@gmail.com

N. Kern
e-mail: norbert.kern@univ-montp2.fr

A. Parmeggiani
DIMNP, UMR 5235 CNRS, Université Montpellier 2 et Montpellier 1,
Montpellier Cedex 5, France

M. Wakayama et al. (eds.), *The Impact of Applications on Mathematics*,
Mathematics for Industry 1, DOI: 10.1007/978-4-431-54907-9_1,

1 Introduction

Biological cells critically rely on controlling fluxes of matter to ensure their viability. Indeed, delivering biological cargoes at the right time to the right place in the cell is necessary to control its internal organization, and plays a role in many crucial cellular functions such as growth, mitosis, motility, differentiation, and intracellular signalling [1]. On the contrary, perturbations or anomalies in this transport lead to pathologies such as neurodegenerative diseases and defects in embryonic development [2–4].

Understanding the organization and regulation of the required intracellular transport of biological cargoes is a formidably complex issue. However, simple first principles reveal quickly that the cell must actively overcome fundamental constraints. In particular, diffusion driven by the Brownian (thermal) motion of cargoes is clearly inefficient for transport over large distances or in very crowded environments of the cytoplasm. A striking example is transport along axons (which can be meters long in large mammals). The Stokes-Einstein relation tells us that the time required to diffuse a distance of 1 m is $\tau_{diff} = (1\ \text{m})^2/6D$, where the diffusion coefficient of the cytoplasm can be estimated as $D \sim 10\ \mu\text{m}^2/s$ for a rather small cargo (of the order of 10 nm in size). It would thus take the cargo $\tau_{diff} \sim 1.7 \cdot 10^{10}$ s, i.e. more than five hundreds years, to cover this distance!

In order to overcome such entropic barriers the cell therefore requires an active transport system, despite the consumption of energy this implies, in order to ensure efficient delivery of cargoes. In eukaryotic cells this task (as well as force production) is accomplished at molecular scales, via nanoscopic engines called *motor proteins* [5]. Driven by ATP hydrolysis, these molecular motors transduce chemical energy into work in order to transport cargoes. They run along tracks, which are biological filaments, which are furthermore interlinked by accessory proteins to form a cellular scaffold called *cytoskeleton*. Recent progress now gives experimental access to these transport mechanisms, and allows quantitative studies of molecular transport from cellular tissues down to the nanometric size of single molecules and single cargoes [6–10].

This review article aims to present some of the progress we recently made on the modelling of active transport on large networks and their emergent properties [11–13], as well as the links to other fields concerned by this interdisciplinary topic.

2 Interdisciplinary Context

Biologists can account for many details of the complexity of this active cytoskeletal transport based on molecular motors. For example, there are several types of filamentous fibers (*filamentous actin*, *microtubules* and so-called *intermediate filaments*), each of them made up from a different type of protein unit, leading to different microscopic structures and different macroscopic mechanical properties etc. Molecular motors come in three families (*myosins*, *kinesins* and *dyneins*), each of them specific to one or several types of filaments, and differing in stepping direction on a given filament, stepping speed, step size, energy consumption etc. One might add the variety in the cargoes, the way in which they are bound to the motors, the complexity of the interactions between several molecular motors. And this is before one even considers molecular crowding in the cellular environment or other crucial issues such as the regulation in the traffic and targeting of cargoes for delivery to specific places in the cell.

Understanding the full complexity of cytoskeletal transport in living cells therefore is a daunting task. It is also clearly an interdisciplinary problem, the facets of which can appeal to a large number of scientific communities.

On experimental grounds, researchers from different disciplines (biology, physics, chemistry, mathematics, electronics and computer sciences) work in interdisciplinary teams to develop powerful techniques in microscopy and videomicroscopy. On more applied grounds, such studies may prove relevant for understanding diseases and pathologies associated to transport defects as well as for the application of modern single molecule therapies [14, 15] and in personalized medicine [16]. On theoretical grounds, an important aspect is to identify the mechanisms underlying cytoskeletal transport as well as the ways in which they can be combined, enhanced, compensated or regulated in order to achieve the required delivery. In addition, specific questions, arising in the context of cytoskeletal transport, in turn pose challenges and provide a stimulus to several scientific disciplines.

To a physicist, intracellular transport driven by motor proteins is an example *par excellence* of a genuine out-of-equilibrium stochastic process, and thus raises challenges as to the foundations of statistical mechanics and theoretical physics [17]. It is also an example of a system with a large number of interacting systems (motors, cargoes etc., which we will refer to as *particles*). The dominating interaction may be expected to be steric exclusion, also known as *excluded volume*, simply due to the fact that no particle can enter the space occupied by another one. The problem is therefore also related to the physics of colloids (on a micrometric scale) and to granular media (on a macroscopic scale). We will review the idea of using *exclusion processes* (EP) to model the many of the essential features of this active transport.

A theoretical physicist may in particular single out the fact that motor proteins step in general along linear and periodic substrates of protein binding sites. This quasi-one-dimensional motion along a lattice is a key feature which provides a link to the physics of low-dimensional systems. Analysing models of cytoskeletal transport may thus also contribute to developing phenomenological approaches and advanced methods in statistical mechanics and mathematical physics (e.g.: low-dimensional field theory and integrable systems) [18–22], and probability theory [23] (see e.g. the recent applications of large deviation functionals to study EP [24, 25]). Other fundamental questions arise naturally: the laws of energy transduction and thermodynamics at molecular scales, self-organized phenomena overcoming entropic barriers in active matter and the role of fluctuations to control the organization of small systems are just a few of a large number of topics which are nowadays inspired by the phenomenology of biological systems involving motor proteins.

To a researcher in *non-linear sciences* the transport problem is interesting precisely due to the interactions between particles. As we will review below, the excluded volume interaction between particles induces a mutual hindrance, such that the current is not monotonous in the density of particles. The transport problem is therefore much richer than for example that of electrical currents in circuits considered by Kirchhoff [26]. The non-linearities lead to highly interesting features, such as *shock waves*.

To a mathematician, it is natural to formalize the problem in terms of (partial) differential equations governing the transport. This approach leads to (generalizations of) the *Burgers equation* for transport on a cytoskeletal filament, which provides valuable insight (see below). But here the problem is in fact one of many such equations, which are coupled due to the interconnected network. In particular biological processes which have to be accounted for require to generalize the problem to a non-conserved particle number, which introduces significant difficulties. A mathematician might also be interested in the *topology* of the cytoskeletal network, describing the way the filaments are interlinked in terms of a (directed) graph.

Finally, contact can be made naturally with a number of other domains, such as traffic phenomena of aeroplanes, trains, cars, pedestrians (e.g. for logistics, traffic control regulation, smart city organization) [27–31], but also in the context of growing interdisciplinary communities working on man-made or ecological systems (e.g. foraging animals, migratory events, collective predator-prey processes) [32]. Some of these transport problems connect in turn to long-standing challenging topics (such as heat conduction anomalies in low-dimensional systems (see references in [24])), as well as to technological applications (like power distribution on grids, routing information on networks, electronic transport on q-dot linear chains [33], spintronics [34] and ionic conductivity [35]).

In the following we will discuss the use of EP to capture many fundamental aspects of cytoskeletal transport driven by motor proteins. We will review how they can be applied to a whole network of filaments, and discuss strategies to solve the transport problem on a network, as well as ways to visualise and interpret the resulting behaviour. We will then discuss perspectives and interdisciplinary aspects of the approach developed.

3 Modelling

We now summarize a simplified picture, based on those biological facts which appear as key features for a generic model based on simple physical processes. Then, we will introduce the theoretical models which describe mathematically the generic properties of motor protein dynamics along cytoskeletal filaments.

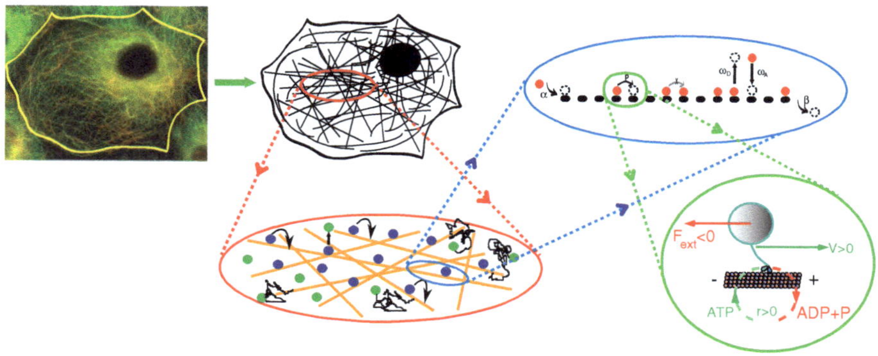

Fig. 1 Motor proteins are molecular machines that hydrolyse ATP to convert chemical energy into force and motion inside the cells. They walk along linear crystalline assemblies of protein filaments, interconnected to form the cytoskeleton, the internal scaffold of the cell. Along this complex network of filaments motor proteins alternate stochastic processive walks with Brownian diffusion. Exclusion processes like TASEP and TASEP-LK (see *upper right* oval) are paradigmatic lattice gas models used to describe biological active transport driven by molecular machines. In the context of this research topic, we have recently developed a multiscale framework to study cytoskeletal transport on networks

3.1 Essential Properties of Motor Protein Transport

First of all, motor proteins are enzymes which operate out of thermodynamic equilibrium. They typically measure around ten up to several tens of nanometers, and have a rather specific structure, of which one usually distinguishes two parts, the so-called *head* and the *tail* domains. Through complex mechanochemical processes [5, 36] motors can bind cargoes to their tail region. The head domain (or domains) bind to the *cytoskeletal filaments*, which are polymer chains made of identical protein complexes (referred here as "monomers"). These biopolymers thus provide a long chain of periodically arranged *binding sites* for certain molecular motors.

In their head domain, motor proteins can cyclically hydrolyse the common fuel of biological cells, ATP (*Adenosin-Tri-Phosphate*), by breaking one phosphate bond to obtain by-products such as ADP and an inorganic phosphate. This process liberates free energy, about $10 - 20k_BT \approx 40 - 80 \cdot 10^{-21}\ J$ per molecule of ATP, which becomes available to push the motor protein well out of thermodynamics equilibrium and induce a cycle of conformational changes of the protein.[1] As a result, the motor moves forward along the filament. This motion can in principle take place towards either side, but is heavily biased to one direction along the filament, which is set by the *polarity* (spatial asymmetry) of the monomer structure.

Motors can thus perform directed (over-damped) motion along these substrates, producing also mechanical forces of several picoNewtons ($1\ pN = 10^{-12}\ N$). These

[1] The typical energy scale is $1\ k_BT \sim 4 \times 10^{-21}$ J with $k_B \simeq 1,38 \times 10^{-23}$ J/K is the Boltzmann constant and $T \simeq 300$ K is the absolute temperature in Kelvin units.

forces may appear very small, but are in reality remarkable for a molecular machine working in contact with a thermally fluctuating environment [37–40]. The resulting step size depends on the type of motors and the underlying filaments, but is typically of the order of tens of nanometers.

In physiological conditions, motor proteins can walk on the filament with speeds ranging from tens of nm/s up to several μm/s, either as single steppers or coordinated with other motors attached to the same cargo. In such conditions, the simplest motor-motor interaction has a steric nature: no more than one motor can occupy the same filament binding site.

It is also important to note that motors have what is called a *finite processivity*, i.e. any given motor will ultimately loose its affinity to the filament due to the strength of thermal fluctuations at nanometric scales. After this detachment, which occurs stochastically, the motor can then re-attach to the filament at a different location, after having undergone Brownian diffusion (often limited to the proximity of the filament by crowding effects). The (idealized) limiting case of a motor which never detaches from a filament would be a *perfect processivity*.

3.2 Exclusion Processes as Models for Motor Protein Collective Transport

Starting from these properties, it is natural to model the stochastic stepping of motors via the dynamics of a gas of particles moving along a linear lattice. Particles cannot occupy the same lattice site and move with stochastic jumps, either totally or partially biased in one direction. In the first case, we reproduce the directed motion in a preferential direction of one motor or its cargo, whereas in the second case we represent the *bidirectionality* of the motor or its cargo [41].

The finite processivity of motors can be incorporated into the model through binding and unbinding events, leading to a stochastic exchange between the lattice and a particle reservoir. Such a reservoir represents either the intracellular cytoplasmic medium or the buffer solution surrounding the filament. In the case of perfect processivity, motors enter at one end of the filament and exit from the other, without ever detaching from the filament. All these microscopic events are coded in terms of rates (probabilities per unit of time) which represent the microscopic parameters of the lattice gas model. Although stochasticity is a requirement for describing the nanoscopic motion of molecular machines, deterministic variants of these models (depending on the kind of update rules to change the occupation of each site in time) [42] have also been studied.

Due to exclusion interactions the average stationary flux $j(x)$ flowing in the lattice is a non-linear function of the average stationary density of particles $\rho(x)$: this condition reads generically as

$$j(x) = \rho(x)\, v_m(\rho(x)). \tag{1}$$

The average speed v_m of the particles therefore depends instantaneously (i.e. without memory or delay effects) on the local density $\rho(x)$ [43].

The rules presented above thus provide the definition of EP, like the TASEP, ASEP or the TASEP-LK (also called PFF in some publications) [17]. These models, despite their simplicity, display a very rich phenomenology, intrinsically out-of-equilibrium and of great interest for probability theory and statistical mechanics studies.

EP are widely studied one-dimensional lattice gas models with many deep relations with other very important classes of models in theoretical physics, like spin-chain models, for example. In equilibrium or out-of-equilibrium, EP allow to write master equations as well as to derive dynamic equations from a field theory operator formalism and to apply various exact or approximated methods to understand the rich phenomenology which emerges [19]. These methods range from the analysis of ordinary and partial differential equations through matrix methods to the application of large deviation functionals from probability theory.

The Totally Asymmetric Simple Exclusion Process (TASEP) is probably the simplest formulation of a non-equilibrium lattice gas of self-propelled particles with mutual interaction. Interestingly, TASEP was proposed several decades ago, already in the biological context, for mRNA translation and protein synthesis by the motion of a collection of ribosomal machines along a messenger RNA [44, 45]. Since the model by MacDonald and collaborators, TASEP, and EP more in general, have become to non-equilibrium statistical mechanics what the Ising model is to equilibrium systems [17]. More recently, TASEP have even been exploited to map ribosome traffic on mRNAs to *gene ontology* bioinformatic concepts [46].

TASEP-LK [47, 48] has been developed much more recently. The model is inspired by the study of motor protein collective transport like TASEP along cytoskeletal filaments coupled with the Langmuir Kinetics (LK) for binding/unbinding of the motors between the filament and the surrounding environment. Contrary to TASEP, in TASEP-LK the flux and particle conservations on the lattice are no more guaranteed.

3.2.1 Mean Field Approach

The simplest approach to studying these models is to neglect correlations between particles in the so-called *Mean Field approach*, in the limit of large lattice sizes ($L \to \infty$ lattices sites, i.e. in the *hydrodynamic* or *continuum limit*) [48, 49]. In this approximation, one can establish and solve a Burgers partial differential equation [50, 51] on the average density $\rho(x)$ (bounded between 0 and 1 due to the excluded volume interaction) with source and sink terms [48]:

$$\partial_t \rho = p \frac{\varepsilon^2}{2} \partial_x^2 \rho - \varepsilon \partial_\rho j \partial_x \rho + \mathscr{S}_A - \mathscr{S}_D. \tag{2}$$

where p represents the microscopic jump rate of each particle. The small (regularization) parameter $\varepsilon = 1/L$ is necessary to match both boundary conditions once the *mesoscopic limit* is taken (see next paragraph). The boundary conditions, in the stationary state, depend on the entry and exit rates α and β through

$$\rho(0) = \alpha/p, \quad \rho(1) = 1 - \beta/p, \tag{3}$$

with the normalized position $x = i/L \in [0, 1]$.

We remark that the values of the density at the boundaries can be interpreted as if the segment was in contact with a reservoir of density α/p at the filament entry and with a reservoir of density $1 - \beta/p$ at the filament end. Note that Eq. (2) can be generalized to many different situations like the case of transport made by extended particles [52, 53]. We will review in the following chapter how the notion of entrance/exit rates for segments can be exploited for analysing transport on networks, where the vertices may be thought of as reservoirs supplying/receiving particles from the segments.

3.2.2 Current-Density Fundamental Relation and "Mesoscopic" Limit

The *current-density fundamental relation* $j\,[\rho(x)]$ in the stationary state reads

$$j(x) = p\,\rho(x)\,(1 - \rho(x)). \tag{4}$$

This relation summarizes, through its simple parabolic form, the exclusion interactions between particles moving in only one direction, which lead to a maximal value of the current $j_{max} = p/4$ for $\rho = 1/2$.

The "source" and "sink" terms $\mathscr{S}_A$ and $\mathscr{S}_D$ in Eq. (2) reflect the binding and unbinding of motors, via the Langmuir process, between the filament and the surrounding medium, respecting the exclusion interaction for binding:

$$\mathscr{S}_A = \omega_A\,(1 - \rho(x)), \quad \mathscr{S}_D = \omega_D\,\rho(x). \tag{5}$$

In the case of perfectly processive motors as in TASEP, source and sink contributions can simply be put to zero. In TASEP-LK on the other hand, the most relevant regime (which actually corresponds the conditions of processive motor proteins), arises when considering an appropriate scaling of the rates ω_A and ω_D with the filament length L (expressed with respect to the number of sites):

$$\omega_A = \frac{p\,\Omega_A}{L} = p\,\varepsilon\,\Omega_A \quad , \quad \omega_D = \frac{p\,\Omega_D}{L} = p\,\varepsilon\,\Omega_D. \tag{6}$$

This scaling, also called *mesoscopic limit* ensures that fluxes at the boundaries, which are controlled by the rates α and β to inject and extract particles, are in competition with the Langmuir kinetics in the filament bulk, controlled by the rescaled rates Ω_A and Ω_D [47, 48]. In this limit, after rescaling the time t by $\tau = p\,\varepsilon\,t$, one can write the (dimensionless) Burgers equation:

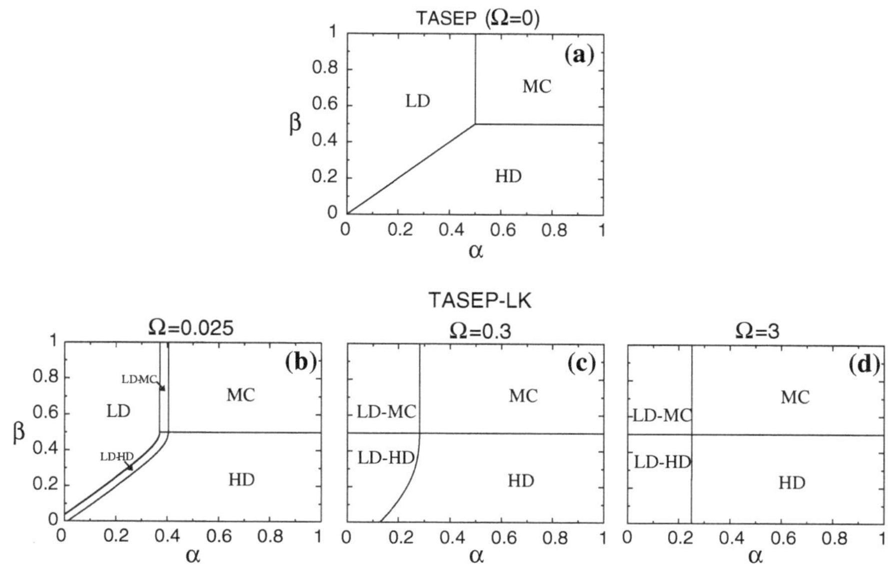

Fig. 2 Examples of phase diagrams for (**a**) TASEP and (**b, c, d**) TASEP-LK in the plane (α, β). For TASEP-LK, the phase diagrams have been obtained for different values of the exchange parameter $\Omega = (\Omega_A + \Omega_D)/2$

$$\partial_\tau \rho = \frac{\varepsilon}{2}\partial_x^2 \rho + (2\rho - 1)\partial_x \rho + \Omega_A(1-\rho) - \Omega_D \rho \tag{7}$$

with rescaled boundary conditions, $\rho(0) = \alpha$ and $\rho(1) = 1 - \beta$, for $\alpha, \beta \in [0, 1]$.

3.2.3 Phase Diagrams

The properties of both models can be represented in terms of phase diagrams on the microscopic parameters controlling the systems. In TASEP, for example, the entry and exit rates from the lattice, α and β respectively, control the boundaries, which sets the average density profile $\rho(x)$ and the average current flow of particles $j(x)$ throughout the one-dimensional lattice. A specific phase diagram in these variables (α, β) for one single filament can then be computed via exact (e.g. Bethe Ansatz and Matrix Ansatz) [19–21, 54–56] or approximate methods (e.g. mean field methods [54] or through so-called *domain wall* approaches [57, 58]). TASEP-LK phase diagrams can be constructed by also considering two additional parameters, such as the binding and unbinding rates, Ω_A and Ω_D, by using approximation techniques [47, 48, 59].

Examples of the phase diagrams for TASEP and TASEP-LK are provided in Fig. 2. In general, in the Low (High) Density region *LD* (*HD*) the normalized density $\rho(x)$ is homogeneous and smaller (larger) than 1/2. In TASEP, in the Maximal Current (*MC*) region, according to the relation (4), the system attains the highest value of the flux $j = p/4$, while the density in the bulk $\rho(x)$ is equal to 1/2. In TASEP-LK the corresponding *MC* phase is described by a specific solution of Eq. (7) expressed in terms of a Lambert-W special function [48].

In the (α, β)-diagram, the phase coexistence between different homogeneous phases arises at the boundaries between phases (see, e.g., the line $\alpha = \beta$ in TASEP), or even throughout extended regions (see, e.g., the $LD - HD$ and $LD - MC$ phases in TASEP-LK). Indeed, whenever the flux j is no longer a monotonic function of the density ρ, different values for the density ρ can coexist since they lead to the same current j. Coexistence thus reflects the emergence of traffic jams of particles along the lattice, where shocks (in TASEP) or domain-walls (in TASEP-LK) match two different density profiles in the system bulk. In TASEP, density jumps characterize a discontinuous (first order) phase transition. On the contrary, transitions to the *MC* phase are continuous in the density (*LD* to *MC*, for example). In these cases, it has been possible to characterize the strong and collective fluctuations of the system and, in the case of continuous transitions, a critical behaviour with diverging correlation lengths like in second order phase transitions [17, 55, 60, 61].

Remarkably, and despite their simplicity, these models can also provide qualitative and quantitative knowledge for biological transport processes. For example, it has been shown recently that TASEP-LK is consistent with experimental observations for kinesin traffic along microtubules in *in-vitro* experiments [9]. In particular, the model predicts the presence of *localized* domain walls representing motor protein traffic jams along the filament. This feature, particular to TASEP-LK and contrary to the stochastic displacement of shocks in the whole lattice as in TASEP, is in good agreement between the theoretical and experimental stationary density profiles of motors along a microtubule [9, 49, 50].

4 Transport on Networks

Complex systems (a cell, a city, a computer processor, internet) are usually spatially distributed, and thus need transport and logistics processes to organize and deliver matter, information and energy, both inside and outside their boundaries. This makes it necessary to dispose of a traffic system like a road network in order to organize transport. In a biological cell this role is played by the cytoskeleton, i.e. the internal cellular scaffold of filaments which are used by motor proteins to transport cargoes [62]. It is surprising to realize the extension and structural complexity of this cytoskeletal network, as they are revealed by the following rough estimate.

For a typical cell of 50 μm in size, its cytoskeletal "road" system (the total length of polymerized actin or microtubules filaments) amounts to several tens of centimeters. To appreciate this length, we rescale the typical step size of a motor protein (about 10 nm) to the human step (about 1 m). In this comparison, while the typical cell size corresponds to a few kilometers in size, the entire cell-contained cytoskeletal network would correspond to several times the total French high-speed train ("TGV") rail network. The situation can be even more impressive for particular cases such as neurons, for which the cytoskeleton can be up to several meters long.

Another estimate provides a lower bound for the number of filaments (10^3), and the number of crossings between filaments (from 10^3 to 10^6). These numbers highlight the structural complexity of the filamentous network, and therefore its connectivity.

4.1 Network Models

In the perspective of studying motor protein based cytoskeletal transport, we thus face the challenging problem of describing transport on large networks. We consider them as consisting of directed segments (mimicking the polar filaments) which interconnect at vertices v, which are points where filaments cross or branch. We restrict the discussion here to a closed network, i.e. we do not consider any loose ends.[2]

The connectivity of the resulting graph may be expected to play an important role for transport. We investigated this role by considering model topologies which allow to distinguish *ordered* (lattice-like) and *random* networks. Such networks can be characterized by the statistical distribution of the vertex connectivity, i.e. of the distributions of the number c_{in} of incoming segments and the number c_{out} of outgoing segments. A first characterization is given by the *average connectivity* c, which states how many segments connect on average to a vertex. When all vertices are connected in the same way ($c = \langle c_{in} \rangle = \langle c_{out} \rangle$ is constant), we call the network *regular* (or a *Bethe* network). In this case, the distribution of the degree k (total number of connecting segments) reads $p_{deg}(k) = \delta_{k,c}$ for each vertex. On the contrary, for *irregular* networks (also called *Poisson* networks) the number of incoming and outgoing vertices can differ, and vary from one vertex to another. In this case, the distribution of the vertex connectivity follows a Poisson distribution $p_{deg}(k) = e^{-c}\, c^k/k!$ with an average connectivity $c = \langle k \rangle$.

We have studied EP like TASEP, ASEP and TASEP-LK on these kinds of structures [11–13] (Fig. 3).

4.2 General Mathematical Framework for Studying the Problem

Our goal is thus to generalize the analysis of transport phenomena from a single segment (single cytoskeletal filament) to a large network (the entire cytoskeleton). The above estimates for the size of the network strongly suggest that we are in trouble, given that the cytoskeleton is indeed a huge network. In this case, finding the solution of Eq. (2) related to TASEP, but now on a whole network, must be expected

[2] Note that in TASEP-LK such a condition is relaxed since particles can enter and leave the system at any site due to the binding/unbinding Langmuir process.

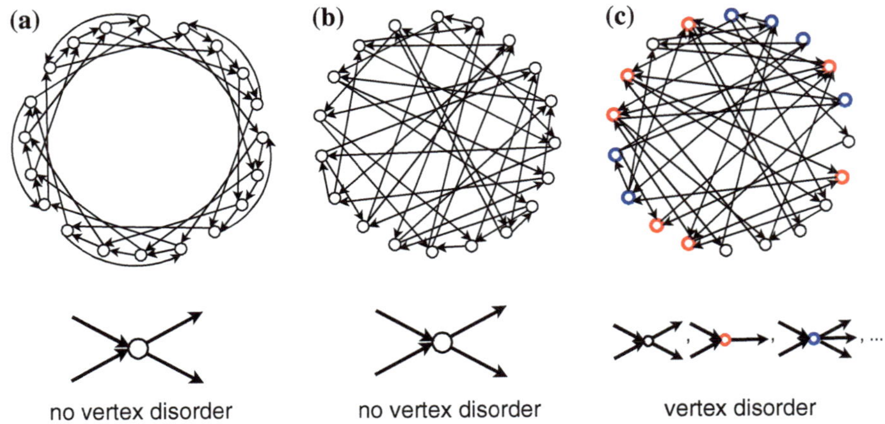

Fig. 3 Examples of regular (Bethe-like) and irregular (Poisson-like) networks: (**a**) a regular lattice with fixed connectivity $c = 2$; (**b**) a random regular graph with fixed connectivity $c = 2$; (**c**) irregular network with average connectivity $c = \langle k \rangle = 2$ (see main text). We consider the networks as consisting of directed segments (mimicking the polar filaments) which interconnect at vertices v, which are the points where filaments cross or branch. We restrict the discussion here to a closed network, i.e. we do not consider any loose ends, although this condition can be relaxed in the case of TASEP-LK. This representation only retains topological connectivity, and does not imply any particular spatial distribution of the vertices/nodes

to be an exceedingly heavy computational task. Indeed, one must in principle study a large number of non-linear differential Burgers equations, coupled with one another, and possibly (as in TASEP) with a global constraint on the conservation of the total number of particles. Unfortunately, advanced methods using integrability properties and related exact solutions are not yet available, even on very simple graphs. Fortunately it is possible, as we will show now, to devise approximate methods, which are already very useful to obtain insight into the emergence of new physical phenomena.

Generalising the transport equations (2) or (7) for a given random graph composed of S directed segments between V vertices v, one would obtain the following problem to solve for each directed segment between vertex v and v':

$$\partial_t \rho_{v\to v'} = p\,\frac{\varepsilon^2}{2}\partial_x^2 \rho_{v\to v'} - \varepsilon\,\partial_\rho j_{v\to v'}\,\partial_x \rho_{v\to v'} + \mathscr{S}_{A,v\to v'} - \mathscr{S}_{D,v\to v'} \tag{8}$$

with

$$j_{v\to v'} = p\,\rho_{v\to v'}(x)\,(1-\rho_{v\to v'}(x)). \tag{9}$$

Depending on whether TASEP or TASEP-LK is considered, one must specify the source and sink contributions. For TASEP-LK, these can be written as

$$\mathscr{S}_{A,v\to v'} = \omega_A\,(1-\rho_{v\to v'}), \quad \mathscr{S}_{D,v\to v'} = \omega_D\,\rho_{v\to v'}, \tag{10}$$

where ω_A and ω_D are specified as in Eq. (6). In general, by following the same rescaling procedure used for the theory on a single segment, see Sect. 3.2, one can write the following system of (dimensionless) Burgers equations:

$$\partial_\tau \rho_{v\to v'} = \frac{\varepsilon}{2}\partial_x^2 \rho_{v\to v'} + (2\rho_{v\to v'} - 1)\partial_x \rho_{v\to v'} + \Omega_A(1-\rho_{v\to v'}) - \Omega_D \rho_{v\to v'}. \quad (11)$$

In general, to solve Eq. (8) [or also Eq. (11)] in the stationary state, one also needs to know the boundary values for the density ρ at each vertex v in the same conditions. The density ρ_v in the stationary state can be computed by the equation for the current conservation for the total in-flux and out-flux at the vertex v (continuity equation):

$$\partial_t \rho_v = \sum_{v'\to v} j_{v'\to v} - \sum_{v'\leftarrow v} j_{v'\leftarrow v}. \quad (12)$$

When the overall average density of particles $\bar{\rho}$ is exactly conserved as in TASEP (i.e. we work in a kind of Canonical ensemble description of statistical mechanics) one has to consider the global constraint on the total number of particles on the lattice N_{tot}:

$$N_{tot} = L\sum_{s=1}^{S}\int_0^1 \rho_s(x)dx + \sum_{v=1}^{V}\rho_v \quad (13)$$

where $\rho_s(x)$ is the average density for the segment s between the vertex v and v' and ρ_v is the local density at the vertex v. From this expression we can define the overall density of particles on the network $\bar{\rho}$:

$$\bar{\rho} = \frac{N_{tot}}{LS+V} \approx \frac{1}{S}\sum_{s=1}^{S}\int_0^1 \rho_s(x)dx \quad (14)$$

where the approximation holds for segments with large size L.

In TASEP-LK the overall density $\bar{\rho}$ is conserved only on average (i.e. we work in a kind of Grand Canonical ensemble). Interestingly, in this case it can be proved that the overall density of particles is equal to the Langmuir isotherm density $\rho_l = \omega_A/(\omega_A+\omega_D)$. Thus the corresponding relation to Eq. (14) for TASEP-LK holds for

$$\bar{\rho} = \rho_l = \frac{\omega_A}{\omega_A+\omega_D} = \frac{\Omega_A}{\Omega_A+\Omega_D} \quad (15)$$

in the large system size limit $L\to\infty$.

In general, the non-linear problem to solve presents S differential Burgers-like Eq. (8), coupled via the vertex densities ρ_v by the current conservation (12) and submitted to a non-local constraint on the overall density $\bar{\rho}$ by Eq. (14) or Eq. (15). For networks comparable to the cytoskeleton, composed of, as we saw, many thousands of interconnected segments, we must find approximative methods in order to simplify the computational problem, which is prohibitive in its exact form.

5 Decomposing the Network via the Effective Rate Approximation

5.1 Generic Characteristics of the Solution

A way to overcome this problem is to look at the physico-mathematical properties of the system in the large size limit $L \to \infty$ [i.e. $\varepsilon \to 0$ in Eq. (8) or Eq. (11)]. As mentioned previously, the segments are low-dimensional systems. In presence of short range interactions, the density and current properties are controlled by the entrance/exit rates at the segment boundaries, and this also holds for TASEP-LK, i.e. in presence of non zero attachment/detachment rates. Indeed, in this case the *LD*, *HD* and *MC* densities and their currents are no longer constant throughout a segment (like in TASEP), but they do show density profiles which vary smoothly in space.

Our mean-field approach essentially consists in decomposing the problem into independent segments [63, 64]. The coupling then arises through effective boundary rates α and β, quantifying the particle exchange with the vertices, which play the role of reservoirs. Indeed, for any segment, with boundary rates α and β, it is these rates which set the density profile along the segment.

In this context, it is useful to consider the characteristics of the general solutions one can build in a given segment of the network. We therefore call ρ_α and ρ_β the corresponding solutions of Eq. (8), as they would be set by the input rate or the output rate, respectively. Both in TASEP and TASEP-LK, despite their qualitative and quantitative differences, three possibilities arise. Solutions can depend either:

i on both boundary rates, α and β
ii on one of them (either α or β)
iii on none of them.

In the first case (*i*), for large L, the solution on the whole segment can be constructed by connecting the two boundary-dependent solutions ρ_α and ρ_β of Eq. (8) [or Eq. (11)] via a shock or a domain wall. In this case, a variation of one of the two boundary rates will change only one branch ρ_α or ρ_β. This corresponds to the motion of the domain wall position within the segment. For example, by increasing the left boundary rate α, the domain wall moves toward the left due to the increasing number of ingoing particles in the lattice. Boundary values of both densities are thus *independent*.

In the second case (*ii*), one of the two solutions, ρ_α or ρ_β, occupies the whole lattice bulk. In this case there no longer is a domain wall in the segment (except for a *boundary layer* confined to one of the segment boundaries). A change of one of the boundary rates will affect the whole bulk density, and eventually the value of the density at the opposite boundary. In this case, the boundary values of both densities are thus *dependent*.

In the third case (*iii*), the density in the bulk becomes boundary rate independent. This occurs for the *MC* phase (where $\rho = 1/2$) or for the case where the Langmuir kinetics dominates for sufficiently high Ω_A and Ω_D rates (where $\rho \simeq \rho_l = \Omega_A/(\Omega_A + \Omega_D)$).

These three different situations will be key in order to rationalize different regimes for the particle distribution along the network.

5.2 The Effective Rate Approximation: An Efficient Algorithm to Explore Large Networks

With this knowledge, it is useful to represent differently the physical behaviour of the network with respect to the previous Sect. 4.2. Since the solutions of each Eq. (8) [or (11)] are known and essentially defined by the cases (*i*), (*ii*) or (*iii*), it is worth considering only Eq. (12) at each vertex v

$$\partial_t \rho_v = \sum_{v' \to v} j_{v' \to v} - \sum_{v' \leftarrow v} j_{v' \leftarrow v} \tag{16}$$

to compute the appropriate boundary conditions via the density at the vertex ρ_v. In general, due to the coupling between all incoming and outgoing currents sharing a given vertex of the network, one must still expect it to be difficult to solve Eq. (16).

However, by considering the physical properties of the system, it turns out that the relations (16) can be solved [11, 13, 64]. Indeed, one can consider that each vertex v represents a reservoir of particles of density ρ_v, at the entry of the segments leaving the vertex v and, reciprocally, at the exit of the segment entering the same vertex.

In general, each segment is therefore characterized by effective entry and exit rates, $\alpha_{v,eff}$ and $\beta_{v,eff}$ respectively[3]:

$$\alpha_{v,eff} = p\frac{\rho_v}{k_v^{out}}, \quad \beta_{v,eff} = p(1 - \rho_v) \tag{17}$$

where k_v^{out} is the local number of outgoing segments from the vertex v. These relations capture the fact that, for a given vertex v, an outgoing particle has to choose one of k_v^{out} segments, whereas particles from all k_v^{in} incoming segments compete for the space on the same vertex.

By using this approximation, all current contributions between the generic segments v and v' can be written self-consistently in terms of the local densities ρ_v and $\rho_{v'}$:

[3] In the effective rates $\alpha_{v,eff}$ and $\beta_{v,eff}$ we keep the explicit dependence on the jump rate p.

$$\partial_t \rho_v = \sum_{v' \to v} j_{v' \to v} \left[x = 1; \frac{\rho_{v'}}{k_{v'}^{out}}, 1 - \rho_v, \Omega_A, \Omega_D \right]$$
$$- \sum_{v' \leftarrow v} j_{v' \leftarrow v} \left[x = 0; \frac{\rho_v}{k_v^{out}}, 1 - \rho_{v'}, \Omega_A, \Omega_D \right], \qquad (18)$$

where we have considered also the dependence on the control parameters Ω_A and Ω_D for the Langmuir process.

In this case the complexity of the problem is (considerably) reduced to computing the stationary densities ρ_v at each vertex v only. Once the vertex densities ρ_v are known, one can use the corresponding solutions of the Burgers equations for given boundary conditions to compute the density $\rho_{v \to v'}$ and the current profiles $j_{v \to v'}$ for each segment $v \to v'$ of the network. This is equivalent to using the model phase diagram for each segment, as a function of the effective entry and exit rates $\alpha_{v,eff} = p\rho_v / k_v^{out}$ and $\beta_{v',eff} = p(1 - \rho_{v'})$ for the given segment.

The possibility of decomposing the network into segments and vertices, the assumption that each vertex can be considered as an effective reservoir for the filaments and the knowledge of the phase diagram for a single segment make it possible to build an efficient algorithm that provides interesting knowledge on the properties of the system [11–13], as we will see in the following.

5.3 A Useful Tool: The Effective Rates Plot

According to the method described above, it is then possible to represent the density and current states of each segment in the entire network on the phase diagram obtained for a single segment. We call this representation the *Effective Rates Plot* (ERP) [13]. The state of the density of each segment, and therefore the stationary density $\rho_{v \to v'}$ (and thus the current $j_{v \to v'}$), are controlled by the microscopic parameters of the model and the network connectivity. In this case one can represent each segment $v \to v'$ by a point $(\alpha_v/p, \beta_{v'}/p)$, on the $(\alpha/p, \beta/p)$ phase diagram of the model considered (Fig. 2), corresponding to its entry and exit rates computed via Eq. (18). The ensemble of these points $(\alpha_v/p, \beta_{v'}/p)$ on the phase diagram will then characterize the state of each segment in the network.

The distribution of the rate points on the ERP diagram of the model immediately identify the distribution functions of the segment densities $W(\rho_s)$, and of the currents $W(j_s)$, throughout the network, where the index s stands for the generic segment $v \to v'$. For regular random networks like the Bethe case, all vertices being equivalent, only one point should appear in the corresponding ERP. This is contrary to irregular networks, as in the Poisson case, where the scattered points on the phase diagram inform on the global organization of densities (and currents) throughout the irregular network.

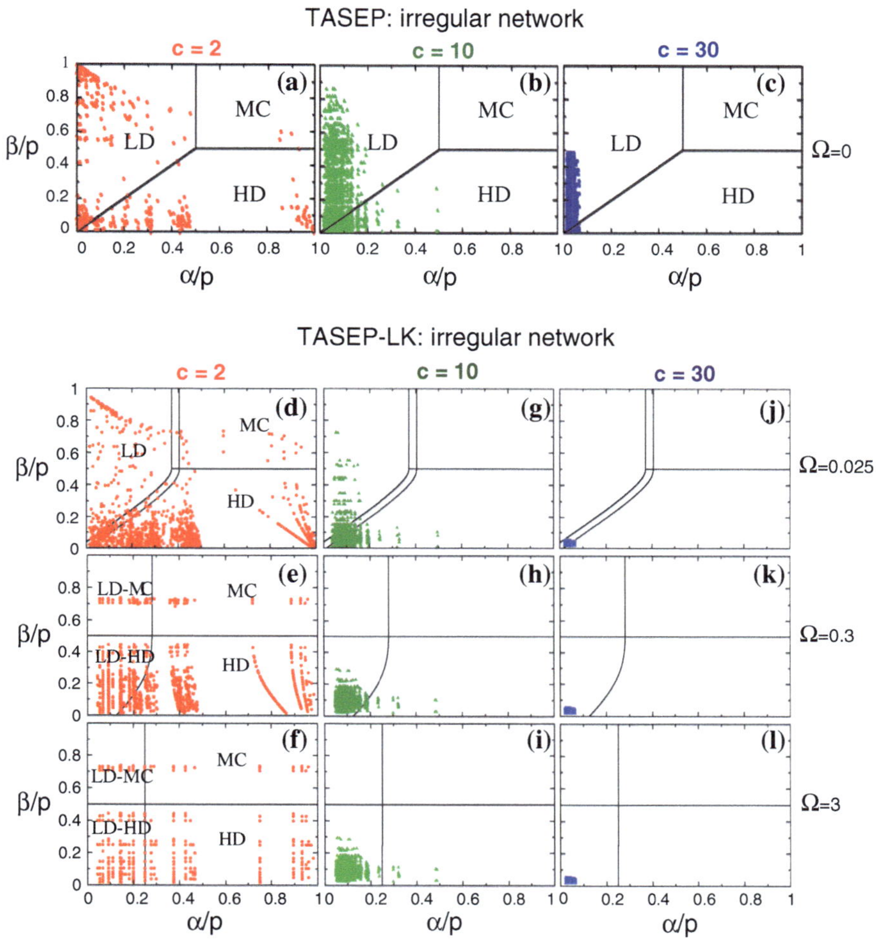

Fig. 4 Effective rate plots for (**a–c**) TASEP and (**d–l**) TASEP-LK for three different values of the network average connectivity $c = 2, 10, 30$. In TASEP, due to the $LD - HD$ coexistence line $\alpha = \beta$, one can find a network (phase separation) regime with segments either in *LD* or *HD*. In contrast, in TASEP-LK, both the average connectivity c and the local binding and unbinding rates, Ω_A and Ω_D, defining the Langmuir kinetics can control the transition from the "network regime" to the "segment regime" (see the figures above and the main text below)

Focusing on the distribution of the scattered points on the ERP, one can see immediately whether new phenomena of phase separation emerge, and also how the dependence of the phase diagram on the microscopic parameters influences the density and current distributions $W(\rho_s)$ and $W(j_s)$ in the network. Indeed, it is clear from this representation that the topology of the TASEP phase diagram implies necessarily a phase separation between *LD* and *HD* density phases in the case of irregular graphs,

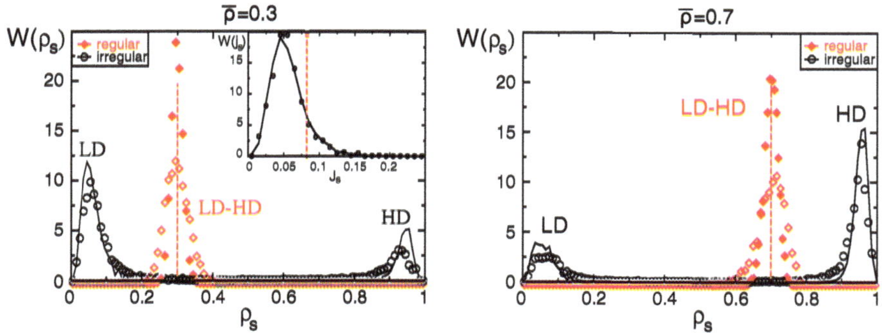

Fig. 5 Distribution function of the densities $W(\rho_s)$ along each network segment of regular and irregular networks, for TASEP on large networks with average connectivity $c = 10$, juxtaposed for two values of the overall densities $\bar{\rho} = 0.3, 0.7$. Bimodality occurs only for irregular networks. For regular networks, the change of the overall density $\bar{\rho}$ induces a shift in the density ρ_s of the center of the monomodal distribution. For irregular networks the bimodal distribution changes the relative contributions of the *LD* and *HD* phases, respectively. The inset in the left figure shows the distribution function for the current j_s, $W(j_s)$

while for TASEP-LK the dependence of the graph on the renormalized binding and unbinding rates Ω_A and Ω_D is more subtle.

For TASEP, the irregular network with an average connectivity[4] c will produce clouds of points with coordinates $(\alpha/p = \rho_v/k_v^{out}, \beta/p = 1 - \rho_{v'})$, separated by the $LD - HD$ coexistence line $\alpha = \beta$.

By increasing the network average connectivity, points gather near the origin, while points in the *MC* disappear since, due to the steric interactions, a maximal current cannot flow from a vertex with more than two outgoing segments. In this situation, segments will have either a *LD* or a *HD* density, and since the $LD - HD$ coexistence line is topologically connected to the origin, phase separation in *LD* and *HD* segments will occur at any given value of the average connectivity c. The multimodality of the density distribution, discovered in [11], is a new phenomenon of phase separation of interacting particles evolving out of thermodynamic equilibrium conditions.

For TASEP-LK, on the contrary, an increase in connectivity can naturally induce a transition from a multimodal distribution of *LD* and *HD* densities to a monomodal $LD - HD$ distribution.[5] The same kind of change of regime occurs when the binding and unbinding rates Ω_A and Ω_D increase while the average connectivity c is kept fixed.

We therefore discover a very interesting behaviour of stationary transport on irregular networks in absence of particle conservation: density inhomogeneities can

[4] A detailed analysis, for large values of the connectivity can be found in the supplementary material of ref. [11].

[5] The current distribution $W(j_s)$ remains monomodal (see the inset in Fig. 5) since the current is globally invariant under the particle-hole symmetry.

be controlled either by global properties such as the network connectivity c, or by more molecular parameters like the binding rate Ω_A (which depends on the local concentration of motors near the filament and their specific binding properties) or the kinetic rate Ω_D (which reflects the particle processivity) (Fig. 4).

Another interesting point emerges from this analysis. In TASEP-LK, the presence of a homogeneous reservoir of particles in contact with a network leads to a competition between two opposing mechanisms. The reservoir tends to homogenize the particle density on the network via the Langmuir kinetics, whereas the directed transport on the network (regular or irregular) pushes the system to build up two different kinds of density inhomogeneities. Phase separation can occur either at the scale of a single segment when this is in the $LD - HD$ or in the $LD - MC$ phase, or at the scale of the network when some of the segments are in LD phase while the others are in the HD phase. It turns out that this behaviour is general, and that even in presence of the homogenizing reservoir the new phenomenon of phase separation on the whole network still occurs [12, 13] for finite rates Ω_A and Ω_D.

6 Some Important Consequences

6.1 Physical and Mathematical Implications

Overall, the analysis of EP such as TASEP and TASEP-LK (and also ASEP, not discussed here) highlights the existence of specific regimes for the organisation of particle density along a network, that we term *network*, *segment* or *site regimes* (see just below).

In the network regime all vertices of the network are coupled to each other: indeed, the bulk density of each segment s depends on a single boundary (at the entry or at the exit of the segment). This is precisely the case (*ii*) of the possible solutions which we discussed in the beginning of Sect. 5, thus implying that entry and exit rates α_v and $\beta_{v'}$ become mutually dependent. This regime occurs in TASEP, and also in TASEP-LK when the exchange rates Ω_A and Ω_D are sufficiently small.

Differently from TASEP, TASEP-LK then has the possibility to leave this regime by increasing the exchange rates. Beyond a critical value of the binding and unbinding rates Ω_A and Ω_D, a domain wall enters the bulk of the generic segment s of the network. This is the regime (i) described above, where the bulk density is built by connecting the solutions imposed by both boundary conditions via a domain wall. Interestingly, this regime occurs independently of the network characteristics such as its regular or irregular connectivity. Contrary to the network regime, in this case the relevant physical and mathematical scale occurs on the single segment length. The properties of the traffic on the network are then dependent on the single segment behaviour. The structural properties of the network are therefore no longer necessary to determine the local behaviour of the density and current of particles, as was the case in the network regime.

Finally, when the Langmuir kinetics dominates on the transport TASEP process, the reservoir imposes that the density is essentially fixed locally to the Langmuir density ρ_l, as if the system were decoupled into single and independent sites in contact with the reservoir. This is the case for the *site regime*.

This classification of the different regimes of density inhomogeneities along the network provides a direct interpretation in terms of the complexity and level of coupling of the differential equations of Eq. (8) throughout the entire network.

Importantly, these results are robust. They can be interesting not only in a biological context, but also for man-made systems where the role of the range of particle-particle interactions would be interesting to investigate further.

We also note that the results we have obtained actually represent a generalization of the classical Kirchhoff's law [26] for linear electrical circuits to non-linear current cases reflecting the interaction of the transported objects. Indeed, if one suppresses the exclusion interactions in the previous models, all phase separation phenomena of the network and segment regimes disappear.

We emphasize that this approach is very powerful and can be generalized to many other lattice gas models. It is also well suited to incorporating and exploiting, eventually, all exact results as they may become available by advanced physico-mathematical tools on a single one-dimensional lattice.

6.2 Biological Implications

From a biological point of view the results obtained open various significant perspectives.

First, the model shows that the internal redistribution of motor proteins can be strongly dependent on the cytoskeleton topology/connectivity. Since motor proteins are key elements involved in the reorganization of cellular compartments, the inhomogeneous distribution of motor proteins along the cytoskeleton is an important qualitative feature for future theoretical and experimental studies. This is of particular interest since the cell cytoskeleton has a very rich regulatory machinery [1]. This theoretical study indeed suggests that the cytoskeleleton structure is a first determinant for the intracellular compartmentalization. This aspect, although intuitive, has to our knowledge never been approached systematically so far by physico-mathematical models.

Second, depending on the level of intracellular crowding, motors can move in two possible regimes of speed v_m along the cytoskeleton. This is due to the non-linear current-density fundamental relation, $j = \rho\, v_m(\rho) = p\, \rho\, (1 - \rho)$, relating the local average current j and density ρ: a given current j may be achieved by many motors moving slowly in a *HD* region (small v_m, large ρ), or by few motors fast in a *LD* region (large v_m, small ρ). Here it has become apparent from the different kinds of phase separation phenomena found that this may have an impact for transport on the scale of the network. Indeed, the presence of a phase separation in *LD* and *HD* on the network scale suggests the possibility that the cytoskeleton can be organized to provide slow lanes as pathways for massive transportation (i.e. forming long queues of motors along the filaments as in a *HD* phase), but at the same time reserve a kind of preferential "high-speed lanes", which might serve for specific targeting or for the logistics of individual cargoes inside the cell.

Third, the phase separation found is a robust phenomenon. Both the network regime and the segment regime occur even when the network of filaments is in contact with a homogeneous reservoir of particles. We expect that, in the presence of diffusion of motors in the cytoplasmic environment [40, 59], such phase separation phenomena will be even enhanced.

Fourth, the overall analysis emphasizes different multiscale mechanisms by which the cell can actively distribute matter, such as motor proteins and all related cargoes. Furthermore, the different regimes of density inhomogeneities can be selected:

1. either by controlling the unbinding kinetics of the motor to the filament (i.e. its processivity);
2. or by the local cytoplasmic organization for the binding kinetics;
3. or, at a completely different scale, by organizing the cytoskeleton topology/connectivity, as can be achieved via the complex cytoskeletal machinery, including the motor protein machinery involved in cytoskeletal filament regulation [65].

7 Perspectives

The study of cytoskeletal transport driven by motor proteins inspires many interesting problems and questions concerning non-equilibrium, collective and non-linear phenomena. The models developed provide tools that can help to explore active systems at different scales: from the behaviour of single molecules up to the organization of large networks with a size comparable to the whole cytoskeleton, for example via the behaviour of motors at junctions between filaments [66, 67] and their mechanochemical complexity [68, 69]. The framework developed so far is very promising for its multiscale nature. It has already provided insight into how collective traffic phenomena arise and has allowed to understand how the biological complexity of many intracellular processes works at different length and time scales.

Theoretical results nowadays challenge the available quantitative experimental techniques in order to understand biological processes. They also suggest to investigate the laws governing the behaviour of complex systems with spatially and temporally distributed degrees of freedom. The study of *active* matter for example is one of the most important ongoing topics in condensed matter and statistical physics [70].

Our comprehension of the real complexity of a cell, or other systems like for example a crowd of pedestrians, is still very incomplete. But even work in progress can provide new important findings, and it hints at a large variety of phenomena, extending beyond the biological realm. Important questions remain to be addressed. One of them is the role of the cytoskeletal network [11–13, 71, 72] and of its dynamics [73–77] on transport. Another one concerns the impact of fluctuations and correlations in transport phenomena, and in particular for transport in contact with the cytoplasm [40, 78]. Achieving these goals will certainly require exploiting a large variety of powerful mathematical and physical methods.

Acknowledgments The authors thank the European Molecular Biology Organization (EMBO), the Centre National de la Recherche Scientifique (CNRS), the National Agency for Research (ANR), the Program of Laboratory of Excellence (LabEx) NUMEV and the Scientific Council of the University of Montpellier 2 for the financial support obtained during these years. We acknowledge the interesting discussions we had on these topics in these years with A. Abrieu, C. Appert, P. Arndt, B. Bassetti, P. Benetatos, M. Bornens, C. Braun-Breton, S. Camalet, G. Cappello, L. Ciandrini, M. Cosentino-Lagomarsino, N. Crampe, O. Dauloudet, S. Diez, J. Dorignac, A. Dunn, B. Embley, M. Evans, T. Franosch F. Geniet, J. Howard, K. Kroy, J.F. Joanny, F. Jülicher, K. Kruse, C. Leduc, M. Lefranc, V. Lorman, P. Malgaretti, K. Mallick, P. Montcourrier, B. Mulder, I. Pagonabarraga, A. Parmeggiani, P. Pierobon, J. Prost, O. Radulescu, A. Raguin, C.M. Romano, E. Sackmann, J. Santos, K. Sasaki, C.F. Schmidt, G.M. Schütz, K. Sekimoto, J. Spudich, F. Turci, C. Vanderzande and M. Wakayama. N.K. and A.P. would like to thank B. Embley for his contribution in the study of TASEP on small networks. In particular, A.P. specially thanks E. Frey for the opportunity to work on exclusion processes for motor protein transport and for the many important and insightful discussions they had on the topic, together with T. Franosch, some years ago.

References

1. Alberts, B., Johnson, A., Walter, P., Lewis, J., Raff, M., Roberts, K.: Molecular Biology of the Cell, 15th edn. Garland, New York (2008)
2. Aridor, M., Hannan, L.A.: Traffic jams II: an update of diseases of intracellular transport. Traffic **3**(11), 781–790 (2002)
3. Hirokawa, N., Takemura, R.: Molecular motors in neuronal development, intracellular transport and diseases. Curr. Opin. Neurobiol. **14**(5), 564–573 (2004)
4. Gunawardena, S., Goldstein, L.S.: Cargo—carrying motor vehicles on the neuronal highway: transport pathways and neurodegenerative disease. J. Neurobiol. **58**(2), 258–271 (2004)
5. Howard, Mechanics of motor proteins and the cytoskeleton, J. Sinauer Associates, Sunderland, MA, (2001)
6. Ross, J.L., Shuman, H., Holzbaur, E.L., Goldman, Y.E.: Kinesin and dynein-dynactin at intersecting microtubules: motor density affects dynein function. Biophys. J. **94**(8), 3115–3125 (2008)
7. Cai, D., McEwen, D.P., Martens, J.R., Meyhofer, E., Verhey, K.J.: Single molecule imaging reveals differences in microtubule track selection between Kinesin motors. PLoS Biol. **7**(10), e1000216 (2009)
8. Pierobon, P., Achouri, S., Courty, S., Dunn, A.R., Spudich, J.A., Dahan, M., Cappello, G.: Velocity, processivity, and individual steps of single myosin V molecules in live cells. Biophys. J. **96**(10), 4268–4275 (2009)
9. Leduc, C., Padberg-Gehle, K., Varga, V., Helbing, D., Diez, S., Howard, J.: Molecular crowding creates traffic jams of kinesin motors on microtubules. Proc. Nat. Acad. Sci. **109**(16), 6100–6105 (2012)
10. Bálint, S., Vilanova, I.V., lvarez, A.S., Lakadamyali, M.: Correlative live-cell and superresolution microscopy reveals cargo transport dynamics at microtubule intersections. Pro. Nat. Acad. Sci. **110**(9), 3375–3380 (2013)
11. Neri, I., Kern, N., Parmeggiani, A.: Totally asymmetric simple exclusion process on networks. Phys. Rev. Lett. **107**(6), 068702 (2011)
12. Neri, I., Kern, N., Parmeggiani, A.: Modeling cytoskeletal traffic: an interplay between passive diffusion and active transport. Phys. Rev. Lett. **110**(9), 098102 (2013)

13. Neri, I., Kern, N., Parmeggiani, A.: Exclusion processes on networks as models for cytoskeletal transport. N. J. Phys. **15**(8), 085005 (2013)
14. Tada, H., Higuchi, H., Wanatabe, T.M., Ohuchi, N.: In vivo real-time tracking of single quantum dots conjugated with monoclonal anti-HER2 antibody in tumors of mice. Cancer Res. **67**(3), 1138–1144 (2007)
15. Farina, F., Pierobon, P., Delevoye, C., Monnet, J., Dingli, F., Loew, D., Quanz, M., Dutreix, M., Cappello, G.: Kinesin KIFC1 actively transports bare double-stranded DNA. Nucl. Acids Res. **41**(9), 4926–4937 (2013)
16. Hamburg, M.A., Collins, F.S.: The path to personalized medicine. N. Eng. J. Med. **363**(4), 301–304 (2010)
17. Chou, T., Mallick, K., Zia, R.K.P.: Non-equilibrium statistical mechanics: from a paradigmatic model to biological transport. Rep Prog. Phys. **74**(11), 116601 (2011)
18. Aghababaie, Y., Menon, G.I., Plischke, M.: Universal properties of interacting Brownian motors. Phys. Rev. E. **59**(3), 2578 (1999)
19. Schütz, G.M.: Exactly solvable models for many-body systems far from equilibrium. Phase Transitions Crit. Phenom. **19**, 1–251 (2001)
20. Blythe, R.A., Evans, M.R.: Nonequilibrium steady states of matrix-product form: a solver's guide. J. Phys. A Math. Theor. **40**(46), R333 (2007)
21. Mallick, K.: Some exact results for the exclusion process. J. Stat. Mech. Theory Exp. **2011**(01), P01024 (2011)
22. Mottishaw, P., Waclaw, B., Evans, M.R.: An exclusion process on a tree with constant aggregate hopping rate. J. Phys. A Math. Theor. **46**(40), 405003 (2013)
23. Liggett, T.M.: Interacting Particle Systems, 276th edn. Springer, Berlin (1985)
24. Derrida, B.: Non-equilibrium steady states: fluctuations and large deviations of the density and of the current. J. Stat. Mech. Theory Exp. **2007**(07), P07023 (2007)
25. Bertini, L., De Sole, A., Gabrielli, D., Jona-Lasinio, G., Landim, C.: Stochastic interacting particle systems out of equilibrium. J. Stat. Mech. Theory Exp. **2007**(07), P07014 (2007)
26. Kirchhoff, G.: Ueber die Auflösung der gleichungen, auf welche man bei der untersuchung der linearen Vertheilung galvanischer Ströme geführt wird Ann. Phys. Chem. **148**, 497–508 (1847)
27. Appert, C., Santen, L.: Modélisation du trafic routier par des automates cellulaires. Actes INRETS, 100, (2002)
28. Kirchner, A., Nishinari, K., Schadschneider, A.: Friction effects and clogging in a cellular automaton model for pedestrian dynamics. Phys. Rev. E. **67**(5), 056122 (2003)
29. Chowdhury, D., Schadschneider, A., Nishinari, K.: Physics of transport and traffic phenomena in biology: from molecular motors and cells to organisms. Phys. Life Rev. **2**(4), 318–352 (2005)
30. Lipowsky, R., Chai, Y., Klumpp, S., Liepelt, S., Müller, M.J.: Molecular motor traffic: From biological nanomachines to macroscopic transport. Phys. A Stat. Mech. Appl. **372**(1), 34–51 (2006)
31. Moussaid, M., Guillot, E.G., Moreau, M., Fehrenbach, J., Chabiron, O., Lemercier, S., Pettr, J., Appert-Rolland, C., Degond, P., Theraulaz, G.: Traffic instabilities in self-organized pedestrian crowds. PLoS computational biology **8**(3), (2012)
32. Nishinari, K., Sugawara, K., Kazama, T., Schadschneider, A., Chowdhury, D.: Modelling of self-driven particles: foraging ants and pedestrians. Phys. A Stat. Mech. Appl. **372**(1), 132–141 (2006)
33. Karzig, T., von Oppen, F.: Signatures of critical full counting statistics in a quantum-dot chain. Phys. Rev. B **81**(4), 045317 (2010)
34. Reichenbach, T., Franosch, T., Frey, E.: Exclusion processes with internal states. Phys. Rev. Lett. **97**(5), 050603 (2006)
35. Chou, T.: An interacting spinflip model for one-dimensional proton conduction. J. Phy. A Math. Gen. **35**(21), 4515 (2002)
36. Karcher, R.L., Deacon, S.W., Gelfand, V.I.: Motor-cargo interactions: the key to transport specificity. Trends Cell Biol. **12**(1), 21–27 (2002)

37. Jülicher, F., Ajdari, A., Prost, J.: Modeling molecular motors. Rev. Mod. Phys. **69**(4), 1269 (1997)
38. Parmeggiani, A., Schmidt, C. F.: Micromechanics of molecular motors: experiments and theory. In: Function and Regulation of Cellular Systems (pp. 151–176). Birkhäuser Basel (2004)
39. Lipowsky, R., Klumpp, S.: Life is motion: multiscale motility of molecular motors. Phys. A Stat. Mech. Appl. **352**(1), 53–112 (2005)
40. Parmeggiani, A.: Non-equilibrium collective transport on molecular highways. In: Traffic and granular flow 07 (pp. 667–677). Springer, Berlin Heidelberg (2009)
41. Gross, S.P.: Hither and yon: a review of bi-directional microtubule-based transport. Phys. Biol. **1**(2), R1 (2004)
42. Nagel, K., Herrmann, H.J.: Deterministic models for traffic jams. Phys. A Stat. Mech. Appl. **199**(2), 254–269 (1993)
43. Schreckenberg, M., Schadschneider, A., Nagel, K., Ito, N.: Discrete stochastic models for traffic flow. Phys. Rev. E **51**(4), 2939 (1995)
44. MacDonald, C.T., Gibbs, J.H., Pipkin, A.C.: Kinetics of biopolymerization on nucleic acid templates. Biopolymers **6**(1), 1–25 (1968)
45. MacDonald, C.T., Gibbs, J.H.: Concerning the kinetics of polypeptide synthesis on polyribosomes. Biopolymers **7**(5), 707–725 (1969)
46. Ciandrini, L., Stansfield, I., Romano, M.C.: Ribosome traffic on mRNAs maps to gene ontology: genome-wide quantification of translation initiation rates and polysome size regulation. PLoS Comput. Biol. **9**(1), e1002866 (2013)
47. Parmeggiani, A., Franosch, T., Frey, E.: Phase coexistence in driven one-dimensional transport. Phys. Rev. Lett. **90**(8), 086601 (2003)
48. Parmeggiani, A., Franosch, T., Frey, E.: Totally asymmetric simple exclusion process with Langmuir kinetics. Phys. Rev. E **70**(4), 046101 (2004)
49. Popkov, V., Rkos, A., Willmann, R.D., Kolomeisky, A.B., Schütz, G.M.: Localization of shocks in driven diffusive systems without particle number conservation. Phys. Rev. E **67**(6), 066117 (2003)
50. Burgers, J.: Kon. Nde. Akad. Wet. Verh. (Eerste Sectie) **17**, 1 (1939)
51. Burgers, J. M.: Mathematical examples illustrating relations occurring in the theory of turbulent fluid motion (pp. 281–334). Springer, Netherlands (1995)
52. Shaw, L.B., Zia, R.K.P., Lee, K.H.: Totally asymmetric exclusion process with extended objects: a model for protein synthesis. Phys. Rev. E **68**(2), 021910 (2003)
53. Pierobon, P., Frey, E., Franosch, T.: Driven lattice gas of dimers coupled to a bulk reservoir. Phys. Rev. E **74**(3), 031920 (2006)
54. Derrida, B., Domany, E., Mukamel, D.: An exact solution of a one-dimensional asymmetric exclusion model with open boundaries. J. Stat. Phys. **69**(3–4), 667–687 (1992)
55. Schütz, G., Domany, E.: Phase transitions in an exactly soluble one-dimensional exclusion process. J. Stat. Phys. **72**(1–2), 277–296 (1993)
56. Derrida, B., Evans, M.R., Hakim, V., Pasquier, V.: Exact solution of a 1D asymmetric exclusion model using a matrix formulation. J. Phys. A Math. Gen. **26**(7), 1493 (1993)
57. Kolomeisky, A.B., Schütz, G.M., Kolomeisky, E.B., Straley, J.P.: Phase diagram of one-dimensional driven lattice gases with open boundaries. J. Phys. A Math. Gen. **31**(33), 6911 (1998)
58. Santen, L., Appert, C.: The asymmetric exclusion process revisited: fluctuations and dynamics in the domain wall picture. J. Stat. Phys. **106**(1–2), 187–199 (2002)
59. Klumpp, S., Lipowsky, R.: Traffic of molecular motors through tube-like compartments. J. Stat. Phys. **113**(1–2), 233–268 (2003)
60. Krug, J.: Boundary-induced phase transitions in driven diffusive systems. Phys. Rev. Lett. **67**(14), 1882 (1991)
61. Pierobon, P., Parmeggiani, A., von Oppen, F., Frey, E.: Dynamic correlation functions and Boltzmann-Langevin approach for driven one-dimensional lattice gas. Phys. Rev. E. **72**(3), 036123 (2005)
62. Luby-Phelps, K.: Physical properties of cytoplasm. Curr. Opin. Cell Biol. **6**(1), 3–9 (1994)

63. Brankov, J., Pesheva, N., Bunzarova, N.: Totally asymmetric exclusion process on chains with a double-chain section in the middle: computer simulations and a simple theory. Phys. Rev. E **69**(6), 066128 (2004)
64. Embley, B., Parmeggiani, A., Kern, N.: Understanding totally asymmetric simple-exclusion-process transport on networks: generic analysis via effective rates and explicit vertices. Phys. Rev. E **80**(4), 041128 (2009)
65. Varga, V., Leduc, C., Bormuth, V., Diez, S., Howard, J.: Kinesin-8 motors act cooperatively to mediate length-dependent microtubule depolymerization. Cell **138**(6), 1174–1183 (2009)
66. Raguin, A., Parmeggiani, A., Kern, N.: Role of network junctions for the totally asymmetric simple exclusion process. Phys. Rev. E **88**(4), 042104 (2013)
67. Embley, B., Parmeggiani, A., Kern, N.: HEX-TASEP: dynamics of pinned domains for TASEP transport on a periodic lattice of hexagonal topology. J. Phys. Condens. Matter **20**(29), 295213 (2008)
68. Nishinari, K., Okada, Y., Schadschneider, A., Chowdhury, D.: Intracellular transport of single-headed molecular motors KIF1A. Phys. Rev. Lett. **95**(11), 118101 (2005)
69. Ciandrini, L., Stansfield, I.C.R.M., Romano, M.C.: Role of the particles stepping cycle in an asymmetric exclusion process: a model of mRNA translation. Phys. Rev. E **81**(5), 051904 (2010)
70. Marchetti, M.C., Joanny, J.F., Ramaswamy, S., Liverpool, T.B., Prost, J., Rao, M., Simha, R.A.: Hydrodynamics of soft active matter. Rev. Mod. Phys. **85**(3), 1143 (2013)
71. Greulich, P., Santen, L.: Active transport and cluster formation on 2D networks. Europ. Phys. J. E **32**(2), 191–208 (2010)
72. Ezaki, T., Nishinari, K.: A balance network for the asymmetric simple exclusion process. J. Stat. Mech. Theory Exp. **2012**(11), P11002 (2012)
73. Kruse, K., Sekimoto, K.: Growth of fingerlike protrusions driven by molecular motors. Phys. Rev. E **66**(3), 031904 (2002)
74. Klein, G.A., Kruse, K., Cuniberti, G., Jülicher, F.: Filament depolymerization by motor molecules. Phys. Rev. Lett. **94**(10), 108102 (2005)
75. Reese, L., Melbinger, A., Frey, E.: Crowding of molecular motors determines microtubule depolymerization. Biophys. J. **101**(9), 2190–2200 (2011)
76. Johann, D., Erlenkmper, C., Kruse, K.: Length regulation of active biopolymers by molecular motors. Phys. Rev. Lett. **108**(25), 258103 (2012)
77. Melbinger, A., Reese, L., Frey, E.: Microtubule length regulation by molecular motors. Phys. Rev. Lett. **108**(25), 258104 (2012)
78. Klumpp, S., Nieuwenhuizen, T.M., Lipowsky, R.: Self-organized density patterns of molecular motors in arrays of cytoskeletal filaments. Biophys. J. **88**(5), 3118–3132 (2005)

Tumour Cell Biology and Some New Non-local Calculus

Graeme Wake, Ali A. Zaidi and Bruce van-Brunt

Abstract Living cell populations which are simultaneously growing and dividing are usually structured by size, which can be, for example, mass, volume, or DNA content. The evolution of the number density $n(x, t)$ of cells by size x, in an unperturbed situation, is observed experimentally to exhibit the attribute of that of an asymptotic "Steady-Size-Distribution" (SSD). That is, $n(x, t) \sim$ scaled (by time t) multiple of a constant shape $y(x)$ as $t \to \infty$, and $y(x)$ is then the SSD distribution, with constant shape for large time. A model describing this is given, enabling parameters to be evaluated. The model involves a linear non-local partial differential equation. Similar to the well-known pantograph equation, the solution gives rise to an unusual first order singular eigenvalue problem. Some results and conjectures are given on the spectrum of this problem. The principal eigenfunction gives the steady-size distribution and serves to provide verification of the observation about the asymptotic growth of the size-distribution.

Keywords Cell-division · Eigenvalue problems · Survival thresholds

1 Introduction and Model

Non-local equations occur quite frequently in applications, yet are rarely included in the curriculum of university courses. This is a pity in view of the richness these problems present in their solutions. The most familiar example of this is the differential **time**-delay equation

---

G. Wake (✉) · A. A. Zaidi
Institute of Natural and Mathematical Sciences, Massey University, Auckland, New Zealand
e-mail: G.C.Wake@massey.ac.nz

A. A. Zaidi
e-mail: A.A.Zaidi@massey.ac.nz

B. van-Brunt
Institute of Fundamental Sciences, Massey University, Palmerston North, New Zealand
e-mail: B.vanBrunt@massey.ac.nz

M. Wakayama et al. (eds.), *The Impact of Applications on Mathematics*,
Mathematics for Industry 1, DOI: 10.1007/978-4-431-54907-9_2,

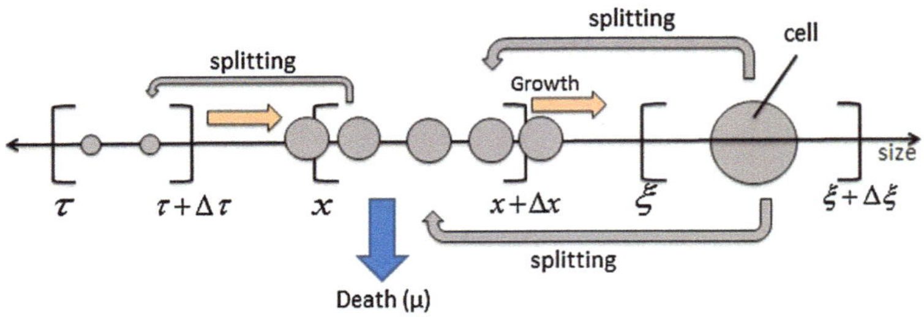

Fig. 1 Cell growth and division for the cohort

$$\begin{aligned} u'(t) &= u(t-T), & t &\geq 0 \\ u(t) &= u_0(t), & -T &\leq t < 0. \end{aligned}$$

The solution to this is well-known (see Bellman and Cooke [1]) and we note that the solution of the differential equation has a countable infinity of linearly independent solutions, whereas when $T = 0$ there is only one. Extensions to multidimensional systems with delays are given in Wake and Byrne [2].

We proceed now to introduce our model. Firstly, we consider the symmetrical case, where cells of size $\xi = \alpha x$ $(\alpha > 1)$ are splitting to give α cells of size x, with frequency b and simultaneously growing at a rate g units per time. Here α can be, in principle, any number greater than one, but is most usually two (binary division). It need not be an integer, amoeba cells show this, say 10 cells can aggregate and simultaneously divide to give 11 cells thereby giving $\alpha = 1.1$. The key requirement is that volume is preserved at the point of division: a cell of size αx is producing α cells of size x. This is shown schematically in Fig 1.

With a per-capita death rate μ, the equation describing this process, see Hall and Wake [3], is

$$\frac{\partial n}{\partial t} + \underbrace{\frac{\partial (gn)}{\partial x}}_{\text{growth}} = \underbrace{b\alpha^2 n(\alpha x, t)}_{\substack{\text{division of}\\ \text{larger cells}}} - \underbrace{bn(x,t)}_{\substack{\text{division into}\\ \text{smaller cells}}} - \underbrace{\mu n(x,t)}_{\text{cell-death}} \tag{1}$$

in the first quadrant $(x, t > 0)$ of the plane. This is complemented by the boundary conditions

$$n(0,t) = 0, \quad n(x,t) \to 0 \quad \text{as} \quad x \to \infty, \tag{2}$$

and initial condition

$$n(x,0) = n_0(x). \tag{3}$$

Cells dividing asymmetrically are essential for generating diverse cell types during development. The capacity for symmetric stem-cell self-renewal may confer developmental plasticity, increased growth and **enhance regenerative** capacity; however,

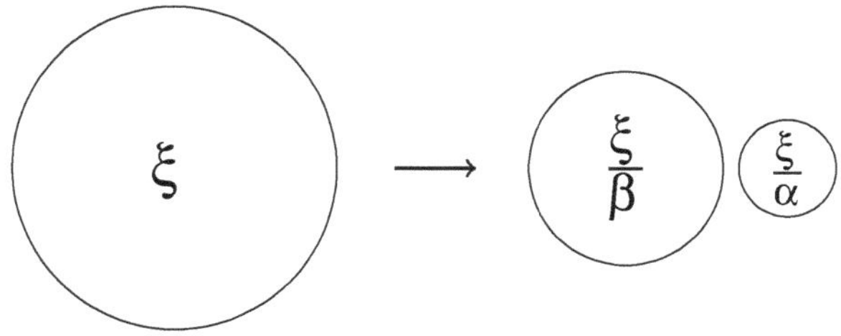

Fig. 2 Schematic representation for binary asymmetrical cell division with $\alpha > \beta > 1$

it may also confer an inherent risk of cancer. When the machinery that regulates asymmetric divisions is disrupted, however, these cells begin dividing symmetrically and form tumours. This needs underpinning rigour to understand the dynamics of cancer-cell growth and regulation of cell-growth. The context is developed in the paper by Basse et al. [4].

A new model is therefore needed of cell-growth with asymmetrical division [two or more daughter cells of different sizes (usually DNA content)] from a single "division-event". This model must capture the key features from earlier models with symmetrical cell-division, where the cell-size distribution tends asymptotically to one of constant shape where the cohort is not disturbed; this being a well-known observation. That is, $n(x,t) \sim T(t)y(x)$ as $t \to \infty$. The function $y(x)$ is also still called a steady-size-distribution (SSD). We consider for simplicity binary asymmetrical cell-division. This is the case where a cell of size ξ divides into two daughter cells of different sizes $\frac{\xi}{\beta}$ and $\frac{\xi}{\alpha}$, shown in Fig 2.

The asymmetry requires the introduction of a transfer rate $W(x,\xi)$ which is the number of cells of size x produced by a cell of size ξ, from its division. This amends Eq. (1) to

$$\frac{\partial n}{\partial t} + \underbrace{\frac{\partial (gn)}{\partial x}}_{\text{growth}} = \underbrace{\int_x^\infty bW(x,\xi)n(\xi,t)d\xi}_{\text{division of larger cells}} - \underbrace{\left(\int_0^x W(\tau,\xi)\frac{\tau}{x}d\tau\right) bn(x,t)}_{\text{loss of cells by division}} - \underbrace{\mu n(x,t)}_{\text{cell-death}} .$$

The second term recognises that there are $W(\tau,x)$ smaller cells produced by the division of a cell of size $x(>\tau)$. Further mass conservation for division requires

$$\int_0^x W(\tau,\xi)\tau d\tau = x, \tag{4}$$

and so we get a new non-local equation

$$\frac{\partial n}{\partial t} + \frac{\partial (gn)}{\partial x} = \int_x^\infty bW(x,\xi)n(\xi,t)d\xi - bn(x,t) - \mu n(x,t). \tag{5}$$

The problem (2), (3), (4) is expected to be well-posed for suitable W.

The two cases above require the following W:

1. Symmetrical division
$$W(x,\xi) = \alpha\delta\left(\frac{\xi}{\alpha} - x\right),$$
where δ is the Dirac-delta function;
2. Binary asymmetrical division
$$W(x,\xi) = \delta\left(\frac{\xi}{\alpha} - x\right) + \delta\left(\frac{\xi}{\beta} - x\right),$$
and mass conservation requires $\frac{1}{\alpha} + \frac{1}{\beta} = 1$ from (4).

Cases (2) and (1) coincide when $\alpha = \beta = 2$.

Using the W in case (2) in Eq. (4) gives a new equation

$$\frac{\partial n}{\partial t} + \frac{\partial(gn)}{\partial x} = b\alpha n(\alpha x,t) + b\beta n(\beta x,t) - (b+\mu)n(x,t). \tag{6}$$

2 Preliminary Results

We illustrate the ideas with the simplifying assumptions that b, g, and μ are constant. We use the more general Eq. (6).

SSD behaviour suggests that there are separable solutions of the form $n(x,t) = T(t)y(x)$ which in Eq. (6) gives $\frac{T'(t)}{T(t)} = \text{constant}(= -\lambda)$ and so

$$n(x,t) \sim e^{-\lambda t}y(x), \tag{7}$$

for some λ, which then gives the interesting non-local ode

$$gy'(x) = b\alpha y(\alpha x) + b\beta y(\beta x) - (b+\mu-\lambda)y(x), \tag{8}$$
$$\text{with } y(0) = y(\infty) = 0.$$

If $y(x)$ is a probability density function this requires (by integration)

$$\lambda = \mu - b, \tag{9}$$

which determines the growth ($b > \mu$) or decay ($b < \mu$) of the solution in Eq. (7) with constant shape of $y(x)$.

Of course Eq. (8) is an eigenvalue problem with a spectrum which satisfies the boundary value problem

$$gy'(x) = b\alpha y(\alpha x) + b\beta y(\beta x) - 2by(x), \quad y(0) = y(\infty) = 0. \tag{10}$$

This has a solution in the form of a double Dirichlet series, scaled so as to be a probability density function

$$y(x) = \sum_{m,n=0}^{\infty} c_{m,n} e^{-(\alpha^m + \beta^n)x}, \tag{11}$$

where $c_{m,n}$ satisfy a complicated recurrence relation, which enables the $c_{m,n}$ to be calculated recursively.

There are, of course, other eigenvalues $\{\lambda_n\}$ and eigenfunctions, which are necessary to fit the initial condition (3).

We expect

$$n(x,t) = \sum_{n=0}^{\infty} a_n y_n(x) e^{-\lambda_n t},$$

with λ_0 given by Eq. (9), and $y_0(x)$ given by Eq. (11).

The nature of these is as yet partially unknown. We would expect $\lambda_n > \lambda_0$, for $n > 0$ and that the set $\{y_n(x)\}$ is complete in some norm. The eigenfunctions are the non-trivial solutions of Eq. (8) when $\lambda = \lambda_n$ with the boundary conditions stated. Usually these are "normalised" in some way (see later).

Some higher eigenvalues and eigenfunctions come from use of the Mellin transform

$$M[y; s] = \int_0^{\infty} x^{s-1} y(x) dx.$$

We obtain the higher eigenvalues and eigenfunctions by using, when $\lambda = \lambda_n$, $y = y_n$, $n \geq 1$,

$$M[y_n; k] = \begin{cases} 0, & k = 1, \ldots, n \\ 1, & k = n+1. \end{cases}$$

Taking transforms of (8), when $k = n$, gives

$$\lambda_n = b + \mu - b\left(\frac{1}{\alpha^n} + \frac{1}{\beta^n}\right) \tag{12}$$

and $y_n(x)$ will be another Dirichlet series.

Clearly:

$n = 0$ in Eq. (12) gives the result in Eq. (9),

$n = 1$ gives $\lambda_1 = \mu$;

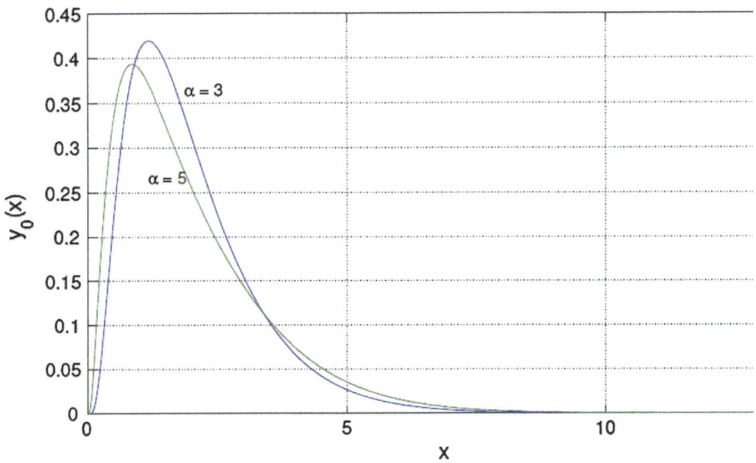

Fig. 3 SSD's for binary asymmetrical division $x \to (\frac{x}{\alpha}, \frac{x}{\beta}) : \frac{1}{\alpha} + \frac{1}{\beta} = 1$

and (λ_n) is monotonic increasing in n, with $(\lambda_n) \to (b + \mu)$ as $n \to \infty$.

We have "normalised" the eigenfunctions by requiring $M[y_n; n + 1] = 1$.

Of course, we have yet to establish whether or not these eigenfunctions are a complete set or even if there are other solutions, as λ_0 is the smallest eigenvalue, clearly $y_0(x)$ is the SSD.

However, the SSD's for various α, β can be computed from Eq. (10) and some are shown in Fig. 3.

3 Concluding Remarks

Cells dividing asymmetrically are essential for generating diverse cell types during development. The capacity for symmetric stem-cell self-renewal may confer developmental plasticity, increased growth and **enhance regenerative** capacity; however, it may also confer an inherent risk of cancer. When the machinery that regulates asymmetric divisions is disrupted, however, these cells begin dividing symmetrically and form tumours. This needs underpinning rigour to understand the dynamics of cancer-cell growth and regulation of cell-growth. This work is relevant to the underlying understanding of cell tumour growth. The application is stimulating new mathematics, for example the spectral theory of non-local singular eigenvalue problems.

Acknowledgments Funding support from *Gravida*; National Centre for Growth and Development, Auckland, New Zealand is gratefully acknowledged (GCW). We also thank the referee for useful suggestions which improved the paper and Miss Babylon for editorial assistance.

References

1. Bellman, R., Cooke, K.: Differential Delay Equations. Academic Press, New York (1963)
2. Wake, G.C., Byrne, H.M.: Calculus from the past: multiple delay systems arising in cancer cell modelling. ANZIAM J. **54**, 117–126 (2013)
3. Hall, A., Wake, G.C.: A functional differential equation arising in modelling cell growth. J. Aust. Math. Soc. Ser. B **30**, 425–434 (1989)
4. Basse, B., Baguley, B.C., Marshall, E., Wake, G.C., Wall, D.J.N.: Modelling cell population growth with applications to cancer therapy in human cell lines. Prog. Biophys. Mol. Biol. **85**, 353–368 (2004)

Industrial Mathematics in Europe

Wil Schilders

Abstract In this paper, we give an overview of the development of industrial mathematics in Europe. The advent of activities is in the 1970s, when, especially in Oxford, the potential of applications of mathematics was realized by Alan Tayler and co-workers, and the very successful study groups with industry were started. It led to discussions about European organisations such as ECMI, started in 1987, to a number of reports on mathematics in industry, to commercial institutes exploiting mathematics for industrial applications, and, finally, to a new organisation that was recently founded, EU-MATHS-IN. It is felt that it is important to share these experiences and activities with colleagues, anticipating that mathematics in industry will be a key enabling technology leading, in many respects, to a better world.

Keywords Industrial mathematics · Mathematical sciences · Virtual design environments · Data science · Computational science · Studygroups mathematics with industry

1 Introduction

The mathematical sciences play a vital part in all aspects of modern society. Without research and training in mathematics, there would be no engineering, economics or computer science; no smart phones, MRI scanners, bank accounts or PIN numbers. Mathematics is playing a key role in tackling the modern-day challenge of cyber security and in predicting the consequences of climate change, as well as in the manufacturing sectors of the automotive and aerospace industries through the utilization of superior virtual design processes. Likewise, the life sciences sector, with significant potential for economic growth, would not be in such a strong position without

---

W. Schilders (✉)
Eindhoven University of Technology, 5612 AZ Eindhoven, The Netherlands
e-mail: w.h.a.schilders@TUE.nl

M. Wakayama et al. (eds.), *The Impact of Applications on Mathematics*,
Mathematics for Industry 1, DOI: 10.1007/978-4-431-54907-9_3,

mathematics research and training, providing the expertise integral to the development of areas such as personalised healthcare and pharmaceuticals and of related medical technologies. The emergence of truly massive datasets across most fields of science and engineering increases the need for new tools from the mathematical sciences.

One of the classic ways in which mathematical science research plays a role in the economy is through collecting data towards understanding it, by using tools and techniques, enabling the discovery of new relationships or models. Modelling of physical phenomena already dates back several centuries, and well-known systems of equations with the names of Maxwell, Navier-Stokes, Korteweg-de Vries and more recently the Schrödinger equation plus many others are now well established. But, it was not until the advent of computers in the middle of the previous century and the development of sophisticated computational methods (like iterative solution methods for large sparse linear systems) that this could be taken to a higher level, by performing computations using these models. Software tools with advanced computational mathematical techniques for the solution of the aforementioned systems of equations have become common place, and are heavily used by engineers and scientists.

Mirroring this activity is increased awareness by society and industry that mathematical simulation is ubiquitous to address the challenging problems of our times. Industrial processes, economic models and critical events like floods, power failures or epidemics have become so complicated that their realistic description does not require the simulation of a single model, but rather the co-simulation of various models. Better scientific understanding of the factors governing these will provide routes to greater innovation power and economic well-being across an increasingly complex networked world with its competitive and strongly interacting agents. Industry, but also science, is highly dependent on the development of virtual environments that can handle the complex problems that we face today, and in the future.

For example, if the origins of life are to be explained, biologists and mathematicians need to work together, and most of the time spent will be on evaluating and simulating the mathematical models. Using the mathematics of evolutionary dynamics, the change from no life to life (referring to the self-replicating molecules dominating early Earth) can be explained. Another example is the electronics industry, which all of us rely on for new developments in virtually every aspect of our everyday life. Innovations in this branch of industry are impossible without the use of virtual design environments that enable engineers to develop and test their complex designs in front of a computer screen, without ever having to go into the time-consuming (several months) process of prototyping.

Principles of computational science and engineering rooted in modern applied mathematics are at the core of these developments, and represent subjects that are set to undergo a renaissance in the 21st century. Indeed, no less a figure than Stephen Hawking is on record as saying that the 21st century will be the century of complexity. Another great figure, yet young, is Fields medallist Terence Tao, who was a major contributor to the recently published document entitled "The mathematical sciences in 2025" [1], stating: "Mathematical sciences work is becoming an increasingly integral and essential component of a growing array of areas of investigation in biology, medicine, social sciences, business, advanced design, climate, finance, advanced materials, and many more—crucial to economic growth and societal well-being".

Growing computing power, nowadays including multicore architectures and GPU's, does not provide the solution to the ever growing demand for more complex and more realistic simulations. In fact, it has been demonstrated that Moore's Law, describing the advances in computing power over the last 40 years, equally holds for mathematical algorithms. Hence, it is important to develop both faster computers and faster algorithms, at the same time. This is essential if we wish to keep up with the growing demands by science and technology for more complex simulations.

Given the above developments, Europe has launched many initiatives to convince industry, society and policy makers that the time is ripe for change. After the OECD report, initiated and chaired by Willi Jaeger from Heidelberg, the European Mathematical Society and the European Science Foundation funded a so-called Forward Look project on "Mathematics in Industry". The result of this project was a report with recommendations to policy makers, industry and the mathematics community, and a very nice book "European success stories in industrial mathematics", containing more than 100 industrial cases in which mathematics played a decisive role. In 2012, this was followed by a report by Deloitte (accountants and advisers) on "The value of the mathematical sciences for industry and society in the UK", revealing that 38 be attributed to results of mathematical sciences research, in a direct, indirect or induced way. A similar study is currently being undertaken in The Netherlands. Germany has published a book entitled "Mathematics, engine of the economy" with more than 20 accounts by captains-of-industry, emphasizing the importance of mathematics. This shows that Europe is putting a lot of effort in to demonstrating the necessity of mathematics for industry and society, and it is anticipated that this will have a very positive influence on the funding situation for mathematics. In this paper, these initiatives will be discussed, as well as the strategy adopted in Europe. All of these efforts have culminated into the formation of a new foundation termed EU-MATHS-IN, that aims at collecting all national and European initiatives in the area of industrial mathematics, so as to learn from each other, to share best practice, and to benefit from a unified approach.

In addition, we will give an overview of some of the activities that have taken place in Europe, and of the results that have thereby been generated, especially in view of the theme of the conference of which this book forms the proceedings, namely "The Impact of Applications on Mathematics".

2 The European Consortium for Mathematics in Industry

From a historical point of view, ECMI, the European Consortium for Mathematics in Industry [1], was one of the first organisations that was founded to foster the potential of applications of mathematics in industry. It celebrated its 25th anniversary in 2012.

Back when it started, in the middle of the 1980s, mathematics was dominated by mathematicians mainly interested in pure mathematics, in algebra, topology, geometry, analysis and so on. Only a small group of people focussed their attention on cooperation with industry. In 1985, this led to the first conference, called the European Symposium for Mathematics in Industry, and abbreviated ESMI. After this successful symposium, it was felt that it would be good to start a European organization, and hence, in 1987, ECMI was founded. The goal was to promote and further the effective use of mathematics and closely related knowledge and expertise in industrial and management settings. More specifically, concerning research, to see what is needed by industry and commerce, to assess what is available, and to discuss what can be done to fill the gaps. Also, it was important to encourage the participating organizations to have joint research ventures. From an educational point of view, the focus was on the creation, organization and quality control of a 2-year postgraduate course on industrial and possibly management mathematics. It was also decided to have an annual conference. Quite quickly this became a biennial conference focussing on applications of mathematics in industry. The ECMI Newsletter of October 2012 [2] contains a very nice account by one of the founding fathers of ECMI, Helmut Neunzert, about the start and the first 10 years of ECMI. Below is a copy of the official list of signatures on the founding document (Fig. 1).

ECMI is now a mature organization with 25 years of experience in the area of mathematics for industry. Its mission can be summarized as follows:

Mathematics, as the universal language of the sciences, plays a key role in technology, economics and life sciences. European industry is increasingly dependent on mathematical expertise in both research and development to keep its world-leading role for high technology innovations and to comply with the EU 2020 agenda for smart, sustainable and inclusive growth. The major objectives to respond to these needs of European industry may be summarized as follows:

- *ECMI advocates the use of mathematical models in industry*
- *ECMI stimulates the education of young scientists to meet the growing demands of industry*
- *ECMI promotes European collaboration, interaction and exchange within academia and industry.*

One of the most successful enterprises of ECMI is in the educational area. Its Educational Committee consists of many experts that meet regularly and discuss curricula for master's degrees in industrial mathematics, as well as keeping a close eye on the quality of such curricula in member universities. This quality control is performed on a regular basis, and new members can apply for the status of qualified node and are then visited by a team of experts evaluating the curriculum and the means used. Nowadays, the ECMI Educational Committee oversees more than 20 high standard master's programs in industrial and econo-mathematics. Students that have graduated from an ECMI centre are awarded an ECMI certificate.

Another one of ECMIs success stories are the annual European Modelling weeks that started 1988 in Bari (Italy) with 30 students working on six projects. Each project

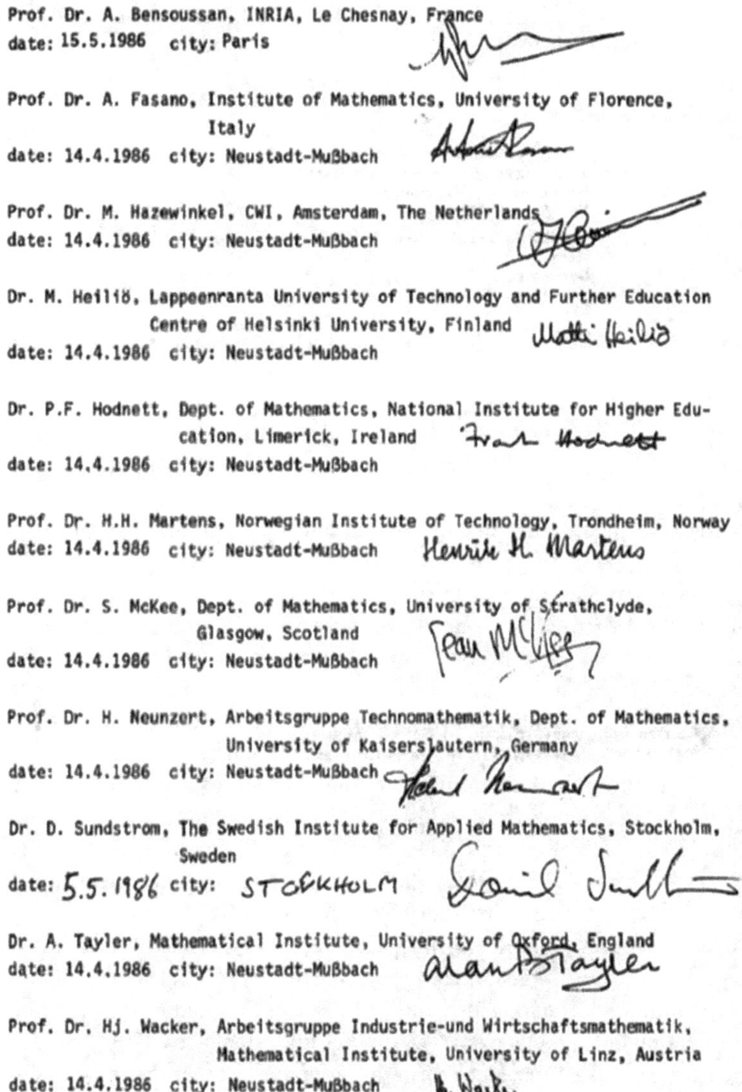

Prof. Dr. A. Bensoussan, INRIA, Le Chesnay, France
date: 15.5.1986 city: Paris

Prof. Dr. A. Fasano, Institute of Mathematics, University of Florence, Italy
date: 14.4.1986 city: Neustadt-Mußbach

Prof. Dr. M. Hazewinkel, CWI, Amsterdam, The Netherlands
date: 14.4.1986 city: Neustadt-Mußbach

Dr. M. Heiliö, Lappeenranta University of Technology and Further Education Centre of Helsinki University, Finland
date: 14.4.1986 city: Neustadt-Mußbach

Dr. P.F. Hodnett, Dept. of Mathematics, National Institute for Higher Education, Limerick, Ireland
date: 14.4.1986 city: Neustadt-Mußbach

Prof. Dr. H.H. Martens, Norwegian Institute of Technology, Trondheim, Norway
date: 14.4.1986 city: Neustadt-Mußbach

Prof. Dr. S. McKee, Dept. of Mathematics, University of Strathclyde, Glasgow, Scotland
date: 14.4.1986 city: Neustadt-Mußbach

Prof. Dr. H. Neunzert, Arbeitsgruppe Technomathematik, Dept. of Mathematics, University of Kaiserslautern, Germany
date: 14.4.1986 city: Neustadt-Mußbach

Dr. D. Sundstrom, The Swedish Institute for Applied Mathematics, Stockholm, Sweden
date: 5.5.1986 city: STOCKHOLM

Dr. A. Tayler, Mathematical Institute, University of Oxford, England
date: 14.4.1986 city: Neustadt-Mußbach

Prof. Dr. Hj. Wacker, Arbeitsgruppe Industrie-und Wirtschaftsmathematik, Mathematical Institute, University of Linz, Austria
date: 14.4.1986 city: Neustadt-Mußbach

Fig. 1 The founding fathers of ECMI

in a modelling week originates from a real life problem and an international student group, supervised by an ECMI instructor, works collaboratively for one week towards a solution. In 2011, the 25th modelling week took place in Milano (Italy), where 75 students worked on 12 projects. Alongside the modelling week, ECMI organizes also its summer school, where lecturers both from ECMI and from its industrial partners give courses in various topics of applied and industrial mathematics.

Fig. 2 The 25th anniversary of ECMI was celebrated at the conference in Lund, July 2012

An important aspect of the ECMI education network is to the organization and the broadening of the exchange of students among the ECMI centres. The strong coherence within the network and the synchronized local master programs taught in English allow for a smooth relocation from one centre to another and for an easy transfer of credits gained at a foreign centre.

ECMI also undertakes many activities in the research area. The Research and Innovation Committee focuses on strategies to increase the interaction between industry and academia, to foster both academic research and industrial innovation. The committee is multidisciplinary. It marshals the power of mathematics, scientific computing and engineering for industrial modelling and simulation. It also fosters special interest groups (SIGs) that focus on a special theme which is either application oriented or methodology based. A SIG identifies a group of experts and has a strong participation by or interest from industry. The SIGs organize regular meetings and workshops. Examples of active SIGs are "Scientific Computing in the Electronics Industry" and "Shape and Size in Medicine, Biotechnology and Material Sciences". The SIGs provide a unique opportunity for cooperation on the European level, (such as the framework programs and the Marie Curie actions—What this means is unclear.) Two very successful examples of past projects are MACSI-net (Mathematics, Computing and Simulation for Industry) and COMSON (Coupled Multiscale Simulation and Optimization in Nanoelectronics). Another important activity is the organization of the biennial ECMI conferences, bringing together academic applied mathematicians and industrial scientists. They provide a forum for scientific experts and young researchers to exchange recent ideas about research and innovation. Special Industry Days allow participants to explore new and relevant industrial areas where mathematics plays an important role. The 17th biennial conference marked the 25th anniversary of ECMI, and was organized in Lund.

In recent years, ECMI has been part of several European initiatives to foster the application of mathematics in industry, and these are discussed in the next few sections (Fig. 2).

3 MACSI-Net

ECMI provided the cradle for a very successful European network, initiated by Prof. Bob Mattheij at TU Eindhoven, one of the founding fathers of ECMI. MACSI-net [3], brief for Mathematics, Computing and Simulation for Industry, was in fact a cooperative venture between ECMI and ECCOMAS [4]. ECCOMAS is a scientific organization grouping together European associations with interests in the development and applications of computational methods in science and technology. The **Mission of ECCOMAS** is to promote joint efforts of European universities, research institutes and industries which are active in the broader field of numerical methods and computer simulation in Engineering and Applied Sciences and to address critical societal and technological problems with particular emphasis on multidisciplinary applications.

When MACSI-net was started, around the change of the century, it was apparent that industry was grappling with ever more challenging problems, which should be solved by using state of the art mathematical and computational tools. Academic institutions often had the knowledge and expertise to be of great help in this. However, enterprises often did not know how to find the proper academic partners, in particular in mathematical areas. On the other hand, academic institutes were still not sufficiently aware of the importance of taking up their role in joint endeavours with both smaller and larger problems that may help Europe's industry to maintain or achieve a competitive edge in a variety of areas. MACSI-net was therefore set up as a network where both enterprises and university institutions could co-operate on the solution of such problems, to their mutual benefit. In particular, the network focused on strategies to increase the interaction between industry and academia in order to help industry (in particular SME) with advanced mathematical and computational tools, and to increase awareness of academia of industrial needs. The network was multidisciplinary, combining the power of mathematics, scientific computing and engineering, for modelling and simulation activities. The network aimed at achieving its goals through

- Strategic meetings with industries about well specified topics
- Summer courses
- Workshops
- Visits of experts
- Foundation of special (interest) groups
- Funding and appointment of post docs
- Activity committees who actively look for funded proposals from EU or other bodies.

The various nodes in the MACSI-net network were each fostering general and specific expertise in areas of mathematics and computing. The role of industrial nodes was somewhat complementary to the academic ones. The network was aiming at the dissemination of ideas, models and algorithms to their mutual benefit, leading to joint research efforts and forging of (often thinly spread) local initiatives. In particular, joint research proposals were expected to make this network attractive for all involved.

MACSI-net was very successful during its 4 years of existence. In the end, there were 17 working groups concentrating on a large variety of topics. Some of these working groups are still active, in a different form, but the researchers have remained in contact. An example of this is working group 2 on Coupled problems and Model Order Reduction. It actually split into two different communities, one of these active within ECCOMAS and organizing biennial conferences on coupled problems (see, for example, [5]). The other group remained concentrated on model order reduction, and has recently been awarded a European COST Action that can be used to coordinate all research in the area that is taking place in Europe [6].

At the end of its lifetime, in 2004, MACSI-net issued an important document which was one of the first reports on industrial mathematics with guidelines and recommendations. Some quotes from this document:

- *Mathematics should be regarded as a technology in its own right. Its crucial role in many industrial problems requires the active participation of mathematicians. Truly multidisciplinary projects will benefit significantly from the involvement of mathematical modellers and this should be encouraged by future funding programmes. Consideration should be given to making the participation of mathematicians in appropriate multidisciplinary projects a condition of project funding.*
- *There is a need for positive action to promote the increased use of mathematics by European industry. The success of local initiatives where mathematicians are working on industrially relevant problems is clear evidence that they are already making a significant contribution to the development of the knowledge-based economy. However, more needs to be done to encourage companies, especially Small and Medium-sized Enterprises (SMEs), to make use of mathematics and mathematicians. Consideration should be given to creating a programme funding projects that will enable companies, especially SMEs, to explore areas where mathematics can make a contribution to their improved competitiveness.*
- *There is an urgent need for more training in the area of industrial mathematics. It is essential to attract bright students to this area and to convey the challenge and the excitement of solving practical problems. Consideration should be given to specific funding for training programmes in industrial mathematics across Europe.*

MACSI-net ended in 2004, but the acronym is still in use in fact. At Limerick University, Prof. Stephen O'Brien attracted funding from the Science Foundation Ireland and is running a project termed MACSI [7], which is a network of mathematical modellers and scientific computational analysts based in Ireland. Its aim is to foster new collaborative research, in particular on problems that arise in industry, in order to produce world-class publications on mathematical modelling.

4 A Renowned Institute for Mathematics and Industry

One of the founding fathers of ECMI, mentioned already, was Prof. Helmut Neunzert, who was also the key driving force behind the creation and subsequent success of the Fraunhofer Institute for Industrial Mathematics (ITWM) that started in Kaiserslautern in the middle of the 1990s. On their website [8], one can find the following remark which encapsulates the essence of the role and importance of industrial mathematics: "The core competence of ITWM is mathematics: the language used by scientists and engineers to formulate models for technical systems. In our time it is particularly important, as it provides efficient algorithms to compute and analyse such models. The ITWM 's mission is to develop this technology to give innovative impulses and put them into practice together with industry partners .Since its foundation in 1995 the ITWM has shown great success in building mathematical bridges between applied sciences and concrete application. Clients are large international companies as well as small and medium regional enterprises. Fraunhofer ITWM focuses on the development of mathematical applications for industry, technology and economy. Mathematical approaches to practical challenges are the specific competences of the institute and complement knowledge in engineering and economics in an optimal way. In 2001 ITWM became the first mathematical oriented institute of the Fraunhofer Gesellschaft. The main emphases are surface quality inspection, financial mathematics, visualization of large data sets, and optimization of production processes, virtual material design and analysis of 3D models of microstructures."

ITWM is an example of how mathematics can successfully be turned into a business. Since ITWM's foundation, its budget has increased by more than five times: beginning with 1,64 million € in 1995, it reached 21 million € in 2012. Nearly 75 % of the operating budget stem from the institute's own profits. At present, ITWM's personnel consist of more than 250 employees, of which 60 are PhD students. They are supported by about 200 research assistants. The proportion of women involved has increased significantly in the last years. It is now at 13 % for the scientists and 26 % for the PhD students.

The success of ITWM has also been observed by others, and by now there are various, small and larger, companies that obtain their business from the application of mathematics to industrial problems. An example is the Laboratory for Industrial Mathematics Eindhoven (LIME) [9] in the Netherlands that originally started at Eindhoven University of Technology, but soon after became an independent company.

5 Captains of Industry Reporting on Mathematics

The book *Mathematik-Motor der Wirtschaft* [10] came about in close cooperation between the Oberwolfach Foundation and the Mathematisches Forschungsinstitut Oberwolfach and features articles by renowned business figures. It was launched by the German Federal Minister of Education and Research, Annette Schavan, at a gala event. Oberwolfach is well known for its workshops on mathematics, but this event was a very special one, involving many captains of industry outlining their opinion about mathematics and its utilization in their companies.

In their articles, various heads of major German companies—Allianz, Daimler, Lufthansa, Linde, and TUI, to name but a few—sum it up in a nutshell: Mathematics is everywhere, and our economy would not work without it. SAP's CEO, Henning

Kagermann, puts it like this: "Corporate management without mathematics is like space travel without physics. Numbers aren't the be all and end all in business life. But without mathematics, we would be nothing."

6 The OECD Report on Mathematics and Industry

While ECMI continued to attract new members and spread its activities further across Europe, including also countries in the eastern part, in Heidelberg the idea came up to use the experience gained in several countries to start a series of discussions and produce a report on mathematics for industry. To this end, the initiator, Prof. Willi Jäger who heads the institute IWR (Interdisciplinary Center for Scientific Computing) [11], suggested the idea to the OECD.

Recognising the importance of mathematics in an industrial context, the delegates to the Global Science Forum (GSF) of the Organization for Economic Co-operation and Development (OECD) agreed to sponsor an international consultation to assess the present state of this interface in the participating countries and to identify mechanisms for strengthening the connection between mathematics and industry (The interaction between mathematics and other sciences was left for future consideration.)

A workshop on "Mathematics in Industry" was then held in Heidelberg in early 2007. The objectives of the workshop were to

- analyse the relationship between the mathematical sciences and industry in the participating countries;
- identify significant trends in research in the mathematical sciences in academia and the mathematical challenges faced by industry in the globalised economic environment, and to analyse the implications of the trends for the relationships between mathematical scientists in academia and industry;
- identify and analyse major challenges and opportunities for a mutually beneficial partnership between industry and academia; and
- formulate action-oriented practical recommendations for the main stakeholders: the community of mathematical scientists, participating industries, and governments.

The report [12] summarised the deliberations, and presented the findings and recommendations of the workshop which will involve further consultations among the participants. The recommendations involved the participation of the academic community, governmental and other funding agencies as well as industry. They were designed to stimulate the interaction between mathematics and industry; to enhance the curriculum for students of mathematics; to improve the infrastructure for increased interactions, both in academia and in industry; and to strengthen coordination and cooperation at national and international levels.

As a follow-up, the OECD also supported an activity that was intended as a corollary to the report "Mathematics-in-Industry". It primarily comprises a factual compendium [13] of the ways in which the various mechanisms cited in that report have been implemented around the world. The compendium, which is not comprehensive, has been compiled with the aim of helping governments, industries and academia to see how they may best exploit mathematics as an industrial resource for both research and training.

7 The Forward Look Initiative of ESF and EMS

Although the reports commissioned by the OECD were a very valuable asset to the mathematics community, it was felt that an even more in-depth understanding of the problems was necessary. Indeed, the impact of mathematics in industry and society had been the subject of numerous studies, but it was decided at the end of 2009 to start a Forward Look project on mathematics and industry, evolving from the belief that European Mathematics as a whole has the potential to boost European knowledge-based innovation, which is essential for a globally competitive economy. The project was fostered by the European Science Foundation and the European Mathematical Society, and involved many members of ECMI, as it was felt they had the experience and knowledge to be able to implement such an activity.

The Forward Look at mathematics and industry sprung from the strong belief that European Mathematics has the potential to be an important economic resource for European industry, helping its innovation and hence its capacity of competing on the global market. To fulfil its potential, special attention has to be paid to the reduction of the geographical and scientific fragmentation in the European Research Area. Overcoming this fragmentation will require the involvement of the entire scientific community. Europe needs to combine all experiences and synergies at the interface between mathematics and industry and create strong areas of interaction to turn challenges into new opportunities.

The project started in 2009, and working groups were set up to discuss the main issues identified. An extensive survey was carried out to identify whether the topics worked on in academic circles reflected the needs of industry. In the figure below, this is illustrated. It confirms that, apart from a very small number of exceptions, mathematicians are indeed doing valuable work in areas of importance in industry (Fig. 3).

The project also organized alignment and consensus conferences, involving many researchers and industrialists from all over Europe, so that the conclusions in the final report [14] were broadly supported and adopted. The final recommendations were:

- Recommendation 1: Policy makers and funding organisations should join their efforts to fund mathematics activities through a European Institute of Mathematics for Innovation.
- Recommendation 2: In order to overcome geographical and scientific fragmentation, academic institutions and industry must share and disseminate best practices across Europe and disciplines via networks and digital means.

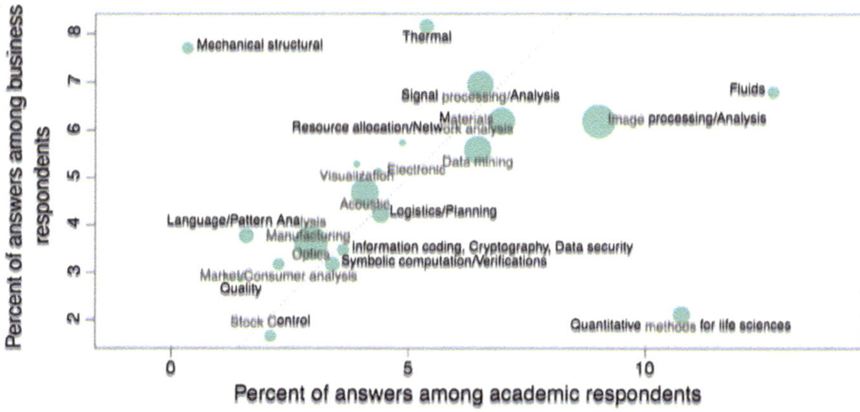

Fig. 3 Main areas of competence available in academia versus major business challenges perceived by industry (size of the bubbles indicates total number of respondents)

- Recommendation 3: Mathematical Societies and academic institutions should create common curricula and educational programmes in mathematics at European level taking into account local expertise and specificity.

Besides these recommendations, the report also gives roadmaps for their implementation.

8 The Network of Networks EU-MATHS-IN

Even though the recommendations from the aforementioned forward look report were widely accepted, it turned out to be quite hard to obtain sufficient support to implement them in practice. Therefore, in 2013, it was decided to take the initiative in our own (mathematical) hands, and start a new organisation in Europe that would enable cross-fertilisation and exchange of best practice. Collaboration provides a much better basis for funding of European organisations. Consequently, at the end of 2013, EU-MATHS-IN was launched in Amsterdam [15]. It is a European service network of mathematics for industry and innovation. As stated on the website: "A new initiative to boost mathematics for industry in Europe. Make the most of our expertise for a more efficient route to innovation!"

EU-MATHS-IN aims to leverage the impact of mathematics on innovations in key technologies through enhanced communication and information exchange between and among the involved stakeholders at a European level. It aims to become a *dedicated one-stop shop* to coordinate and facilitate the required exchanges in the field of application-driven mathematical research and its exploitation for innovations in industry, science and society. For this, it aims to build an *e-infrastructure* that provides tailored access to information and facilitates communication and exchange by

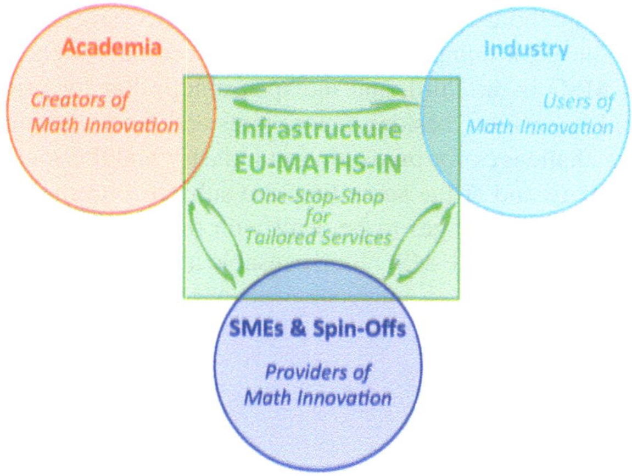

Fig. 4 Graphical illustration of the strategy of EU-MATHS-IN

player-specific sets of *services*. It will *act as facilitator, translator, educator and link* between and among the various players and their communities in Europe. In the figure below, a graphical illustration is given of the intended structure (Fig. 4).

The important features of the long term goals of the organisation are:

- Establish strategic connections among the national networks and centers working in the field of industrial mathematics and mathematics for innovation;
- Create a European service unit that can foster the competitive advantage of the European industry through international cooperation;
- Promote the technological aspects of mathematics raising public awareness;
- Stimulate the cooperation at European level of mathematical research with companies and administrations;
- Establish a one-stop-shop at European level for industrial users of mathematical scientific research results;
- Provide European industry, in particular SMEs, with a competitive advantage taking profit of the scientific excellence of the continent (give Europe the possibility to cash a "scientific dividend");
- Acquire funding for the performance of activities that serve the realisation of the Association's aims.

It is felt that, with EU-MATHS-IN, Europe has a powerful new organisation that will be able to bring together all national initiatives, such as those that have arisen in many European countries, to learn from each other, to share experiences and together form a community that is recognized for its capability of bridging the gap between mathematics and industry.

One of the first initiatives of the organisation is to strongly call for the establishment of Mathematical modelling, simulation and optimization (MSO) as a transversal (universal?) Key Enabling Technology (KET). The arguments are as follows. There is no doubt that continuous multidisciplinary research and novel mathematical and computational methods are needed to provide the necessary tools for industrial innovation and European competitiveness. It has become widely recognized that the approach of MSO is the third, and indispensable, pillar for scientific progress and technological innovation, besides experiments and theory building. Experience shows that future challenges for innovation in industry and the society will involve increasing complexity and at the same time are subject to ever-shorter innovation cycles. The real-world challenges to be dealt with on our way towards innovation exhibit opportunities that make MSO indispensable and at the same time, a far from trivial task.

9 Conclusion

Europe has always been very active in trying to bridge the gap between mathematics and industry. Already since the 1970s, when the Oxford study groups with mathematics started to be held, mathematicians realized the potential for breakthroughs and innovations in industry and for societal problems. In this paper, we have given a chronological, but probably not entirely complete, picture of what has happened in Europe since the 1970s. We observe a very natural and continuous growth of activities, stepping up intensity over the years. In recent years, a strong linking of mathematicians and mathematics with industry personnel and problems has beeb occurring with the importance of industrial mathematics being realized. This is confirmed by recent reports issued by Deloitte, both in the UK [16] and very recently in the Netherlands [17], stating that about 30 % of gross domestic product added can be attributed to the results of research in the mathematical sciences. This is an enormously large percentage, and it will hopefully wake up politicians and policy makers to investing more into mathematics. For example, in the Netherlands, investigations by mathematicians into the need for increasing the heights of dikes have led to an alternative plan saving more than 4 billion Euro as compared to a plan made by a committee without mathematicians. It once again proves that mathematics is everywhere, and invaluable.

The foregoing leads to the natural question: how will our world look like in 2025? And what will be the role of the mathematical sciences in shaping that world? Since the start of the 21st century, it has become clear that the mathematical sciences are gaining a new stature. They are increasingly providing the knowledge to enable innovative breakthroughs and insights in many other disciplines such as biology, healthcare, social sciences and climatology, alongside their traditional role in physics, chemistry and computer science. The importance of the mathematical sciences is also rapidly increasing in the business world, for example in design processes, electronics and finance. All these developments are vital for economic growth and competitive strength, and demand an in-depth review of the overall way we look at the mathematical sciences. This involves the integration of mathematics with statistics, operations research and computational science, and it carries implications for the nature and scale of research funding.

References

1. http://www.ecmi-indmath.org
2. http://www.mafy.lut.fi/EcmiNL/issues.php?issueNumber=52
3. http://www.macsinet.org
4. http://www.eccomas.org
5. http://congress.cimne.com/coupled2013/
6. http://www.cost.eu/td1307
7. http://www.macsi.ul.ie/
8. http://www.itwm.fraunhofer.de/
9. http://www.limebv.nl/
10. http://www.springer.com/mathematics/book/978-3-540-78667-2
11. http://www.iwr.uni-heidelberg.de/
12. http://www.oecd.org/dataoecd/47/1/41019441.pdf July 2008
13. http://www.oecd.org/science/sci-tech/42617645.pdf
14. https://www.ceremade.dauphine.fr/FLMI/
15. http://www.eu-maths-in.eu
16. http://www.epsrc.ac.uk/SiteCollectionDocuments/Publications/reports/DeloitteMeasuringThe EconomicsBenefitsOfMathematicalScienceResearchUKNov2012.pdf
17. http://www.platformwiskunde.nl

Visualizing Multivariate Data Using Singularity Theory

Osamu Saeki, Shigeo Takahashi, Daisuke Sakurai, Hsiang-Yun Wu, Keisuke Kikuchi, Hamish Carr, David Duke and Takahiro Yamamoto

Abstract This is a survey article on recent developments in visualization of large data, especially that of multivariate volume data. We present two essential ingredients. The first one is the mathematical background, especially the singularity theory of differentiable mappings, which enables us to capture topological features of given multivariate data in a mathematically rigorous way. The second one is a new development in computer science, called the joint contour net, which can encode topological

O. Saeki (✉)
Institute of Mathematics for Industry, Kyushu University, Motooka 744, Nishi-ku, Fukuoka 819-0395, Japan
e-mail: saeki@imi.kyushu-u.ac.jp

S. Takahashi · D. Sakurai · H-Y. Wu · K. Kikuchi
Graduate School of Frontier Sciences, The University of Tokyo, 5-1-5 Kashiwanoha, Kashiwa, Chiba 227-8561, Japan
e-mail: shigeo@visual.k.u-tokyo.ac.jp

D. Sakurai
e-mail: sakurai@visual.k.u-tokyo.ac.jp

H-Y. Wu
e-mail: yun@visual.k.u-tokyo.ac.jp

K. Kikuchi
e-mail: kikuchi@visual.k.u-tokyo.ac.jp

H. Carr · D. Duke
School of Computing, University of Leeds, Leeds LS2 9JT, Leeds, UK
e-mail: h.carr@leeds.ac.uk

D. Duke
e-mail: d.j.duke@leeds.ac.uk

T. Yamamoto
Faculty of Engineering, Kyushu Sangyo University, 3-1 Matsukadai 2-chome, Fukuoka 813-8503, Japan
e-mail: yama.t@ip.kyusan-u.ac.jp

M. Wakayama et al. (eds.), *The Impact of Applications on Mathematics*,
Mathematics for Industry 1, DOI: 10.1007/978-4-431-54907-9_4,

structures of a given set of multivariate data in an efficient and robust way. Some applications to real data analysis are also presented.

Keywords Multivariate data · Singular fiber · Differential topology · Joint contour net · Reeb space · Data visualization · Jacobi set · 3-Manifold with boundary

1 Introduction

This is a survey article describing recent developments in visualization of multivariate data, especially from the viewpoint of singularity theory in differential topology. We will also present a new development in computer science which is adapted to visualize multivariate data efficiently.

In scientific situations, a set of large data, obtained by a simulation or an experiment, can often be considered to be a discrete set of sample values of a differentiable mapping between Euclidean spaces, or between manifolds. Therefore, in data visualization, analyzing discrete sample points for mappings between n and m-dimensional spaces has been considered important for a long time and studied extensively. As one of the important ingredients for such studies, we have the method of extracting singularities of mappings using *differential topology*. This kind of an idea started to appear in the literature in the 1990's. Nowadays it is drawing considerable attention as one of the most striking recent innovations in data visualization, and its power of expression has been strengthened remarkably. The research in data visualization so far has been focused on differential topological analysis of 2- or 3-dimensional scalar fields (i.e., $n = 2$ or 3 and $m = 1$); recently its generalization to the case of higher dimensional domains has been developed.

Our main purpose of this article is to analyze multivariate functions, especially in the case $n = 3$ and $m = 2$, from differential topological viewpoints, which will be an important issue from now on. For visual data analysis of such multivariate functions it will be necessary to study the topology of singularities of differentiable mappings using the mathematical theory of singularities.

The paper is organized as follows. In Sect. 2, we recall some basic materials from singularity theory of differentiable mappings. The notion of a fiber and that of a Reeb space will be introduced, which will play essential roles in this article. In Sect. 3, we summarize the existing method for visualizing 3-dimensional scalar function data (i.e., the case of $n = 3$ and $m = 1$), presenting some explicit applications to real data. In Sect. 4, we present a new development in computer science, called the joint contour net, which has been recently developed for visualizing multivariate data in a robust and efficient way. In Sect. 5, based on the singularity theory of differentiable mappings, we classify the topological types of fibers for mappings of 3-dimensional spaces to 2-dimensional ones. This classification can then be used to identify the fiber types for a given set of data, with the help of the joint contour net. Finally in Sect. 6, we give some explicit examples of applications of our techniques, including the case of a simple analytic mapping together with the case of hurricane data. We also mention that this kind of techniques can help to explore or discover new phenomena in singularity theory itself.

2 Preliminaries

Let N be a compact orientable 3-dimensional manifold of class C^∞ possibly with boundary: for example, a closed bounded domain enclosed by a smooth surface in $\mathbf{R}^3$ is such an example. If the reader is not familiar with the theory of differentiable manifolds, then N can be safely assumed to be such a domain in the following.

Let $f : N \to \mathbf{R}^2$ be a differentiable mapping of class C^∞. The manifold N is often called the *spatial domain* (or *domain*) and $\mathbf{R}^2$ the *data domain* (or *range*). Giving such a mapping is equivalent to giving two differentiable functions $f_i : N \to \mathbf{R}$, $i = 1, 2$, so that we have $f = (f_1, f_2)$. In this sense, such a mapping f as above is often called a *multivariate function* or a 2-*variate function.*

We will later assume that data (of the f-values) are given at a discrete set of points in the spatial domain. However, for the moment, we consider a differentiable mapping as above in order to introduce some theoretical concepts.

For a value (or rather, a point) $c \in \mathbf{R}^2$ in the data domain, the set $f^{-1}(c) = \{x \in N \mid f(x) = c\}$ is called a *fiber* (for details, see [17]). It is sometimes called a *level set*, which is commonly used for the case of scalar functions $N \to \mathbf{R}$. In this article, in order to emphasize that we are treating the case of multivariate functions, we use the terminology "fiber" rather than "level set". Generically, a fiber of a mapping $f : N \to \mathbf{R}^2$ constitutes a union of curves (with singularities, in general) and is also called an *isoline*. For the analysis of the multivariate data structure given by the mapping f, it is important to visualize the data in such a way that the structure of the fibers are clearly encoded.

For a point $x \in N$, let $df_x : T_x N \to T_{f(x)}\mathbf{R}^2$ be the differential of f at x, where $T_x N$ and $T_{f(x)}\mathbf{R}^2$ are the tangent spaces of N and $\mathbf{R}^2$ at x and $f(x)$, respectively. In other words, df_x can be identified with the linear mapping $\mathbf{R}^3 \to \mathbf{R}^2$ represented by the Jacobian matrix of f with respect to local coordinates around x for N. A *singular point* is a point $x \in N$ with rank $df_x < 2$. The set of all singular points is denoted by $J(f)$ and is called the *Jacobi set* or *singular point set* of f (see [5]). It is known that generically, $J(f)$ forms a smooth 1-dimensional curve in N without singularities. We can consider the mapping $f_\partial = f|_{\partial N} : \partial N \to \mathbf{R}^2$ obtained by restricting f to the boundary surface of N, and define its Jacobi set $J(f_\partial)$ in a similar way.

Fibers that pass through singular points of f or f_∂ are called *singular fibers* (see Fig. 1). Singular fibers play an important role in extracting topological features of given data.

For extracting topological features or topological structures of a given mapping, the following concept plays an essential role as well.

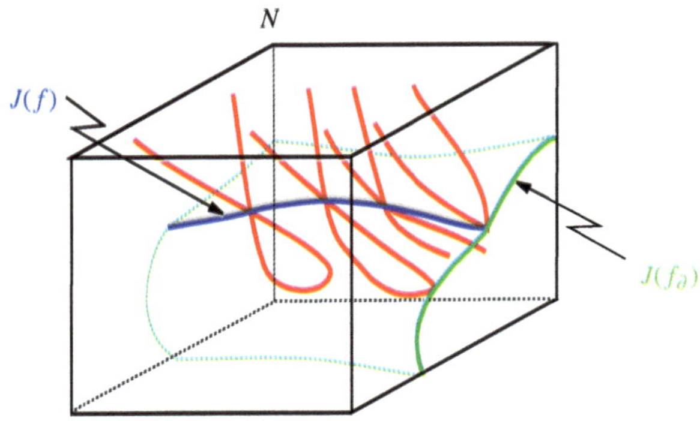

Fig. 1 Example of Jacobi sets and singular fibers: the red curves represent singular fibers, which have singular points along the Jacobi sets $J(f)$ or $J(f_\partial)$

Definition 2.1 Let $f : N \to \mathbf{R}^2$ be a differentiable mapping of a 3-dimensional manifold N into the plane. Two points $x, x' \in N$ are *f-equivalent* if $f(x) = f(x')$ $(= c \in \mathbf{R}^2)$, and x and x' belong to the same connected component of the fiber $f^{-1}(c)$. This defines an equivalence relation on N, and we denote by W_f the quotient space of N with respect to f-equivalence, where W_f is endowed with the quotient topology induced by the quotient mapping $q_f : N \to W_f$. We call the space W_f the *Reeb space* of f (see [6]). It is then easy to see that there exists a unique continuous mapping $\bar{f} : W_f \to \mathbf{R}^2$ that makes the following diagram commutative:

$$\begin{array}{ccc} N & \xrightarrow{\quad f \quad} & \mathbf{R}^2 \\ & {}_{q_f}\searrow \quad \nearrow_{\bar{f}} & \\ & W_f. & \end{array}$$

The above commutative diagram is called the *Stein factorization* of f (see [13]).

For the case of a scalar function $h : N \to \mathbf{R}$, one can define the Reeb space in exactly the same way, and the resulting space is known as the *Reeb graph* of h (see [16]). An example for a 2-dimensional scalar function is shown in Fig. 2. It is known that for a scalar function h, the Reeb space is actually a graph, consisting of vertices and edges, provided that the function h is generic enough.

The Reeb graph is indispensable for visualizing 3-dimensional scalar functions. In fact, its vertices correspond to the critical points of h, and it can encode the topological transition of the level sets around each critical point (for example, see Fig. 3).

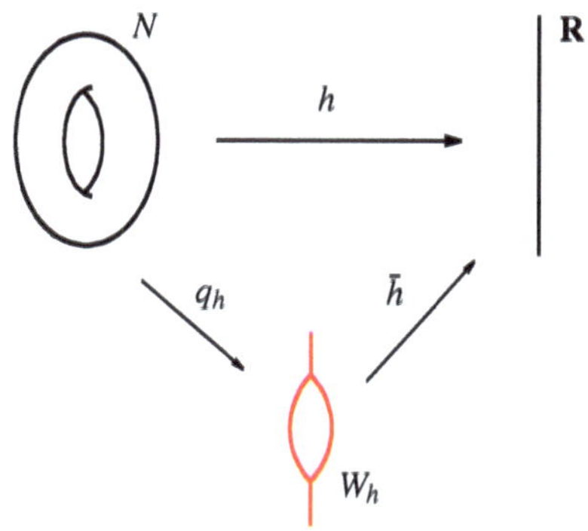

Fig. 2 Example of a Reeb graph: the vertices of the Reeb graph W_h correspond to the critical points of the scalar function h

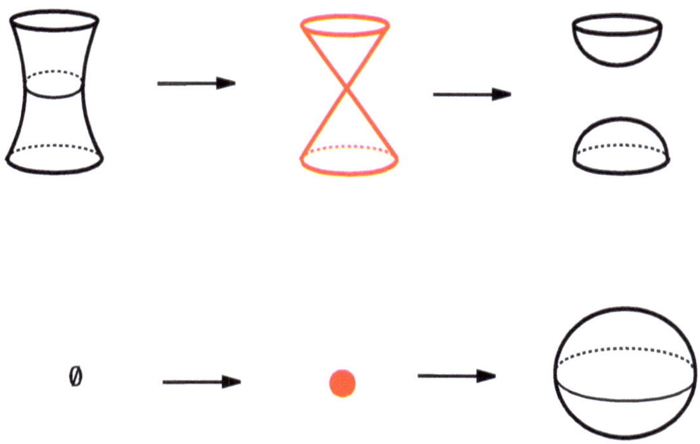

Fig. 3 Examples of level-surface change for a 3-dimensional scalar function: the red figures represent level surfaces containing critical points, and around such surfaces, the topology of level surfaces changes

3 Visualization of Scalar Functions

Before discussing visualization of multivariate data, let us present some explicit examples of application of the Reeb graphs and topological transitions of the level-surfaces to the visualization of real data.

Given a set of volume data, we can analyze it to get the corresponding Reeb graph. Nowadays, we can compute such a graph quite rapidly and in a robust way (for example, see [3, 9, 15]). Then, the resulting graph enables us to design the transfer function automatically in such a way that the function emphasizes the values where topological changes occur. Finally, this transfer function is used to render the visualization result (see Fig. 4). Such a technique is called *direct volume rendering* and is thoroughly studied in [21–23].

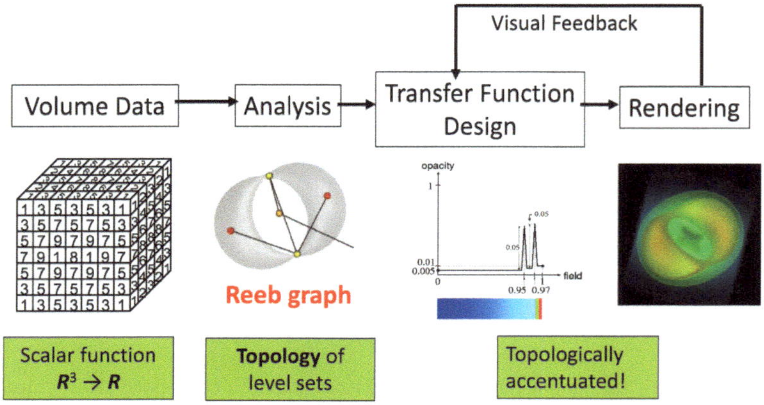

Fig. 4 Direct volume rendering: a given set of large volume data, which usually has complicated structures and may contain noise, can be analyzed using differential topology, and then the transfer function is designed automatically in such a way that topologically important level surfaces are accentuated in the visualization result

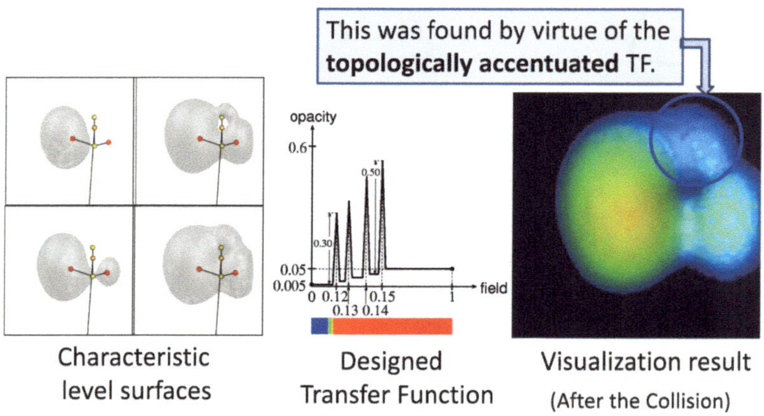

Fig. 5 Proton and hydrogen atom collision: accentuated characteristic surfaces help us to detect a special event

For example, in [8], this technique is applied to visualize a set of simulation data for the electron density function during a proton and hydrogen atom collision. This is a spatio-temporal case and corresponds to dimension four. However, for each fixed time T, one can analyze the volume data, and by varying T, one can extract the Reeb graph change to detect the characteristic time. By this method, we can effectively observe the electron density change during the collision and can extract the features of the data efficiently, much better than a simple movie visualization (see Fig. 5).

4 Joint Contour Net

Let us now turn to the case of multivariate functions. For computational visualizations of multivariate data, several studies have been known. Some of the known results are summarized in the state-of-the-art report [7].

In [1], the concept introduced as a refinement of scatterplots for discrete data values. Using such a technique, one can more or less grasp the curve of the Jacobi set image in the data domain. In [12], a more sophisticated algorithm for detecting the Jacobi set image has been introduced, and the problem of counting the number of singular fiber components has been raised, although they did not give an answer to it. Any way, the authors did not identify the link between their approach and fiber analysis.

In this section, as a new development in computer science for treating the case of multivariate data, we use the concept of a joint contour net (JCN, for short) [2].

In order to visualize the data given by a multivariate function $f : N \to \mathbf{R}^2$ from a 3-dimensional manifold N, it is essential to visualize the following.

(a) Images of $J(f)$ and $J(f_\partial)$ by f as singular curves in the data domain.
(b) Type of the singular fiber associated to each edge and each singular point of the curves (a) above.

The image curves (a) correspond to the loci where the fibers change their topology. The type of the singular fiber (b) tells us how the fibers change their topology around the corresponding edge or singular point.

For these purposes, we use the concept of a joint contour net [2], which decomposes the spatial domain into regions of equivalent behavior. In the following, we assume that f-values are given at a discrete set of points in the spatial domain. The main idea of JCN is that we quantize the f-values. Instead of taking a point $c \in \mathbf{R}^2$, we consider a small pixel $P \subset \mathbf{R}^2$. Instead of a fiber $f^{-1}(c)$, we consider a fattened fiber $f^{-1}(P)$. In this way, we can identify singular fibers in a robust way, since fattened fibers contain essential information on its central fiber.

The main idea of the construction of the JCN is as follows. We take a tetrahedral mesh of the spatial domain and a rectangular mesh of the data domain. Then, we decompose the tetrahedra in the domain into smaller pieces according to their (quantized) f-values. More precisely, for each pixel P in the range, we assign the set $T(P)$ of tetrahedra in the domain which contains a point mapped into P (see Fig. 6).

Then, we unite neighboring pieces that have the same (quantized) f-value. More precisely, at each pixel P, we put a node to each subset of $T(P)$ that constitutes a connected component of the union of the tetrahedra in $T(P)$. Such a node corresponds to a connected component of the inverse image of P, which can be considered to be a component of a fattened fiber (see Fig. 7).

Finally, we encode the adjacency information of the fattened fibers by edges. More precisely, if two nodes are in neighboring pixels and they share a common tetrahedron, then we put an edge that connects the nodes (see Fig. 8).

In this way, we get a graph called *joint contour net*, which describes the adjacency relations among the connected components of fattened fibers. This graph then

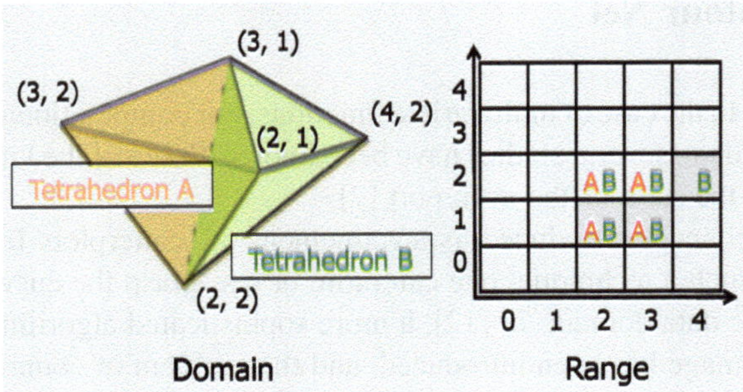

Fig. 6 For each pixel in the range, a set of tetrahedra is assigned

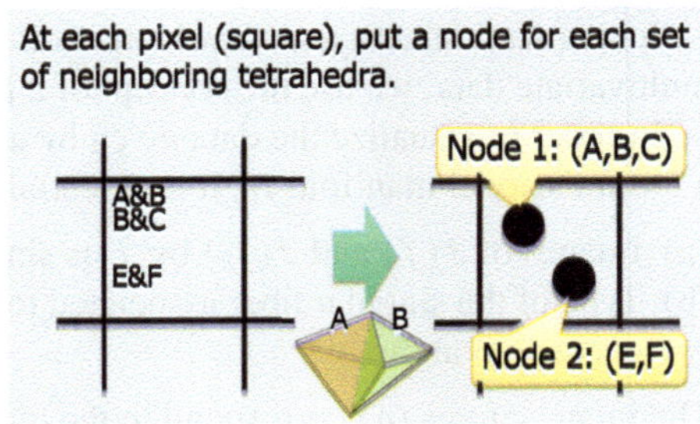

Fig. 7 We put a node corresponding to each component of a fattened fiber

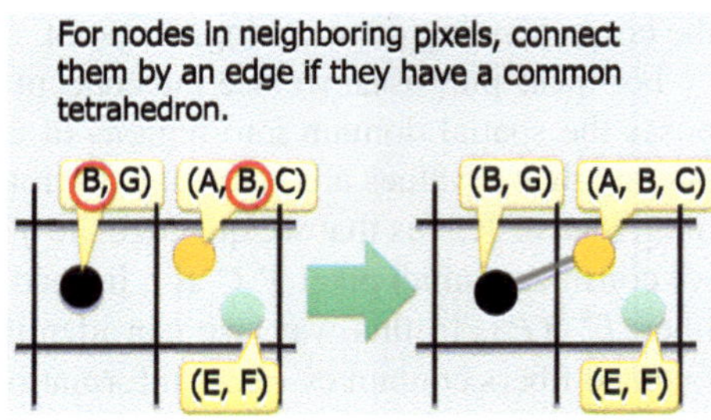

Fig. 8 Adjacent nodes are connected by edges

approximates the Reeb space of f. Therefore, the joint contour net can then be used to detect birth-death or split-merge of fibers (see Fig. 9).

In this way, we can completely extract the information on the number of connected components of fattened fibers, and the adjacency relations among the components.

We can also construct a similar object for the mapping $f|_{\partial N} : \partial N \to \mathbf{R}^2$ on the boundary. The resulting JCN will tell us how the fibers of $f : N \to \mathbf{R}^2$ intersect with the boundary.

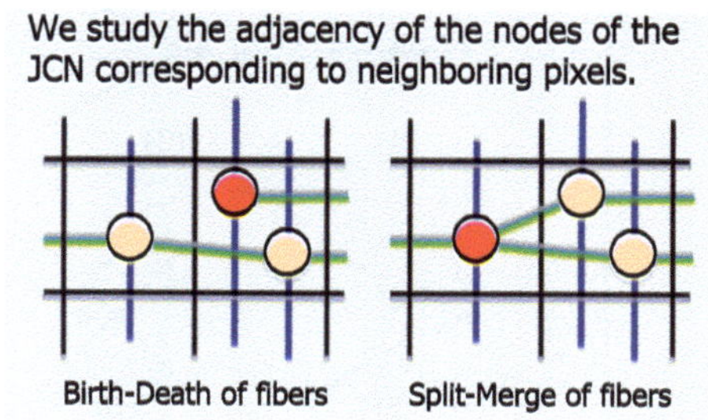

Fig. 9 JCN detects birth-death or split-merge of fibers. Green horizontal edges and blue vertical ones connect adjacent nodes. As a consequence, we obtain yellow nodes that are adjacent to four neighboring nodes, and the remaining red nodes correspond to topological change of fibers

5 Fiber Types

By getting the JCN of a given set of multivariate data, we can identify the Jacobi set image in the range. Then, as we have seen before, our next task is to identify the fiber types for each point in the range. For this purpose, we utilize the theory of singularities of differentiable mappings.

Let N be a 3-dimensional compact manifold possibly with boundary and $f : N \to \mathbf{R}^2$ a generic differentiable mapping, where, in mathematical terminology, "generic" here means "C^∞ stable" (for a precise definition of a C^∞ stable mapping, see [10], for example). By using results of [14, 20], we have the following characterization of C^∞ stable mappings $f : N \to \mathbf{R}^2$.

Proposition 5.1 *Let N be a 3-dimensional compact manifold possibly with boundary. A C^∞ mapping $f : N \to \mathbf{R}^2$ is C^∞ stable if and only if it satisfies the following conditions.*

1. *(Local conditions) In the following, for $p \in \partial N$, we use local coordinates (x, y, z) for N around p such that* Int N *and ∂N correspond to the sets $\{z > 0\}$ and $\{z = 0\}$, respectively.*

   (1a) *Around each point $p \in$* Int N, *there exist appropriate local coordinates such that f is described by one of the following normal forms:*

$$(x, y, z) \mapsto \begin{cases} (x, y), & p\text{: regular point,} \\ (x, y^2 + z^2), & p\text{: definite fold point,} \\ (x, y^2 - z^2), & p\text{: indefinite fold point,} \\ (x, y^3 + xy - z^2), & p\text{: cusp point.} \end{cases}$$

   (1b) *Around each point $p \in \partial N \setminus J(f)$, f can be described by one of the following normal forms:*

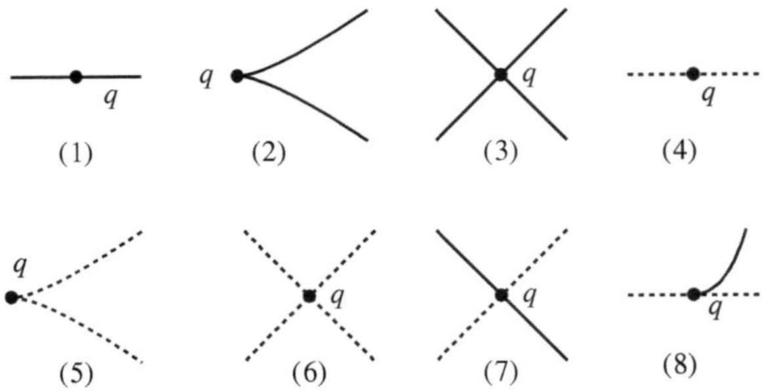

Fig. 10 Local behaviors of the mapping $f|_{J(f)\cup J(f|_{\partial N})}$: the *solid* and *dashed lines* depict the singular value sets $f(J(f))$ and $f(J(f|_{\partial N}))$, respectively

$$(x,y,z) \mapsto \begin{cases} (x,y), & p\text{: regular point of } f|_{\partial N}, \\ (x, y^2+z), & p\text{: boundary definite fold point,} \\ (x, y^2-z), & p\text{: boundary indefinite fold point,} \\ (x, y^3+xy+z), & p\text{: boundary cusp point.} \end{cases}$$

(1c) *Around each point $p \in \partial N \cap J(f)$, f can be described by the normal form*

$$(x,y,z) \mapsto (x, y^2 + xz \pm z^2).$$

2. *(Global conditions) For each $q \in f(J(f)) \cup f(J(f|_{\partial N}))$, the map $f|_{J(f)\cup J(f|_{\partial N})}$ at the points in $f^{-1}(q) \cap (J(f) \cup J(f|_{\partial N}))$ is described by one of the eight mappings as depicted in Fig. 10, where the solid lines correspond to the singular value set $f(J(f))$ and the dashed lines to $f(J(f|_{\partial N}))$:* (1) *corresponds to a single fold point,* (4) *corresponds to a single boundary fold point,* (3), (6) *and* (7) *represent normal crossings of two immersion germs, each of which corresponds to a fold point or a boundary fold point,* (2) *corresponds to a cusp point,* (5) *corresponds to a boundary cusp points, and* (8) *corresponds to a single point in $\partial N \cap J(f)$.*

In singularity theory, the natural equivalence for fibers of differentiable mappings is formulated as follows.

Definition 5.2 Let $f_i \colon N_i \to M_i$ be differentiable mappings between C^∞ manifolds, $i = 0, 1$. For $q_i \in M_i$, $i = 0, 1$, we say that the fibers over q_0 and q_1 are C^∞ *equivalent* if for some open neighborhoods U_i of $q_i \in M_i$, there exist diffeomorphisms $\Phi : f^{-1}(U_0) \to f^{-1}(U_1)$ and $\varphi : U_0 \to U_1$ with $\varphi(q_0) = q_1$ that make the following diagram commutative:

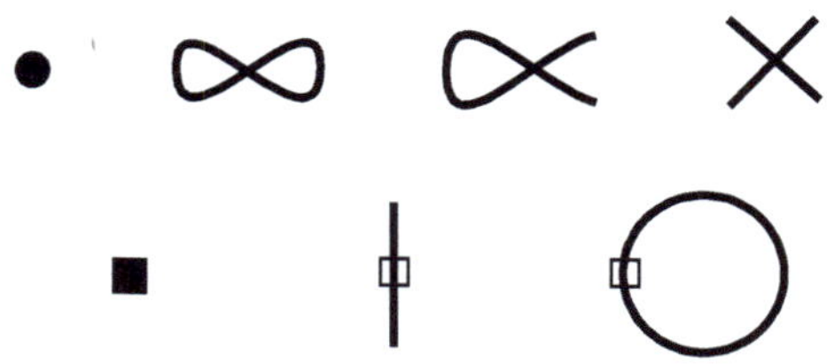

Fig. 11 Singular fibers of $\kappa = 1$: squares correspond to boundary tangency points in ∂N

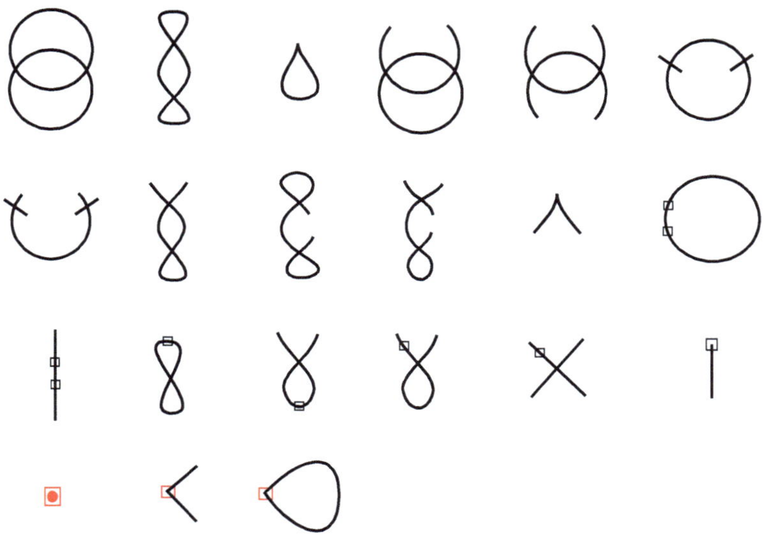

Fig. 12 Singular fibers of $\kappa = 2$: *red points* correspond to $J(f) \cap \partial N = J(f) \cap J(f_\partial)$

$$\begin{array}{ccc} (f_0^{-1}(U_0), f_0^{-1}(q_0)) & \xrightarrow{\Phi} & (f_1^{-1}(U_1), f_1^{-1}(q_1)) \\ {\scriptstyle f_0}\downarrow & & \downarrow{\scriptstyle f_1} \\ (U_0, q_0) & \xrightarrow{\varphi} & (U_1, q_1). \end{array}$$

Then, the fibers of generic differentiable mappings of 3-dimensional manifolds into the plane are classified as follows (for details, see [19]).

Theorem 5.3 *Let $f : N \to \mathbf{R}^2$ be a C^∞ stable mapping of a compact orientable 3-dimensional manifold N with boundary into the plane. Then, every component of a fiber containing a singular point is equivalent to one of the fibers in the lists in Figs. 11 and 12: there are seven fibers of codimension $\kappa = 1$, and twenty one fibers of $\kappa = 2$.*

In Figs. 11 and 12, the *codimension* κ refers to the codimension of the set of points in $\mathbf{R}^2$ whose corresponding fibers are equivalent to the relevant one (see [17]

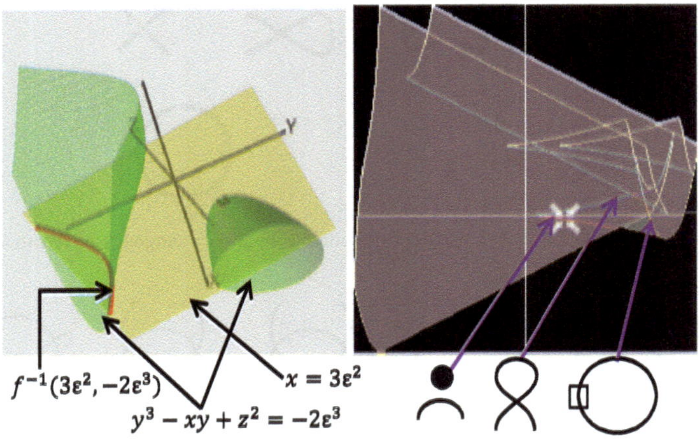

Fig. 13 Analytic multivariate function $f(x, y, z) = (x, y^3 - xy + z^2)$: we have the spatial domain in the *left*, and the data domain in the *right*. A birth-death of fibers can be observed

for details). In the lists, squares correspond to boundary tangency points in ∂N, and *red points* correspond to $J(f) \cap \partial N = J(f) \cap J(f_\partial)$. Note that the figures depict only the components of inverse images: however, mathematically, they represent mappings defined around them.

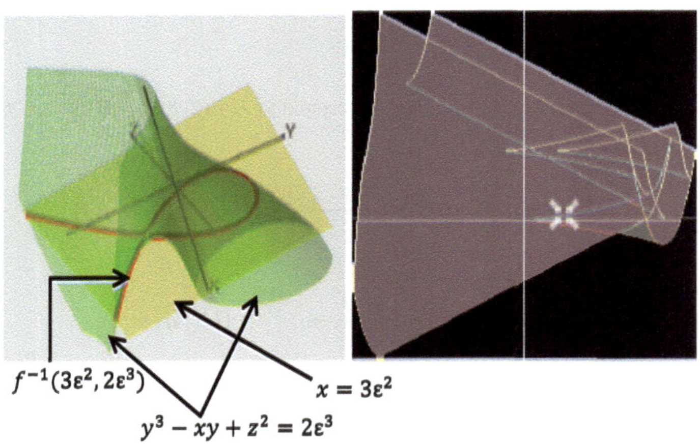

Fig. 14 The same analytic multivariate function: a split-merge of fibers can be observed

Theorem 5.3 is proved by using the relative version of Ehresmann's fibration theorem together with the analysis of the local fiber changes using the normal forms in Proposition 5.1. See [17–19] for details.

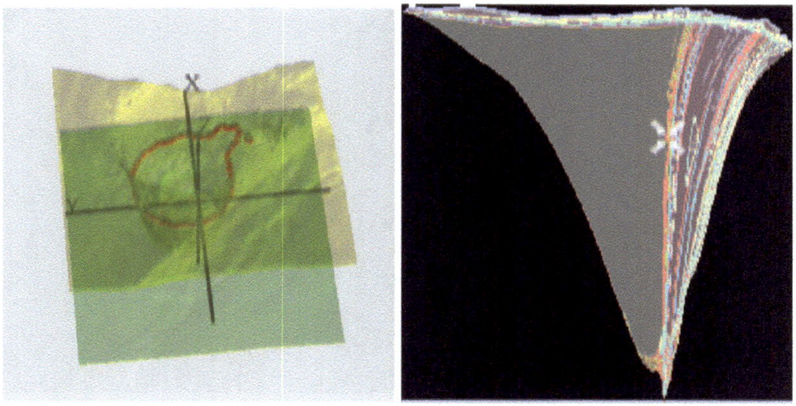

Fig. 15 Hurricane Isabel Data (1): the singular fiber in the *left* corresponds to the crossing in the *right*

6 Examples of Visualization

Using the techniques explained above, we can effectively visualize multivariate data. In this section, we show some explicit examples of applications.

As a simple example, Figs. 13 and 14 show the case of the analytic mapping $f : N \to \mathbf{R}^2$ given by

$$f(x, y, z) = (x, y^3 - xy + z^2),$$

where $N = [-1, 1] \times [-1, 1] \times [-1, 1]$. On the left hand side, we have the spatial domain where the level surfaces of the two coordinate scalar functions are depicted. On the right hand side, the images of the Jacobi set curves are depicted with colors varying according to the corresponding codimension 1 singular fibers lying over them. The black curves (and a point) on the right indicate the fibers which lie over the corresponding points in the range. The left hand side level surfaces correspond to the crossings in the data domain and their intersection red curves indicate the corresponding fibers.

As an example of an application to real data, Figs. 15 and 16 show the visualization of the fibers of the Hurricane Isabel data. We can observe characteristic curves on the right where some special phenomena apparently happen.

Another application to nuclear scission data visualization utilizing the JCN technique can be found in [4].

We can also use these techniques for investigating mathematical theories, as Fig. 17 shows. This shows the impact of our techniques also to the research of mathematics.

Acknowledgments The "Hurricane Isabel" data set was produced by the Weather Research and Forecast (WRF) model, courtesy of NCAR and the U.S. National Science Foundation (NSF). This

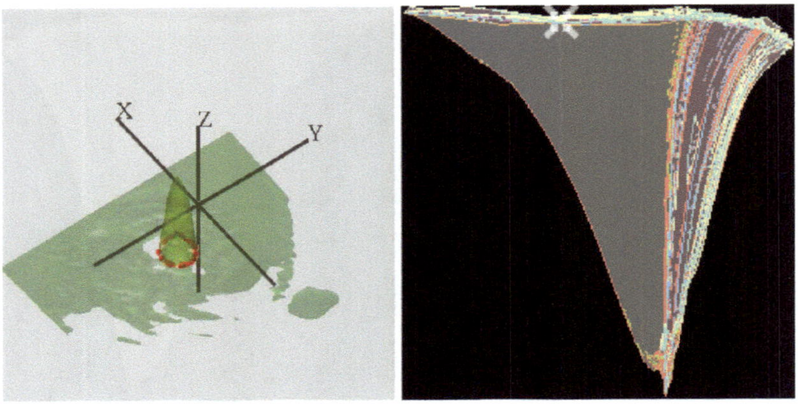

Fig. 16 Hurricane Isabel Data (2)

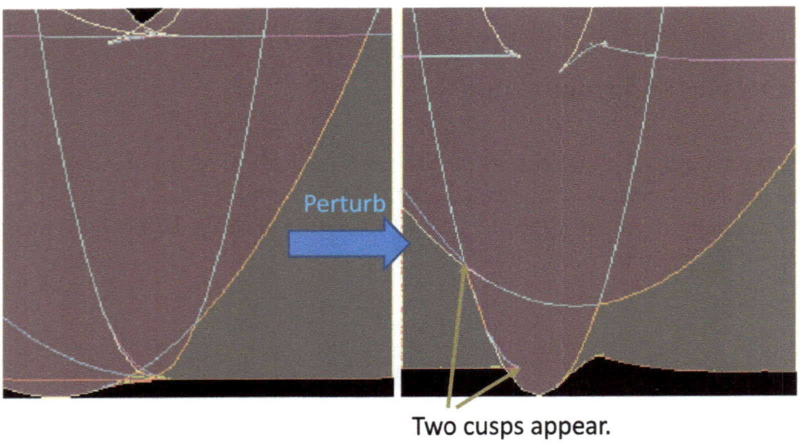

Fig. 17 This supports a theoretical result: the mapping on the *left* hand side is degenerated: however, after a perturbation, two or more cusps appear. This was predicted by a theorem [11] in singularity theory: now it has been visually verified

research has been partially supported by JSPS KAKENHI Grant Number 25540041, and EPSRC EP/J013072/1.

References

1. Bachthaler, S., Weiskopf, D.: Continuous scatterplots. IEEE Trans. Vis. Comput. Graph. **14**(6), 1428–1435 (2008)
2. Carr, H., Duke, D.: Joint contour nets. to appear in IEEE Transactions on Visualization and Computer Graphics (2013)

3. Carr, H., Snoeyink, J., Axen, U.: Computing contour trees in all dimensions. Comput. Geom. Theory Appl. **24**, 75–94 (2003)
4. Duke, D., Carr, H., Knoll, A., Schunck, N., Namh, A., Staszczak, A.: Visualizing nuclear scission through a multifield extension of topological analysis. IEEE Trans. Vis. Comput. Graph. **18**(12), 2033–2040 (2012)
5. Edelsbrunner, H., Harer, J.: Jacobi sets of multiple Morse functions. Foundations of computational mathematics: Minneapolis, 2002, London Mathematical Society Lecture Note Series, vol. 312, pp. 37–57, Cambridge Univ. Press, Cambridge (2004)
6. Edelsbrunner, H., Harer, J., Patel, A.K.: Reeb spaces of piecewise linear mappings. Proceedings of the twenty-fourth annual symposium on computational geometry, pp. 242–250 (2008)
7. Fuchs, R., Hauser, H.: Visualization of multi-variate scientific data. Comput. Graph. Forum **28**(6), 1670–1690 (2009)
8. Fujishiro, I., Otsuka, R., Takahashi, S., Takeshima, Y.: T-Map: A topological approach to visual exploration of time-varying volume data. In: Labarta, J., Joe, K., Sato, T. (eds.) High-Performance Computing. Lecture Notes in Computer Science, vol. 4759, pp. 176–190, Springer, Berlin (2008)
9. Ge, X., Safa, I., Belkin, M., Wang, Y.: Data skeletonization via Reeb graphs. Twenty-Fifth Annual Conference on Neural Information Processing Systems, pp. 837–845 (2011)
10. Golubitsky, M., Guillemin, V.: Stable mappings and their singularities. Graduate Texts in Mathematics, vol. 14, Springer (1973)
11. Ikegami, K., Saeki, O.: Cobordism of Morse maps and its application to map germs. Math. Proc. Camb. Phil. Soc. **147**, 235–254 (2009)
12. Lehmann, D.J., Theisel, H.: Discontinuities in continuous scatterplots. IEEE Trans. Vis. Comput. Graph. **16**(6), 1291–1300 (2010)
13. Levine, H.: Classifying immersions into $\mathbf{R}^4$ over stable maps of 3-manifolds into $\mathbf{R}^2$. Lecture Notes in Math, vol. 1157, Springer, Berlin (1985)
14. Martins, L.F., Nabarro, A.C.: Projections of hypersurfaces in $\mathbf{R}^4$ with boundary to planes. Glasgow Math. J. **56**(1), 149–167 (2014)
15. Pascucci, V., Scorzelli, G., Bremer, P.T., Mascarenhas, A.: Robust on-line computation of Reeb graphs: Simplicity and speed. ACM Trans. Graph. 26, No. 3, (2007), Article 58, 58.1–58.9
16. Reeb, G.: Sur les points singuliers d'une forme de Pfaff complètement intégrable ou d'une fonction numérique. C. R. Acad. Sci. Paris **222**, 847–849 (1946)
17. Saeki, O.: Topology of singular fibers of differentiable maps. Lecture Notes in Math, vol. 1854, Springer (2004)
18. Saeki, O.: Cobordism of Morse functions on surfaces, universal complex of singular fibers, and their application to map germs. Algebr. Geom. Topol. **6**, 539–572 (2006)
19. Saeki, O., Yamamoto, T.: Singular fibers of stable maps of 3-manifolds with boundary into surfaces and their applications. preprint (2014)
20. Shibata, N.: On non-singular stable maps of 3-manifolds with boundary into the plane. Hiroshima Math. J. **30**, 415–435 (2000)
21. Takahashi, S., Takeshima, Y., Fujishiro, I.: Topological volume skeletonization and its application to transfer function design. Graph. Models **66**, 24–49 (2004)
22. Takeshima, Y., Takahashi, S., Fujishiro, I., Nielson, G.M.: Introducing topological attributes for objective-based visualization of simulated datasets. In: Proceedings of the Volume Graphics 2005, pp. 137–145 (2005)
23. Weber, G., Dillard, S., Carr, H., Pascucci, V., Hamann, B.: Topology-controlled volume rendering. IEEE Trans. Vis. Comput. Graph. **13**(2), 330–341 (2007)

Two Applications of Geometric Optimal Control to the Dynamics of Spin Particles

Bernard Bonnard and Monique Chyba

Abstract The purpose of this article is to present the application of methods from geometric optimal control to two problems in the dynamics of spin particles. First, we consider the saturation problem for a single spin system and second, the control of a linear chain of spin particles with Ising couplings. For both problems the minimizers are parameterized using Pontryagin Maximum Principle and the optimal solution is found by a careful analysis of the corresponding equations.

Keywords Optimal control · Bloch equations · Saturation problem · Dynamics of spins particles

1 Introduction

The past few years have witnessed an intense recent research activity on the optimal control of the dynamics of spin systems controlled by a radio frequency magnetic field (RF-magnetic field) with application to Nuclear Magnetic Resonance (NMR) spectroscopy and Medical Resonance Imaging (MRI). A pioneer application of geometric optimal control was the contribution to the saturation problem of a single spin system [20], where the authors replaced the standard inversion recovery sequence by the time minimal solution formed by a sequence of bang RF-pulses with

B. Bonnard (✉)
Institut de Mathématiques de Bourgogne, CNRS 5584, Université de Bourgogne, 9 Avenue Savary, 21078 Dijon, France
e-mail: bernard.bonnard@u-bourgogne.fr

B. Bonnard
INRIA Sophia Antipolis, Route des Lucioles BP 93, 06902 Sophia Antipolis Cedex, France

M. Chyba
University of Hawaii, 2565 McCarthy Mall, Honolulu, HI 96822, USA
e-mail: chyba@hawaii.edu

M. Wakayama et al. (eds.), *The Impact of Applications on Mathematics*, Mathematics for Industry 1, DOI: 10.1007/978-4-431-54907-9_5,

maximal amplitude and pulses with intermediate amplitude corresponding to singular trajectories in optimal control [6]. This result is a consequence of the Pontryagin Maximum Principle [24] after performing a reduction to a two dimensional system which allows a complete analysis of the problem.

The first objective of this article is to present in details the geometric tools to analyze this saturation problem as well as some extensions of it in relation with the contrast problem in MRI [20, 21]. The second objective of this article is to discuss the time optimal control of a linear chain of spin particles with Ising couplings [19] in relation with quantum computing. Restricting to the case of three spins we introduce the geometric framework to analyze the problem and we improve the preliminary results contained in [25, 26]. As for the single spin system, the minimizers are parameterized using the Maximum Principle and the optimal solution is computed using the framework of recent developments of invariant sub-Riemannian geometry (SR-geometry) on SO(3) [1].

2 Preliminary: The Pontryagin Maximum Principle

In this section, we recall the necessary optimality conditions [24] which allows us to parameterize the optimal solutions. For our purpose it is sufficient to formulate them in the time minimum case.

2.1 Necessary Optimality Conditions

Consider the minimum time problem for a smooth control system:

$$\frac{dx}{dt}(t) = f(x(t), u(t)), \tag{1}$$

where $x(t) \in M$, with M a n-dimensional manifold, the control domain U is a subset of R^m and the state variable satisfies the boundary conditions $x(0) \in M_0$ and $x(t_f) \in M_1$, with M_0, M_1 being smooth submanifolds of M.

We introduce the pseudo-Hamiltonian:

$$H(x, p, u) = \langle p, f(x, u) \rangle, \tag{2}$$

where $\langle , \rangle$ is the scalar product and $p \in T^*M$ denotes the adjoint vector. The Maximum Principle states that if the trajectory $t \to x(t)$, $t \in [0, t_f]$ associated to the admissible control $u: [0, t_f] \to U$ is optimal, then there exists $p: [0, t_f] \to T^*M$ non zero and absolutely continuous such that the following equations are satisfied almost everywhere on $[0, t_f]$:

$$\frac{dx}{dt}(t) = \frac{\partial H}{\partial p}(x(t), p(t), u(t)) \tag{3}$$

$$\frac{dp}{dt}(t) = -\frac{\partial H}{\partial x}(x(t), p(t), u(t)), \tag{4}$$

as well as the maximum condition:

$$H(x(t), p(t), u(t)) = M(x(t), p(t)) \tag{5}$$

where $M(x(t), p(t)) = \max_{v \in U} H(x(t), p(t), v)$. Moreover $M(x(t), p(t))$ is constant along the trajectory and the following boundary conditions are satisfied:

$$x(0) \in M_0, \quad x(t_f) \in M_1, \tag{6}$$

$$p(0) \perp T_{x(0)}M_0, \quad p(t_f) \perp T_{x(t_f)}M_1 \text{(Transversality conditions)}. \tag{7}$$

3 The Saturation Problem

In this section, we recall prior results on the single spin system and introduce the model for multiple spins.

3.1 The Case of a Single Spin and Its Extensions to Systems of Spins

The single spin system is modeled by the Bloch equation written in a moving frame and adapted coordinates:

$$\frac{dx}{dt}(t) = -\Gamma x(t) - u_2(t)z(t) \tag{8}$$

$$\frac{dy}{dt}(t) = -\Gamma y(t) + u_1(t)z(t) \tag{9}$$

$$\frac{dz}{dt}(t) = \gamma(1 - z(t)) - u_1(t)y(t) + u_2(t)x(t), \tag{10}$$

where $q = (x, y, z)$ represents the magnetization vector restricted to the Bloch ball: $|q| \leq 1$, (Γ, γ) are the physical parameters which are the signature of the chemical species and satisfies $2\Gamma \geq \gamma > 0$, and $u = (u_1, u_2)$ is the bounded RF-applied magnetic field $|u| \leq m$.

The objective of the saturation problem is to bring the magnetization vector q from the north pole: $N = (0, 0, 1)$ (which is the equilibrium point of the free system) to the center $O = (0, 0, 0)$ of the Bloch ball. The physical interpretation is related to the fact that in MRI, the amplitude $|q|$ corresponds to a grey level, with $|q| = 1$ corresponding to white and $|q| = 0$ to black. A direct generalization of the statement above is to bring the system from a forced equilibrium position (associated to a nonzero fixed constant control) to the center O.

Equations (8)–(10) can be written in a compact form as an affine control system

$$\frac{dq}{dt}(t) = F(q(t)) + u_1(t)G_1(q(t)) + u_2(t)G_2(q(t)) \tag{11}$$

where F represents the dissipation of the system and the G_i's the controlled vector fields:

$$F(q(t)) = \begin{pmatrix} -\Gamma x(t) \\ -\Gamma y(t) \\ \gamma(1 - z(t)) \end{pmatrix}, \quad G_1(q(t)) = \begin{pmatrix} 0 \\ z(t) \\ -y(t) \end{pmatrix}, \quad G_2(q(t)) = \begin{pmatrix} -z(t) \\ 0 \\ x(t) \end{pmatrix}. \tag{12}$$

An extension with application to MRI is to consider an ensemble of N spin systems, associated to the same chemical species (i.e. with the same physical parameters Γ and γ). We denote by $q_s(t)$, $s = 1, \dots, N$, the solutions of the system:

$$\frac{dq_s}{dt}(t) = F(q_s(t)) + (1 - \varepsilon_s)\{u_1(t)G_1(q_s(t)) + u_2(t)(G_2(q_s(t))\} \tag{13}$$

where the control $u = (u_1, u_2)$ satisfies $|u| \leq m$.

The saturation problem for the ensemble of spins is to steer the system from the north pole $N = ((0, 0, 1), \dots, (0, 0, 1))$ to the center $O = ((0, 0, 0), \dots (0, 0, 0))$ of the products of the Bloch balls. This is equivalent to the saturation problem in MRI, where the amplitude $m(1 - \varepsilon_s)$ corresponds to the variation of the applied RF-field induced by the spatial position of the spin s in the image.

3.2 The Saturation Problem in Minimum Time for a Single Spin

First of all, we observe that due to the symmetry of revolution of the problem with respect to the polarizing z-axis we can restrict our analysis to a $2D$- single-input system:

$$\frac{dy}{dt}(t) = -\Gamma y(t) + u(t)z(t) \tag{14}$$

$$\frac{dz}{dt}(t) = \gamma(1 - z(t)) - u(t)y(t), \tag{15}$$

where $|u(t)| \leq m$. The system can be written as $\frac{dq}{dt} = F(q) + uG(q)$, with $q = (y, z)$.

Applying the Maximum Principle, an optimal solution is found by concatenation of regular and singular arcs defined as follows.

Definition 1 For a system of the form $\frac{dq}{dt} = F(q) + uG(q)$, the control is said to be:

- *singular* if $\langle p(t), G(q(t))\rangle \equiv 0$, and is determined by differentiating this implicit equation.
- *regular* if $\langle p(t), G(q(t))\rangle \neq 0$, and is given for a.e. t by $u(t) = m\ \mathrm{sign}\langle p(t), G(q(t))\rangle$.

One denotes by σ_s a singular arc, and by $\sigma_{\pm m}$ bang arcs such that the control is given by $u = \pm m$ and $\sigma_1\sigma_2$ an arc σ_1 followed by an arc σ_2.

The first step is to compute the singular arcs. They satisfy $\langle p(t), G(q(t)\rangle = 0$, and differentiating this equation with respect to time one gets the relations:

$$\langle p(t), G(q(t))\rangle = \langle p(t), [G, F](q(t))\rangle = 0, \tag{16}$$

$$\langle p(t), [[G, F], F](q(t)) + u(t)[[G, F], G](q(t))\rangle = 0, \tag{17}$$

where $[,]$ denotes the Lie bracket. The singular trajectories are therefore located on the set $S = \{q;\ det(G, [G, F])(q) = 0\}$, which is given in our case by $y(-2\delta z+\gamma) = 0$ with the notation $\delta = \gamma - \Gamma$. It is formed by the $z-$ axis of revolution $y = 0$ and the horizontal direction: $z_0 = \gamma/2\delta$.

The singular control is computed as a feedback using (17). We have $u(t) = \langle p(t), [[G, F], F](q(t))/[[G, F], G](q(t))\rangle$ which implies:

- For $y = 0$, the singular control is zero and the corresponding system is $\frac{dz}{dt}(t) = \gamma(1 - z(t))$. It relaxes to the equilibrium state.
- For $z_0 = \gamma/2\delta$, the feedback singular control is $u_s = \gamma(2\Gamma - \gamma)/2\delta y$ and the dynamics is given by:

$$\frac{dy}{dt}(t) = -\Gamma y(t) - \frac{\gamma^2(2\Gamma - \gamma)}{4(\gamma - \Gamma)^2 y(t)}, \tag{18}$$

  and $u_s \to \infty$ when $y \to 0$. The time to steer $(-1, *)$ to $(0, *)$ is given by the integral $T = \int_0^1 \frac{dy}{\Gamma y + \frac{\gamma^2(2\Gamma-\gamma)}{4(\gamma-\Gamma)^2 y}}$ which can be easily computed.

The interesting case in the saturation problem is when the horizontal line $z_0 = \gamma/2\delta$ is such that $-1 < z_0 < 0$ which imposes the following condition on the physical parameters: $2\Gamma > 3\gamma$, and we shall restrict our analysis to this case.

To complete the analysis, we must determine the optimality status of the singular line. It can be small time maximizing (slow) or minimizing (fast). To distinguish the two cases one uses the generalized Legendre-Clebsch condition [6] as follows.

Let $D'' = det(G, F) = \gamma z(z-1) + \Gamma y^2$ and $C = \{D'' = 0\}$ be the collinear set. If $\gamma > 0$, this set is not reduced to a point, but forms an oval curve joining the north pole $(0, 1)$ to the center $(0, 0)$ of the Bloch ball and the intersection with the horizontal singular line is empty. Denoting $D = det(G, [[G, F], G])$, the singular lines are fast displacement directions if $DD'' > 0$ and slow if $DD'' < 0$. From this condition one deduces that the z-axis of revolution is fast if $1 > z > z_0$ and slow if $z_0 > z > -1$, while the horizontal singular line is fast.

From this analysis, we deduce that the standard inversion sequence: $u(t) = m$ to steer $(0, 1)$ to $(0, -*)$ followed by $u(t) = 0$ to relax the system along the z-axis is not time optimal. It has to be replaced by a policy using the horizontal singular line. The complete analysis is not straightforward since $|u_s| \to \infty$ when $y \to 0$ along the singular horizontal line and the singular control saturates the constraint $|u| \leq m$.

The final result is given in Theorem 1, see [20] for more details.

Theorem 1 *In the time minimal saturation problem, the optimal policy is of the form* $\sigma_m \sigma_s \sigma_m \sigma_s$.

3.2.1 Interpretation

The first bang arc is used to move the system from the equilibrium point $(0, 1)$ to the horizontal singular line while the second bang arc σ_m connects the horizontal singular arc to the vertical and this occurs before saturating the singular control.

4 The Geometry of a Linear Three Spin System with Ising Couplings

4.1 Mathematical Model

We restrict ourselves to the optimal control of three coupled spins, but the problem can be generalized to a chain with any number of spins. We follow here the presentation of [25, 26].

We introduce the spin 1/2 matrices σ_α, where α represents the number of the particle carrying spin, related to the Pauli matrices by a 1/2 factor. Such matrices satisfy:

$$[\sigma_x, \sigma_y] = i\sigma_z, \quad \sigma_x^2 = \sigma_y^2 = \sigma_z^2 = 1/4. \tag{19}$$

The Hilbert space of the system consists of a dimensional space formed by the tensorial product of the three two dimensional spin 1/2 Hilbert spaces. The Hamiltonian of the system can be written as follows:

$$H = H_C + H_F, \tag{20}$$

where H_C the Hamiltonian of the free system and H_F the Hamiltonian corresponding to the RF-magnetic field are given by

$$H_C = 2J_{12}\sigma_{1z}\sigma_{2z} + 2J_{23}\sigma_{2z}\sigma_{3z} \tag{21}$$

$$H_F = u(t)\sigma_{2y}, \tag{22}$$

with the coefficients J_{ij} representing the coupling constants between the spins i and j.

We consider the time evolution of the vector $X = (x_1, x_2, x_3, x_4)$ where $x_1 = \langle\sigma_{1x}\rangle, x_2 = \langle 2\sigma_{1y}\sigma_{2z}\rangle, x_3 = \langle 2\sigma_{1y}\sigma_{2x}\rangle, x_4 = \langle 4\sigma_{1y}\sigma_{2y}\sigma_{3z}\rangle$ where $\langle\ \rangle$ denotes here the expectation value. To compute the dynamics, we introduce the density matrix ρ, a 8×8-matrix which satisfies:

$$\frac{d\rho}{dt} = -i[H, \rho]. \tag{23}$$

Using the definition of the expectation value of a given operator:

$$\langle O\rangle = Tr(O\rho), \tag{24}$$

one gets:

$$\frac{d}{dt}\langle\sigma_{1x}\rangle = Tr\left(\sigma_{1x}\frac{d\rho}{dt}\right) = -iTr(\sigma_{1x}[H, \rho]) = -iTr([\sigma_{1x}, H]\rho). \tag{25}$$

Hence we deduce that:

$$\frac{dx_1}{dt} = -J_{12}Tr(2\sigma_{1y}\sigma_{2z}\rho). \tag{26}$$

By rescaling the time by the factor J_{12}, we obtain:

$$\frac{dx_1}{dt} = -x_2. \tag{27}$$

Similar computations lead to the evolution of X given by:

$$\frac{dX}{dt} = \begin{pmatrix} 0 & -1 & 0 & 0 \\ 1 & 0 & -u & 0 \\ 0 & u & 0 & -k \\ 0 & 0 & k & 0 \end{pmatrix} X, \quad k = J_{23}/J_{12}. \tag{28}$$

The optimal control problem is to transfer in minimum time the initial position $(1, 0, 0, 0)$ to the position $(0, 0, 0, 1)$ as an intermediate step to realize the transfer in minimum time from σ_{1x} to σ_{3x}. Indeed it connects the first spin to the third one by controlling the second spin.

Introducing the coordinates

$$r_1 = x_1, r_2 = \sqrt{x_2^2 + x_3^2}, \quad r_3 = x_4, \tag{29}$$

and

$$\tan\alpha = x_3/x_2, \tag{30}$$

we obtain the system:

$$\frac{d}{dt}\begin{bmatrix} r_1 \\ r_2 \\ r_3 \end{bmatrix} = \begin{bmatrix} 0 & -\cos\alpha & 0 \\ \cos\alpha & 0 & -k\sin\alpha \\ 0 & k\sin\alpha & 0 \end{bmatrix}\begin{bmatrix} r_1 \\ r_2 \\ r_3 \end{bmatrix}, \tag{31}$$

where $r = (r_1, r_2, r_3) \in S^2$ (the two dimensional sphere), and $u_1 = -k\sin\alpha$, $u_3 = -\cos\alpha$.

In those coordinates the minimum time problem is equivalent to find the fastest transfer on the sphere from $(1, 0, 0)$ to $(0, 0, 1)$. It can be written:

$$\frac{dr_1}{dt} = u_3 r_2, \quad \frac{dr_2}{dt} = -u_3 r_1 + u_1 r_3, \quad \frac{dr_3}{dt} = -u_1 r_2, \tag{32}$$

$$\min_{u(.)} \int_0^T (I_1 u_1^2 + I_3 u_3^2)dt, \quad k^2 = I_1/I_3. \tag{33}$$

This problem is equivalent to a Riemannian problem on the sphere S^2, with a singularity at the equator $r_2 = 0$, the metric being

$$g = \frac{dr_1^2 + k^2 dr_3^2}{r_2^2}. \tag{34}$$

Introducing the spherical coordinates

$$r_2 = \cos\varphi, \quad r_1 = \sin\varphi\cos\theta, \quad r_3 = \sin\varphi\sin\theta, \tag{35}$$

where $\varphi = \pi/2$ is the equator, the metric g takes the form

$$\begin{aligned} g = {} & (\cos^2\theta + k^2\sin^2\theta)d\varphi^2 + 2(k^2 - 1)\tan\varphi\sin\theta\cos\theta d\varphi d\theta \\ & + \tan^2\varphi(\sin^2\theta + k^2\cos^2\theta)d\theta^2, \end{aligned}$$

with the associated Hamiltonian

$$H = \frac{1}{4k^2}\{p_\varphi^2(\sin^2\theta + k^2\cos^2\theta) + p_\theta^2\text{cotan}^2\varphi(\cos^2\theta + k^2\sin^2\theta)$$
$$- 2(k^2-1)p_\varphi p_\theta \text{cotan}\varphi \sin\theta cos\theta\}.$$

If $k = 1$, the Hamiltonian takes the form $H = \frac{1}{4}(p_\varphi^2 + p_\theta^2\text{cotan}^2\varphi)$ and describes the standard Grushin metric on S^2.

4.2 Connection with Invariant Metrics on SO(3) and Integration

A first approach consists in lifting the problem on SO(3). We introduce the matrix $R(t) = (r_{ij}(t))$ of SO(3) where $r_1 = r_{11}, r_2 = r_{12}, r_3 = r_{13}$ are the components of the first row and we consider the right-invariant control system:

$$\frac{d}{dt}R^t = \begin{bmatrix} 0 & u_3 & 0 \\ -u_3 & 0 & u_1 \\ 0 & -u_1 & 0 \end{bmatrix} R^t. \tag{36}$$

The first column equation corresponds to our problem. Hence our optimal control problem becomes:

$$\min_{u(.)} \int_0^T (I_1u_1^2 + I_3u_3^2)dt \tag{37}$$

for the right invariant control system with the following boundary conditions:

$$R^t(0) = \begin{bmatrix} 1 * * \\ 0 * * \\ 0 * * \end{bmatrix}, \quad R^t(T) = \begin{bmatrix} 0 * * \\ 0 * * \\ 1 * * \end{bmatrix}, \tag{38}$$

and we want to steer the first axis of the frame R^t from e_1 to e_3, where e_i denotes the canonical basis.

Equivalently, it is transformed into a left-invariant control problem to use the computations in [18]:

$$\frac{dR}{dt} = R\begin{bmatrix} 0 & -u_3 & 0 \\ u_3 & 0 & -u_1 \\ 0 & u_1 & 0 \end{bmatrix}, \quad \min_{u(.)} \int_0^T (I_1u_1^2 + I_3u_3^2)dt \tag{39}$$

with the corresponding boundary conditions. This defines a left-invariant SR-problem on SO(3) depending on the parameter $k^2 = I_1/I_3$. Upon an appropriated limit process $I_2 \to +\infty$, it is related to the Euler-Poinsot rigid body motion [23]:

$$\frac{dR}{dt} = R \begin{bmatrix} 0 & -u_3 & u_2 \\ u_3 & 0 & -u_1 \\ -u_2 & u_1 & 0 \end{bmatrix}, \quad \min_{u(.)} \int_0^T (I_1 u_1^2 + I_2 u_2^2 + I_3 u_3^2) dt \tag{40}$$

which is well-known model for invariant metrics on SO(3) depending on 2 parameters, the ratio I_2/I_1 and I_3/I_1. There are two special cases:

1. The bi-invariant case $I_1 = I_2 = I_3$ where the geodesics solutions are the rotations of SO(3).
2. The case of revolution where $I_1 = I_2$.

The optimal solutions can be parameterized by the Pontryagin maximum principle and thanks to the explicit formula given in [18], the solutions can be computed in both the Riemannian and the sub-Riemannian cases using the elliptic functions. See [20] for the details of the computations. We here only sketch the main points.

We introduce the following matrices:

$$A_1 = \begin{pmatrix} 0 & 0 & 0 \\ 0 & 0 & -1 \\ 0 & 1 & 0 \end{pmatrix}, \quad A_2 = \begin{pmatrix} 0 & 0 & 1 \\ 0 & 0 & 0 \\ -1 & 0 & 0 \end{pmatrix}, \quad A_3 = \begin{pmatrix} 0 & -1 & 0 \\ 1 & 0 & 0 \\ 0 & 0 & 0 \end{pmatrix} \tag{41}$$

with the Lie brackets relations:

$$[A_1, A_2] = -A_3, \quad [A_1, A_3] = A_2, \quad [A_2, A_3] = -A_1. \tag{42}$$

The optimal control problem is then stated as:

$$\frac{dR}{dt} = \sum_{i=1}^{3} u_i R A_i, \quad \min_{u(.)} \int_0^T \sum_{i=1}^{3} I_i u_i^2 dt. \tag{43}$$

Applying the Maximum Principle and denoting by H_i the symplectic lifts of the vector fields RA_i leads to introduce the pseudo-Hamiltonian:

$$H = \sum_{i=1}^{3} u_i H_i - \frac{1}{2} \sum_{i=1}^{3} I_i u_i^2. \tag{44}$$

From the maximization condition $\frac{\partial H}{\partial u} = 0$ we deduce that $u_i = H_i/I_i$, $i = 1, 2, 3$. Plugging back this expression for the controls u_i in the Hamiltonian H defines the true Hamiltonian:

$$H_n = \frac{1}{2} \left(\frac{H_1^2}{I_1} + \frac{H_2^2}{I_2} + \frac{H_3^2}{I_3} \right). \tag{45}$$

Then we derive the Euler equation:

$$\frac{dH_i}{dt} = \{H_i, H_n\}, \tag{46}$$

where {, } is the Poisson bracket. Using Lie brackets computations, one obtains:

$$\frac{dH_1}{dt} = H_2 H_3 \left(\frac{1}{I_3} - \frac{1}{I_2}\right), \tag{47}$$

$$\frac{dH_2}{dt} = H_1 H_3 \left(\frac{1}{I_1} - \frac{1}{I_3}\right), \tag{48}$$

$$\frac{dH_3}{dt} = H_1 H_2 \left(\frac{1}{I_2} - \frac{1}{I_1}\right). \tag{49}$$

The SR-case can be derived formally by setting $u_2 = \varepsilon v_2$, $\varepsilon \to 0$ to obtain a bi-input system. Since $u_i = H_i/I_i$, this is equivalent to $I_2 \to +\infty$. The Hamiltonian reduces to:

$$H_n = \frac{1}{2}\left(\frac{H_1^2}{I_1} + \frac{H_3^2}{I_3}\right), \tag{50}$$

with the corresponding Euler equation where the parameter $k^2 = I_1/I_3$ is the invariant classifying the SR-metrics.

To get an uniform integration procedure we use the following result from [18].

Proposition 1 *For each invariant Hamiltonian on SO(3) (the Riemannian and the SR-case) the system is integrable by quadrature using the four first integrals: the Hamiltonian H_n and the Hamiltonian lifts of the right invariant vector fields $A_i R$.*

4.2.1 Integration

The general algorithm consists in integrating the Euler equation, while the remaining quadrature is deduced using the Euler angles Φ_i defined by taking the following decomposition of a matrix R in SO(3):

$$R = (\exp\Phi_1 A_3) \circ (\exp\Phi_2 A_2) \circ (\exp\Phi_3 A_3) \tag{51}$$

while the angles Φ_2, Φ_3 can be found from the relations:

$$H_1 = -|H| \sin\Phi_2 \cos\Phi_3, \quad H_2 = |H| \sin\Phi_2 \sin\Phi_3, \quad H_3 = |H| \cos\Phi_2 \tag{52}$$

and the angle Φ_1 is computed by integrating the differential equation

$$\frac{d\Phi_1}{dt} = \frac{|H|}{(H_2 \sin \Phi_3 - H_1 \cos \Phi_3)} \left(\sin \Phi_3 \frac{\partial H_n}{\partial H_2} - \cos \Phi_3 \frac{\partial H_n}{\partial H_1} \right). \tag{53}$$

In the SR-case, the Euler equation is integrated as follows. We fix the level set of the true Hamiltonian to 1/2: $H_n = 1/2$, and we introduce α such that $\cos \alpha = H_1/\sqrt{I_1}$ and $\sin \alpha = H_3/\sqrt{I_3}$. We obtain the pendulum equation:

$$\frac{d^2\alpha}{dt^2} = \frac{k^2 - 1}{2I_1} \sin 2\alpha, \quad k^2 = I_1/I_3. \tag{54}$$

Details of the parameterizations of the solutions are given in [11]. The computations in the Riemannian case are standard [18, 22].

In conclusion, in the spin time optimal problem the optimal solutions can be found among extremals solutions of the Maximum Principle.

4.3 Direct Integration on the Sphere

We identify S^2 as the homogenous space SO(3)/SO(2). In this interpretation, the Hamiltonian $|H|^2 = H_1^2 + H_2^2 + H_3^2$ corresponds to the bi-invariant case and represents the Casimir function which commutes with the Hamiltonian associated to every invariant Riemannian and sub-Riemannian metrics.

On the homogeneous space this defines the round sphere with constant curvature $+1$, whose metric in spherical coordinates is given by:

$$g = d\varphi^2 + \sin^2 \varphi d\theta^2 \tag{55}$$

and with Hamiltonian

$$F = p_\varphi^2 + \frac{p_\theta^2}{\sin^2 \varphi}. \tag{56}$$

We have the following result.

Lemma 1 *Consider the standard Grushin metric on S^2 with Hamiltonian $H_0 = p_\varphi^2 + p_\theta^2 \text{cotan}^2 \varphi$. Then $\{H_0, F\} = 0$ where F is the Hamiltonian of the round metric on S^2, $F = p_\varphi^2 + \frac{p_\theta^2}{\sin^2 \varphi}$.*

Proof Using the symmetry of revolution of the Grushin metric, we have that p_θ is a constant (Clairaut relation). Hence p_θ^2 is a constant. Therefore $p_\varphi^2 + \frac{\cos^2 \varphi}{sin^2 \varphi} p_\theta^2 + p_\theta^2$ is a constant, which means that F is a first integral.

Up to a constant renormalization the Hamiltonian becomes:

$$H = H_0 + k^* H', \quad k^* = k^2 - 1, \tag{57}$$

where

$$H' = G^2, \quad G = p_\varphi \cos\theta - p_\theta \text{cotan}\varphi \sin\theta. \tag{58}$$

We have the following.

Proposition 2 *The following relations are satisfied:*

$$\{H_0, F\} = \{G, F\} = 0. \tag{59}$$

Therefore, $\{H, F\} = 0$ *for each* k^*.

4.3.1 Integration

To integrate the geodesic flow we use the standard Birkhoff method, see [4, 5] for the details. The Hamiltonian H admits a first integral F which is quadratic in p and corresponds to a Liouville metric on S^2. The metric is written in the isothermal form:

$$g = \lambda(x, y)(dx^2 + dy^2) \tag{60}$$

outside the equator using a rescaling of r_3.

Any diffeomorphism

$$x = \varphi(u, v), \quad y = \psi(u, v) \tag{61}$$

which is preserving the isothermal form and the orientation satisfies the Cauchy-Riemann relation:

$$\varphi_u = \psi_v, \quad \varphi_v = -\psi_u. \tag{62}$$

In the isothermal coordinates, the first integral becomes:

$$F(x, y) = b_1(x, y)p_x^2 + 2b_2(x, y)p_x p_y + b_3(x, y)p_y^2. \tag{63}$$

We introduce:

$$R = (b_1 - b_3) + 2ib_2 \tag{64}$$

which is an holomorphic function of $z = x + iy$ using Birkhoff's relations [4]. Let us denote

$$w = u + iv, \tag{65}$$

and let us introduce the holomorphic change of variables

$$\Phi\colon w \to z. \tag{66}$$

We obtain

$$p_x = D(p_u\psi_v - p_v\psi_u), \quad p_y = D(-p_u\varphi_v + p_v\varphi_u), \tag{67}$$

with

$$D = (\varphi_u\psi_v - \psi_u\varphi_v)^{-1}. \tag{68}$$

Expressing F in the (u, v) coordinates we have:

$$F(u, v) = p_u^2 b_1'(u, v) + 2p_u p_v b_2'(u, v) + p_v^2 b_3'(u, v). \tag{69}$$

An easy computation shows that:

$$S = (b_1' - b_3' + 2ib_2') = D^2(\varphi_u - i\psi_u)^2(b_1 - b_3 + 2ib_2) \tag{70}$$
$$= (\varphi_u + i\psi_u)^{-2}(b_1 - b_3 + 2ib_2). \tag{71}$$

We choose the change of coordinates such that we have the following normalization: $S = 1$. Hence, we must solve the equation

$$(\varphi_u + i\psi_u) = \sqrt{R(z)}. \tag{72}$$

In the new coordinates the metric takes the Liouville normal form:

$$g(u, v) = (f(u) + g(v))(du^2 + dv^2), \tag{73}$$

and the integration is standard see for instance [5].

4.3.2 Grushin Singularity

The family of metrics g is defined as a metric on the distibution $\Delta = \mathrm{Span}\{F_1, F_3\}$ where F_1, F_3 are respectively the vector fields corresponding to rotations with axis e_1 and e_3. By construction such a metric has a Grushin singularity [2] at the equator E where rank Δ is one and the distribution is transverse to the equator. The local normal form near a point of the equator is described in the aforementioned article. Hence our family of metrics defines an almost-Riemannian metric on the sphere with Grushin singularity at the equator.

4.4 The Optimality Problem

We discuss briefly in this section the optimality problem which can be handled using the technical framework developed in [11] combining geometric analysis and numerical techniques.

We use the following concepts. On the almost-Riemannian manifold (S^2, g), the cut point along a geodesic curve γ, projection of an extremal curve solution of the Maximum Principle, emanating from $q_0 \in S^2$ is the first point where it ceases to be minimizing and we denote $C_{cut}(q_0)$ the set of such points forming the cut locus. The first conjugate point is the point where it ceases to be minimizing among the geodesics C^1- close from γ and we denote $C(q_0)$ the set of such points, forming the conjugate locus.

Our optimality problem amounts to transfer with minimum length the point $q_0 = (1, 0, 0)$ given by $\varphi_0 = \pi/2, \theta_0 = 0$ in spherical coordinates to the point $q_1 = (0, 0, 1)$ defined by $\varphi_1 = \pi/2, \theta_1 = \pi/2$. In particular the problem is solved by computing the cut locus $C(q_0)$ of the equatorial point.

First, we have the following lemma.

Lemma 2 *The family of metrics g depending upon the parameter k have a discrete symmetry group defined by the two reflexions:* $H(\varphi, p_\varphi) = H(\pi-\varphi, -p_\varphi)$ *(reflexion with respect to the equator) and* $H(\theta, p_\theta) = H(-\theta, -p_\theta)$ *(reflexion with respect to the meridian* $\theta = 0$*).*

The next step is to use the Grushin singularity resolution described in [7] and the previous symmetries.

Proposition 3 *Near* q_0 *identified to* 0, *the conjugate and cut loci for the metric restricted to a neighborhood of* 0 *can be computed using the local model* $dx^2 + \frac{dy^2}{x^2}$. *The cut locus is a segment* $[-\varepsilon, +\varepsilon]$ *minus* 0 *while the conjugate locus is formed by four symmetric curves of the form* $x = cy^2$ *minus* 0 *and tangential to the meridian* $\theta_0 = 0$ *(although the Gaussian curvature is strictly negative and tends to* $-\infty$ *at the equator).*

In [18] the cut and conjugate loci in the Grushin case $k = 1$ on S^2 of an equatorial point are completely described making an homotopy $g_\lambda = d\varphi^2 + G_\lambda(\varphi)d\theta^2$, $G_\lambda(\varphi) = \frac{\sin^2\varphi}{(1-\lambda\sin^2\varphi)}$, $\lambda \in [0, 1]$ from the round metric to the Grushin case which explains in particular the curvature concentration at the equator in the Grushin case. We get the following result.

Proposition 4 *In the Grushin case* $k = 1$, *the cut locus of the equatorial point* $\varphi_0 = \pi/2, \theta_0 = 0$ *is the whole equator minus this point while the conjugate locus has a double heart shape, with four meridional singularities, two at the origin described previously and two cusps on the opposite meridian.*

The general case is studied in [11] using a continuation method on the conjugate locus starting from a Grushin case (observe also that the cut locus is decribed in [14] using the equatorial symmetry). From the geometric point of view the neat framework is given by recent works to describe the conjugate and cut loci on Liouville surfaces generalizing the ellipsoid case [16, 17].

Note the following result that can be easily proved.

Proposition 5 *For every k^*, the cut locus of the equatorial point is the equator minus the point.*

Proof A simple computation shows that the Gaussian curvature in each hemisphere is strictly negative. Hence there is no conjugate point for a geodesic starting form the equatorial point before returning to the equator. Due to the reflectional symmetry with respect to the equator, two geodesics starting from the equatorial point intersects with same length when returning to the equator. This proves the result.

References

1. Agrachev, A., Sachkov, Y.: Control Theory from the Geometric Viewpoint, vol 87 of Encyclopaedia of Mathematical Sciences. Control Theory and Optimization, II, xiv+412 pp. Springer, Berlin (2004)
2. Agrachev, A., Boscain, U., Sigalotti, M.: A Gauss-Bonnet-like formula on two-dimensional almost-Riemannian manifolds. Discrete Contin. Dyn. Syst. **20**(4), 801–822 (2008)
3. Arnold, V.I.: Mathematical methods of classical mechanics, Translated from the Russian by Vogtmann, K., and Weinstein, A., 2nd edn. Graduate Texts in Mathematics, vol 60, xvi+508 pp. Springer, New York (1989)
4. Birkhoff, G.D.: Dynamical Systems, vol IX. American society colloquium publications (1927)
5. Bolsinov, V., Fomenko, A.T.: Integrable geodesic flows on two-dimensional surfaces. In: Monographs in Contemporary Mathematrics. Kluwer Academic, Dordretcht (2000)
6. Bonnard, B., Chyba, M.: Singular Trajectories and their Role in Control Theory, Vol 40 of Mathématiques & Applications, xvi+357 pp. Springer, Berlin (2003)
7. Bonnard, B., Caillau, J.B., Sinclair, R., Tanaka, M.: Tanaka Conjugate and cut loci of a two-sphere of revolution with application to optimal control. Ann. Inst. H. Poincaré Anal. Non Linéaire **26**(4), 1081–1098 (2009)
8. Bonnard, B., Caillau, J.B., Janin, G.: Conjugate-cut loci and injectivity domains on two-spheres of revolution. ESAIM Control Optim. Calc. Var. **19**(2), 533–554 (2013)
9. Bonnard, B., Chyba, M., Marriott, J.: Singular trajectories and the contrast imaging problem in nuclear magnetic resonance. SIAM J. Control Optim. **51**(2), 1325–1349 (2013)
10. Bonnard, B., Cots, O., Jassionnesse, L.: Geometric and numerical techniques to compute conjugate and cut loci on Riemannian surfaces. In: Stefani, G., Boscain, U., Gauthier, J.-P., Sarychev, A., Sigalotti, M. (eds.) Geometric Control Theory and sub-Riemannian Geometry. Springer INdAMSeries 5, pp. 53–72. Springer International Publishing, Switzerland (2014). doi:10.1007/978-3-319-02132-4_4
11. Bonnard, B., Cots, O., Pomet, J.-B., Shcherbakova, N.: Riemannian metrics on 2d-manifolds related to the Euler-Poinsot rigid body motion. ESAIM Control Optim. Calc. Var. (to appear 2014)
12. Bonnnard, B., Caillau, J.B.: Metrics with equatorial singularities on the sphere. Annali di Matematica (to appear 2014). doi10.1007/s10231-013-0333-y
13. Boscain, U., Charlot, G., Gauthier, J.P., Guérin, S., Jauslin, H.-R.: Optimal control in laser-induced population transfer for two- and three-level quantum systems. J. Math. Phys. **43**(5), 2107–2132 (2002)

14. Boscain, U., Chambrion, T., Charlot, G.: Nonisotropic 3-level quantum systems: complete solutions for minimum time and minimum energy. Discrete Contin. Dyn. Syst. Ser. B **5**(4), 957–990 (electronic, 2005)
15. Helgason, S.: Differential Geometry, Lie Groups, and Symmetric Spaces, vol 80 of Pure and Applied Mathematics, xv+628 pp. Academic Press, Inc., Harcourt Brace Jovanovich Publishers, New York (1978)
16. Itoh, J.-I., Kiyohara, K.: The cut loci and the conjugate loci on ellipsoids. Manuscripta Math. **114**(2), 247–264 (2004)
17. Itoh, J.-I., Jin-ichi, K.: Kiyohara cut loci and conjugate loci on Liouville surfaces. Manuscripta Math. **136**(1–2), 115–141 (2011)
18. Jurdjevic, V.: Geometric Control Theory, vol 52 of Cambridge Studies in Advanced Mathematics, xviii+492 pp. Cambridge University Press, Cambridge (1997)
19. Khaneja, N., Glaser, S.J., Steffen, R.: Brockett sub-Riemannian geometry and time optimal control of three spin systems: quantum gates and coherence transfer. Phys. Rev. A **65**(3), part A, 032301, 11 pp (2002)
20. Lapert, M., Zhang, Y., Braun, M., Glaser, S.J., Sugny, D.: Singular extremals for the time-optimal control of dissipative spin particles. Phys. Rev. Lett. **104**, 083001 (2010)
21. Lapert, M., Zhang, Y., Janich, M., Glaser, S.J., Sugny, D.: Exploring the physical limits of saturation contrast in magnetic resonance imaging. Nature- Sci. Rep. **2**, 589 (2012)
22. Lawden, D.F.: Elliptic Functions and Applications, vol 80 of Applied Mathematical Sciences, xiv+334 pp. Springer, New York (1989)
23. Levitt, M.H.: Spin Dynamics—Basics of Nuclear Magnetic Resonance. Wiley, New York (2001). 686 pp
24. Pontryagin, L.S., Boltyanskii, V.G., Gamkrelidze, R.V., Mishchenko, E.F.: The Mathematical Theory of Optimal Processes, viii+360 pp. Translated from the Russian by Trirogoff, K.N. L. W. Neustadt Interscience Publishers, Wiley, New York (1962)
25. Yuan, H.: Geometry, Optimal Control and Quantum Computing. Harvard Ph.D. thesis (2006)
26. Yuan, H., Zeier, R., Khaneja, N.: Elliptic functions and efficient control of Ising spin chains with unequal couplings. Phys. Rev. A **77**, 032340 (2008)

Cryptographic Technology for Benefiting from Big Data

Keisuke Hakuta and Hisayoshi Sato

Abstract "Big Data" technology is the process of collecting and storing large amounts and wide varieties of data sets, and extracting valuable information and/or knowledge by analyzing them. Big Data analytics plays an important role in improving business services and quality of life. However, Big Data might include personal information such as credit card data, health-related data, purchasing history, geographic location data, etc. In Big Data analytics, the data sets may be accessed not only by the data holders but also by the third parties. This indicates a potential privacy breach. In addition, when public cloud is used as a platform of Big Data analytics, the risk of privacy breach might further increase. To protect against such threats, it is desirable to develop encryption schemes which are as efficient as possible and encryption schemes which allow to perform computations on encrypted data without decrypting it. In this paper, we present some of the latest results of our research related to the above challenge for Big Data security and privacy.

Keywords Big data · Cloud computing · Security · Privacy · Cryptography · Anonymization

1 Introduction

"Big Data" technology is the process of collecting and storing large amounts and wide varieties of data sets, and extracting valuable information and/or knowledge by analyzing the data sets (Fig. 1).

K. Hakuta (✉) · H. Sato
Yokohama Research Laboratory, Hitachi, Ltd., 292, Yoshida-cho, Totsuka-ku, Yokohama, Kanagawa 244-0817, Japan
e-mail: keisuke.hakuta.cw@hitachi.com

H. Sato
e-mail: hisayoshi.sato.th@hitachi.com

M. Wakayama et al. (eds.), *The Impact of Applications on Mathematics*, Mathematics for Industry 1, DOI: 10.1007/978-4-431-54907-9_6,

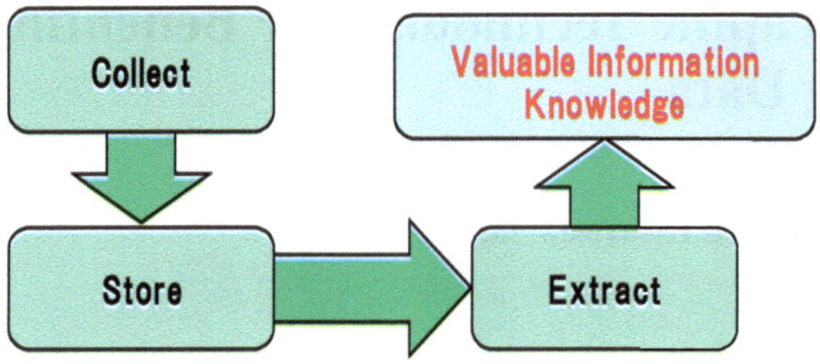

Fig. 1 Process of big data technology

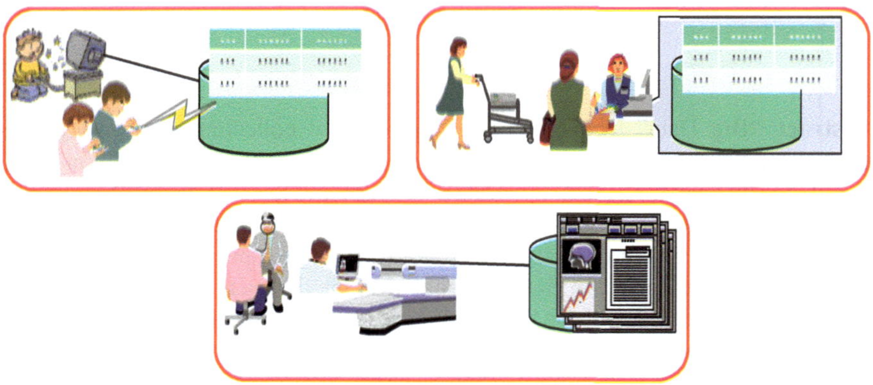

Fig. 2 Examples of benefiting from big data technology

Big data would be useful in many situations such as providing business services and improving quality of life [7]. Figure 2 shows three examples of services which will benefit from Big Data technology.

(1) Online games: Such the game is a game that is played by multiple players across a computer network. A wide range of player behavior or activities is useful to improve online game development and hence to increase company's market penetration.
(2) Purchasing history: It can assist in better understanding of customer purchasing patterns and behaviors through basket analysis.
(3) Cost management of premedical care: It is valuable in improving treatment and reducing the cost of medical care.

What types of computing environments are suitable for big data analytics? These are the so-called "on-premise" and the public cloud.

We show the difference between them in Fig. 3. On-premise is essentially an extension of a traditional datacenter. That is, the company A stores its confidential data locally in a secure manner, but data center management needs high costs. On the other hand, in public cloud, data center can be managed at low cost. However,

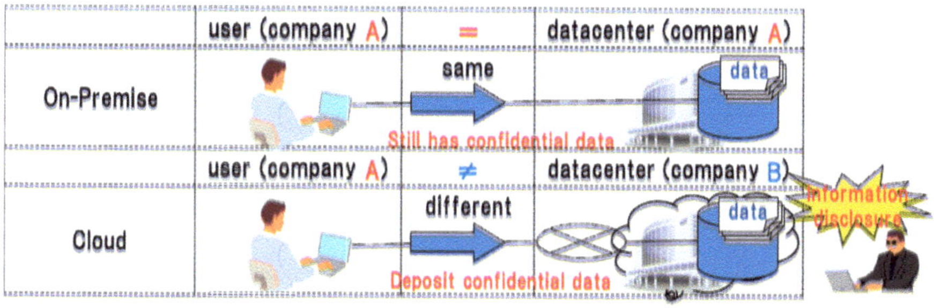

Fig. 3 Difference between on-premise and cloud

Table 1 Security risks in big data analytics

Year	Num. of incidents	Damage per incident	Amount of damage = Num. of incidents × Damage per incident
2008	1,373	216 million yen	29.7 billion yen
2009	1,539	171 million yen	26.4 billion yen
2010	1,679	168 million yen	28.1 billion yen
2011	1,551	189 million yen	29.2 billion yen

company A's user deposits her confidential data there, but they may not be stored securely, hence a certain risk of information disclosure is present [1, 2].

Table 1 indicates the number of incidents and the corresponding amount of damages caused by information disclosure in the recent years.

As we can see, the amount of damage remains quite high. These incidents are include all incidents related to public cloud. However, for companies, data disclosure not only causes financial damages but also destroys the credibility. For individuals, data disclosure may lead to leakage of their identities. In order to prevent data disclosure, one needs to produce new guidelines for data security and to develop security technologies for big data and cloud computing.

In order to properly utilize Big Data, one must address security and privacy issues. In particular, the U.S. Government has released a white paper on privacy [16] and the European Commission has proposed a general data protection regulation [3]. Recently, Japanese government has discussed how to utilize Big Data taking privacy concerns into account. The guidelines [3, 16], in particular, cover data protection and privacy enhancing technologies.

For protecting against data disclosure, we consider the following four techniques to be necessary as privacy enhancing technologies (Fig. 4).

The first technique is highly efficient cryptosystem. Highly efficient public key cryptosystems are suitable when many users have access to a datacenter, because many key exchanges are needed for encryption of data which is sent through communication channel between each user and the datacenter. The second one is searchable

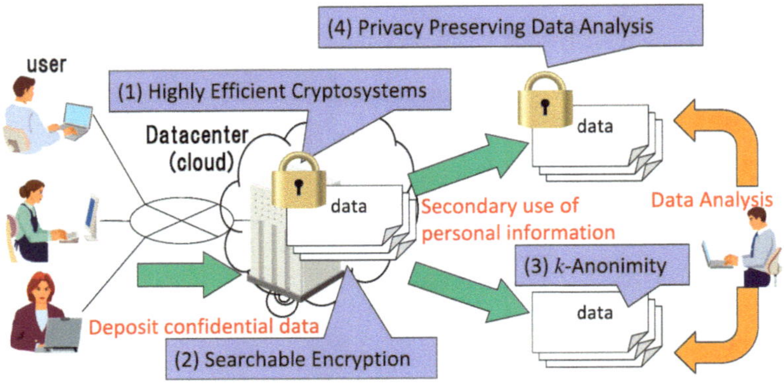

Fig. 4 Security techniques which are necessary as privacy enhancing technologies

encryption. Searchable encryption is suitable when a small number of users operate at a datacenter and only a part of deposited data which is selected by the user can be downloaded. The third and the fourth ones are k-anonymity and privacy preserving data analysis, respectively. These are data protection techniques for secondary use of personal information. k-Anonymity provides a measure of privacy protection by preventing re-identification. Therefore highly accurate and secure data analysis can be done. Privacy preserving data analysis is more secure than k-anonymity, but it is difficult to realize data analysis with similar accuracy as in the case of k-anonymity. Thus these two techniques are complementary.

In this paper, we present some of the latest results of our research related to the above challenge for Big Data security and privacy. The rest of this paper is organized as follows. Section 2 briefly describes the multivariate public key cryptosystems (MPKC) and their security analysis. Section 3 through Sect. 5 present cryptographic schemes which are useful for big data analytics. Section 6 concludes the paper.

2 Highly Efficient Public Key Cryptosystems

First, we explain highly efficient public key cryptosystems. Highly efficient public key cryptosystems are suitable when many users operate this system because many key exchanges are needed for data encryption.

Most widely used public key cryptosystems are divided into two types.

(1) RSA cryptosystem [12]: This cryptosystem is de facto standard, it is based on the difficulty of integer factorization.
(2) Elliptic curve cryptosystems [6, 8]: Elliptic curve cryptosystems provide many advantages, for example, shorter key length and faster computation speed as compared to those of RSA.

The others are the public key cryptosystem for next generation. There are tree main candidates.

(3) Lattice-based cryptosystems: The security of lattice-based cryptosystems depends on the intractability of finding the shortest vector in a lattice. Lattice-based cryptosystems have many applications such as fully homomorphic encryption.
(4) Code-based cryptosystems: Code-based cryptographic schemes build on error correcting codes. The main advantages of code-based schemes are the encryption/decryption speed, and simpler structure of the schemes (thus suitable for light-weight implementation).
(5) Multivariate public key cryptosystem (MPKC for short): The security of MPKC is based on an NP-complete problem, namely the problem of solving a random system of multivariate polynomial equations. Furthermore, encryption/signature schemes based on MPKC are usually much faster than other public key schemes. Some types of MPKC are related to the Jacobian conjecture.

In this paper, we focus on MPKC. In MPKC, a polynomial map (over a field k) is used to construct encryption (resp. signature) schemes. Secret key is a tuple of polynomial maps, and their composition is the corresponding public key F. Trivially, the security of MPKC depends on the difficulty of decomposition of F.

Polynomial automorphism is an invertible polynomial map. The set of polynomial automorphisms is denoted by $\mathrm{Aut}(k, n)$. Similarly, the set of degree one polynomial automorphisms is denoted by $\mathrm{Aff}(k, n)$ and the set of elementary polynomial automorphisms is denoted by $E(k, n)$, where $X_1, \ldots, X_n$ are indeterminates over k, and elementary automorphism is of the form

$$E_{a_i} = (X_1, \ldots, X_{i-1}, X_i + a_i, X_{i+1}, \ldots, X_n),\ a_i \in k[X_1, \ldots, \hat{X_i}, \ldots, X_n].$$

$\mathrm{Aff}(k, n)$ and $E(k, n)$ are subgroups of $\mathrm{Aut}(k, n)$. $T(k, n)$ is a subgroup of $\mathrm{Aut}(k, n)$ generated by $\mathrm{Aff}(k, n)$ and $E(k, n)$. The element of $T(k, n)$ is called tame automorphism and the element of this set is called wild automorphism. Tame generators problem is an open problem in mathematics to decide whether $\mathrm{Aut}(k, n) = T(k, n)$ or not. The equality is true for $n = 1$ and $n = 2$. However, the equality does not hold for $n = 3$. The result is derived from the fact that the Nagata automorphism is wild [13, Corollary 9].

Tame transformation method (TTM for short) is one example of the MPKC scheme [9, 10]. The secret key is a tuple of tame automorphisms and the corresponding public key is their composition. The security of TTM is based on the intractability of the decomposition. An open question is as follows: For a given F, is the decomposition really intractable in general? Our result is that we give the decomposition of some specific tame automorphisms which comprise the public key of an MPKC scheme. For more details of our security analysis, we refer the reader to [4, Main Theorem].

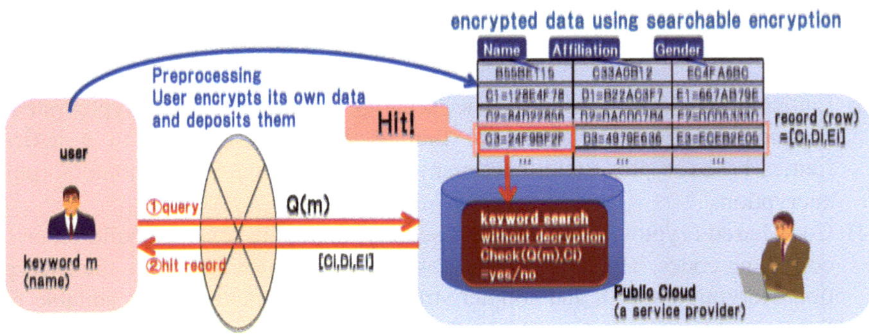

Fig. 5 General outline of searchable encryption

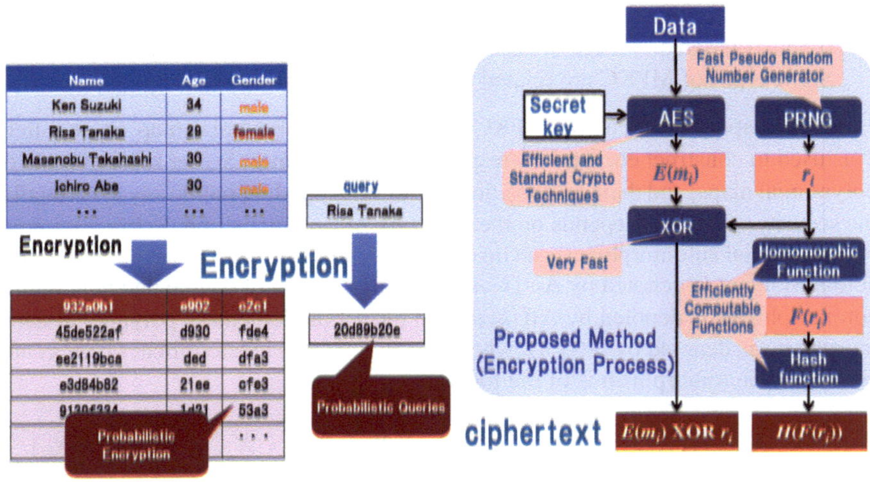

Fig. 6 Brief overview of Hitachi's searchable encryption

3 Searchable Encryption

Searchable encryption is an encryption scheme enabling the search for data in encrypted databases.

Figure 5 shows a brief overview of searchable encryption. First, a user encrypts its own data and deposits them. If the user wants to search a keyword, for example a name, the user encrypts the name, and send the encrypted keyword to the database. Then the database searches the name without decryption, that is the database compares an encrypted data and the encrypted keyword. If the database finds the keyword, the database sends the corresponding record. The database does not learn the keyword. Previously proposed methods are either vulnerable to statistical attacks or inefficient.

Figure 6 shows a brief overview of the Hitachi's searchable encryption scheme. Moreover, the proposed method uses probabilistic algorithms, so Hitachi's searchable encryption scheme is secure against the statistical attacks. The proposed method only uses efficient cryptographic primitives. Therefore Hitachi's searchable encryption scheme is highly efficient. For more details on Hitachi's searchable encryption scheme, we refer the reader to [18].

4 *k*-Anonymity

Next, we explain Hitachi's k-anonymity [5]. k-Anonymity and privacy preserving data analysis are data protection techniques for secondary use of personal information. k-Anonymity provides a measure of privacy protection by preventing re-identification. Hereby, highly accurate and secure data analysis, described in the next section, can be achieved. Privacy preserving data analysis is more secure than k-anonymity, but it is difficult to realize data analyze with similar accuracy as in the case of k-anonymity. Thus these two techniques are complementary.

Traditional anonymitiy algorithms are constructed by deletion/obscuration of personal information. Therefore, traditional anonymity algorithms are not sufficient. On the other hand, k-anonymity [14, Definition 3] requires that each record is indistinguishable from at least $k-1$ other records.

There are two main problems in conventional k-anonymity algorithms. One is the high operational costs for generation of domain generalization hierarchies. The other one is the excessive information loss which is caused by k-anonymity. Hitachi's k-anonymity algorithm has the following two advantages. First, we have succeeded to automate the generation of domain generalization hierarchies tree. Secondly, one can reduce information loss via tree based on Huffman code. For more details of Hitachi's k-anonymity algorithm, we refer the reader to [5].

5 Privacy Preserving Data Analysis

5.1 Enciphered Frequency Analysis

One example of data mining is the machine learning. Machine learning has two phases: learning phase and test phase. Multiple linear regression is the most widely used predictive model in the field of machine learning. In multiple linear regression, the criterion variable (or dependent variable) Y is modeled as the following form (called a multiple linear regression equation) $Y = a_0 + a_1X_1 + \cdots + a_pX_p + \varepsilon$ which consists of a linear combination $a_0 + a_1X_1 + \cdots + a_pX_p$ of the corresponding values of the predictor variables (independent variables) together with an error term ε. In multiple linear regression analysis, data holder collects a data set, namely, a finite number of tuples in the form of a value of the criterion variable and values of the predictor variables $\{(y_i, x_{i,1}, \ldots, x_{i,p})\}_{1 \le i \le N}$. Then the data holder derives a

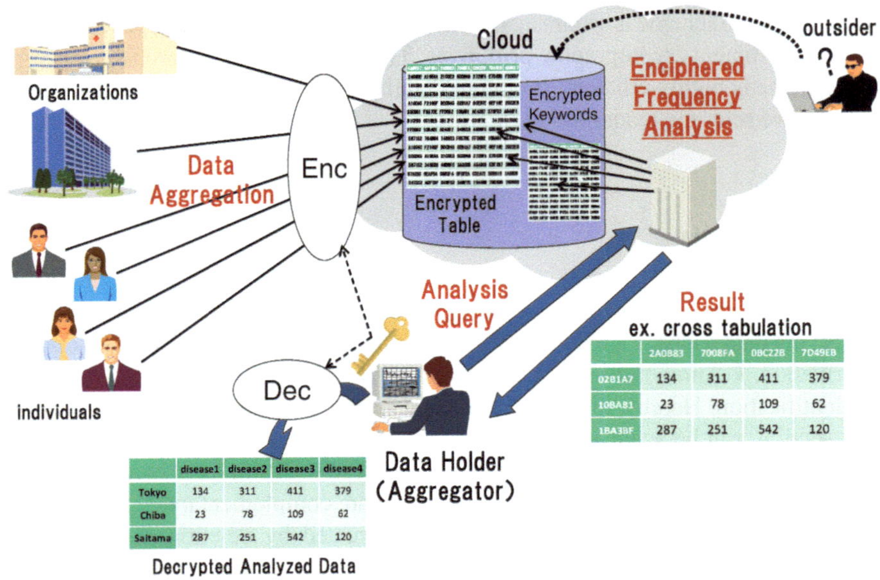

Fig. 7 Typical usecase of enciphered frequency analysis

regression equation (coefficients a_i, $i = 0, 1, \ldots, p$) in the learning phase. Finally the data holder substitutes observed values for the predictor variables X_i in the test phase in order to evaluate the criterion variable Y. Machine learning deals with two types of data: ordered data and nominal scale (i.e., variable that does not really have any evaluative distinction). For each combination of the phase and data type, these techniques have been developed to realize secure data analysis. In this paper, we focus on the case of nominal scale.

One of the most fundamental examples of statistical analysis is frequency analysis. As explained in Sect. 3, searchable encryption is an encryption scheme for searching a data in encrypted databases. By using a searchable encryption scheme, the database can count the number of hits. Namely, the database is able to do the frequency analysis without decryption. We call this technique the *enciphered frequency analysis*. Enciphered frequency analysis is realized using searchable encryption. An authorized agent can obtain some information on the data, but outsiders obtain nothing from the encrypted data. We can realize a secure frequency analysis. In contrast to conventional searchable encryption (SE) schemes, Hitachi's SE has high efficiency. Thus Hitachi's SE can treat big data. For more details of Hitachi's enciphered frequency analysis, we refer the reader to [11].

Figure 7 indicates a typical use case of enciphered frequency analysis. First, individuals or organizations deposit their data to the data holder. Then the data holder aggregates their data and encrypts them. The data holder stores the encrypted table into the data storage in the public cloud. The public cloud is used both to store encrypted table and as a computing environment. If the data holder wants to run

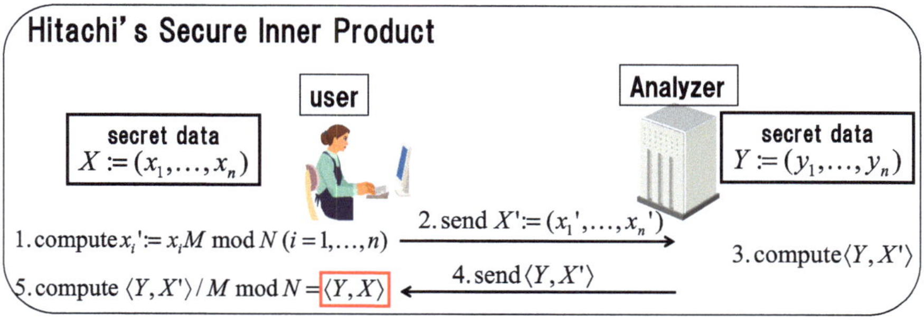

Fig. 8 Example of secure and efficient inner product protocol

frequency analysis securely, she uses enciphered frequency analysis as explained above, and then the cloud returns the result. We remark that the cloud does not know each attribute. Finally, the data holder decrypts each attribute and obtains the analyzed data.

5.2 Secure Inner Product Protocol

Multiple linear regression analysis is as basic as frequency analysis. We assume that there are two parties A and B, and each party has a subset of the data set, namely, party A has $\{(y_i, x_{i,1}, \ldots, x_{i,n})\}_{1 \le i \le N}$ and party B has $\{(x_{i,n+1}, \ldots, x_{i,p})\}_{1 \le i \le N}$. We assume that they would like to analyze the data set using multiple linear regression without disclosing their original subset of the data set to each other. This situation often occurs for some business reason (See [17, Sect. 3.1] for more details).

In such a situation, secure inner product protocol plays a crucial role in realize multiple linear regression securely [17, Sect. 5.2]. Secure inner product protocol is a security protocol to compute the inner product of w and X, where $w := (w_1, \ldots, w_n)$, $X := (X_1, \ldots, X_n) \in \mathbb{R}^n$. Here,"secure" means that the user and the analyzer want to hide their secret data w, X from each other. The following two candidates are known: the one based on homomorphic encryption and the one using orthogonal complementary space (cf. [17, Chap. 5]).

Figure 8 shows an example of secure and efficient inner product protocol [15]. Let N be a random positive integer, and M an element of $\mathbb{Z}_N^*$. Let X and Y be two elements of this set. We assume that x_i and y_i are randomly chosen and the inner product of X and Y is less than N. Note that the user wants to keep X secret and Analyzer wants to keep Y secret. The protocol is as follows: The user computes $x_i' = x_i M \text{mod} N$ for each i $(i = 1, \ldots, n)$ and sends them to Analyzer. The user has to keep N and M secret. Then Analyzer computes the inner product of Y and X' and sends the inner product back to the user. Finally, the user obtains the inner product $\langle Y, X \rangle$ by computing $\langle Y, X' \rangle / M \bmod N$. The protocol is very efficient compared to the conventional methods [17, Sect. 5.2]. A third party (an attacker of this protocol) does not know X, Y, M, and N. Hence it seems that the attacker does not discover any secret information. However we do not know how to prove the security of the protocol.

6 Conclusion

Big data analytics is useful in many situations such as improving business services and quality of life, but there is a need to develop relevant security technologies. In this paper, we presented some of the latest results of our research for big data security and privacy, multivariate public key cryptosystems, searchable encryption, k-anonymity, and privacy preserving data analysis. In information technology, system architecture is changing constantly. By this changing, new security technologies become essential to the new system architecture. For industrial applications, we need not only theoretical results, but also practical solutions.

Acknowledgments The authors would like to thank the anonymous reviewers for their careful reading of our manuscript and their many constructive comments and suggestions to improve the quality of the paper. A part of this work has been supported by Ministry of Internal Affairs and Communications of the Japanese Government.

References

1. Cloud Security Alliance (CSA).: Security guidance for critical areas of focus in cloud computing V3.0. https://downloads.cloudsecurityalliance.org/initiatives/guidance/csaguide.v3.0.pdf. Cited 22 Mar 2014
2. Cloud Security Alliance (CSA).: Top threats to cloud computing v1.0. https://cloudsecurityalliance.org/topthreats/csathreats.v1.0.pdf. Cited 22 Mar 2014
3. European Commission.: Proposal for a regulation of the European parliament and of the council on the protection of individuals with regard to the processing of personal data and on the free movement of such data (General data protection regulation). http://ec.europa.eu/justice/data-protection/document/review2012/com_2012_11_en.pdf. Cited 22 Mar 2014
4. Hakuta, K., Sato, H., Takagi, T.: On tameness of Matsumoto-Imai central maps with three variables, In: Computer security symposium 2011 (3C3-4) (2011)
5. Harada, K., Sato, Y.: k-Anonymization schemes with automatic generation of generalization trees and distortion measuring using information entropy. In: IPSJ SIG technical reports 2010-CSEC-50, no.47, 1–7 (2010)
6. Koblitz, N.: Elliptic curve cryptosystems. Math. Comp. **48**(177), 203–209 (1987)
7. McKinsey Global Institute.: Big data: The next frontier for innovation, competition, and productivity. http://www.mckinsey.com/insights/business_technology/big_data_the_next_frontier_for_innovation Cited 22 Mar 2014
8. Miller, V.: Uses of elliptic curves in cryptography. In: Williams, H.C. (ed.) CRYPTO 1985. LNCS, vol. 218, pp. 417–426. Springer, Heidelberg (1986)
9. Moh, M.M.: A fast public key system with signature and master key functions. Comm. Algebra **27**(5), 2207–2222 (1999)
10. Moh, M.M.: An application of algebraic geometry to encryption: tame transformation method. Rev. Mat. Iberoamericana **19**(2), 667–685 (2003)

11. Naganuma, K., Sato, H., Yoshino, M., Sato, Y.: Privacy preserved data analysis using searchable encryption, In: Symposium on cryptography and information, security 2014 (2C4-1)
12. Rivest, R., Shamir, A., Adleman, L.: A method for obtaining digital signatures and public key cryptosystems. Commun. ACM **21**(2), 361–396 (1978)
13. Shestakov, I.P., Umirbaev, U.U.: The tame and the wild automorphisms of polynomial rings in three variables. J. Amer. Math. Soc. **17**(1), 197–227 (2004)
14. Sweeney, L.: k-Anonymity: a model for protecting privacy. Int. J. Uncertainty, Fuzziness Knowl.-Based Syst. **10**(5), 557–570 (2002)
15. Takahashi, K., Okeya, K., Japan patent 5297688
16. The White House.: Consumer data privacy in a networked world: a framework for protecting privacy and promoting innovation in the global digital economy, Feb 2012. Available at http://www.whitehouse.gov/sites/default/files/privacy-final.pdf
17. Vaidya, J., Clifton, C.W., Zhu, Y.M.: Privacy preserving data mining. vol 19 in advances in information security, springer, New York (2006)
18. Yoshino, M., Naganuma, K., Sato, H.: Symmetric searchable encryption for database applications, In: International Conference on Network-Based, Information Systems, pp. 657–662 (2011)

Secure Cryptographic Module Implementation and Mathematics

Dooho Choi, Yongjae Choi, Yousung Kang and Seungkwang Lee

Abstract Cryptographic Engineering is defined by the discipline of using cryptography to solve human problems (from the Wikipedia [1]). Main focus of the cryptographic engineering is to implement the cryptographic primitives based on mathematics to the real world device as the manner of software or hardware. Therefore, to study the cryptographic engineering field, mathematics backgrounds are needed as well as the computer engineering and computer science. In this article, we briefly review the trend of the cryptographic engineering field for the last decade. After that, side-channel attack for the crypto modules are introduced and several efforts are explained for preventing the side-channel attack in the area of the cryptographic engineering.

Keywords Cryptographic engineering · Side channel attack · SCARF · Side channel analysis resistant framework · WBC · White-box cryptography · Masking · Shuffling

D. Choi (✉)
Cyber Security Research Laboratory at ETRI, University of Science and Technology(UST), Daejeon, South Korea
e-mail: dhchoi@etri.re.kr

Y. Choi · Y. Kang · S. Lee
Cyber Security Research Laboratory at ETRI, Daejeon, South Korea
e-mail: choiyjskwang@etri.re.kr

Y. Kang
e-mail: youskang@etri.re.kr

S. Lee
e-mail: skwang@etri.re.kr

M. Wakayama et al. (eds.), *The Impact of Applications on Mathematics*,
Mathematics for Industry 1, DOI: 10.1007/978-4-431-54907-9_7,

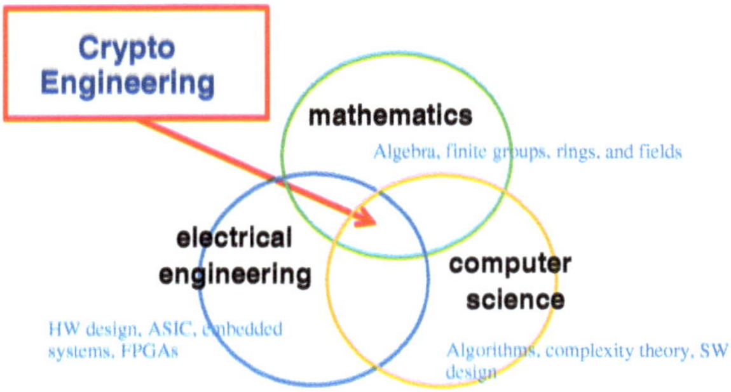

Fig. 1 Concept of cryptographic engineering

1 Introduction

The cryptographic engineering is an emerging bridge research field among mathematics, computer engineering, and computer science. For the research of this field, the background studies are needed for mathematics, electrical engineering, and computer science (see Fig. 1). Hardware design, ASIC, embedded systems, FPGAs, and etc. have to be studied in the electrical engineering. In the computer science field, algorithms, complexity theory, software design, and etc. have to be studied as a background. First of all, the mathematical background is the most difficult part. It is because that almost all cryptographic primitives are based on the mathematical computations (for example, RSA, Diffie-Hellman key sharing, elliptic curve cryptography). Especially, algebra, finite group/rings/fields, elliptic curve theory, and etc. are well studied.

Figure 2 shows the brief history of the crypto engineering. About 12 years ago, at the beginning of this field, the major research topic is an **efficient** hardware and software implementation of the asymmetric key crypto algorithm such as RSA. At that time, one only major purpose of the crypto engineering was the **efficient** crypto implementation.

After the discover of **side-channel attack** (differential power analysis) by Kocher [2], a **secure** implementation was added in the main topic of the crypto engineering. Even if it is mathematically proved that a crypto algorithm is theoretically secure, the secret key can be revealed by analyzing the side-channel information during running the implemented crypto algorithm in the device (for example, power consumption, computation time, electro-magnetic radiation, etc.).

Additionally, one of major issues was emerged in the crypto engineering about 6 or 7 years ago, that is, a **lightweight** implementation. Lightweight means the efficient implementation under the low computing environment (e.g., low power consumption, low memory, and resource constraint computing ability). Recently, one of the emerging research trend is **key hiding** technique. For the key hiding, some researchers and companies are studying and developing the technique based on PUF (Physically Unclonable Function), and some of researchers and companies

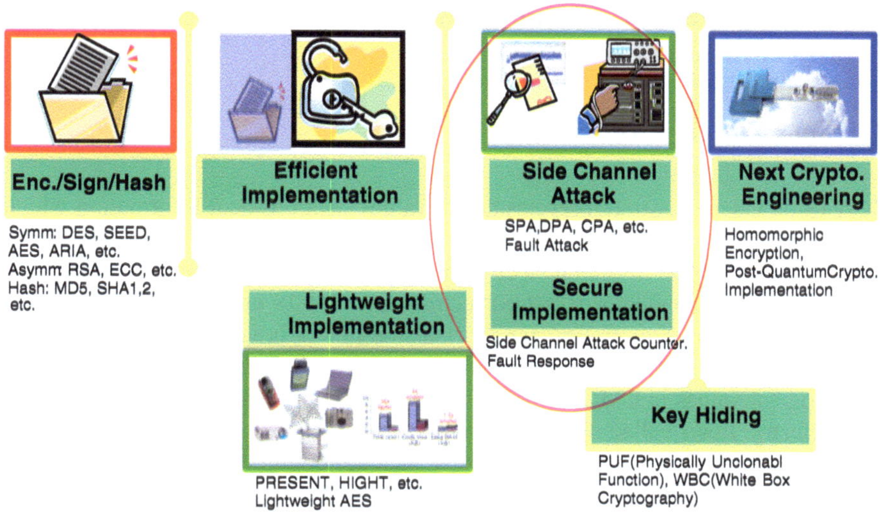

Fig. 2 Trend of cryptographic engineering

are trying to find a software solution based on the white-box cryptography for the key hiding.

In the cryptography field, one of the hot topics is a homomorphic encryption. If an encryption function preserves some operations, it is called a homomorphic encryption. Main application of the homomorphic encryption can be a personal privacy-enhanced cloud service. However, until now, almost all suggested homomorphic encryptions cannot guarantee practical performance. Nevertheless, the next main topic of the crypto engineering will be the efficient implementation for the homomorphic encryption. Another next crypto engineering mainstream will be the implementation of the post-quantum cryptography.

In Sect. 2 we introduce the side-channel attack and relevant project, especially, SCARF (Side Channel Analysis Resistant Framework) project which is the government funded side-channel analysis related project of Korea. In Sect. 3 we consider the principals to protect the side-channel attack for the crypto implementation, and introduce two countermeasure algorithms against the side-channel attack. We address the white-box cryptography as one of the secure crypto implementation solutions in Sect. 4, and then we give some conclusion.

2 Side Channel Attack and Relevant Research Project in Korea

2.1 Side Channel Attack

SCA (Side Channel Attack) reveals secret values with physical information which is leaked from target devices. The leakages are originated from their own physical architecture. Most of digital electric circuits are implemented in complementary metal oxide semiconductor (CMOS) technology. CMOS circuit has high integration property. Furthermore, its standby leakage power is very low and it has the wide noise margin. The efficient structure, however, leaks *important information* when logical values in CMOS circuits are changed. The transitions of internal data make electric flows called dynamic currents. The electric flows are composed with charging currents that is made by load capacitance and saturation currents of transistors. This flows make information related to cryptographic operation. For instance, currents during operation cycles generate power consumptions and electromagnetic radiations. The physical information of the crypto circuit allows an attacker to break crypto system.

Algorithm 1 Left-to-Right Exponentiation Algorithm

Require: $m, d = [d_{n-1}d_{n-2}d_1d_0]$ with binary expression, $N = p \cdot q$ where p, q large primes
Ensure: $c = m^d \mod N$
  $c \leftarrow 1$
  for $k = n - 1$ down to 0 **do**
    $c \leftarrow c \cdot c \mod N$
    if $d_k = 1$ **then**
      $c \leftarrow c \cdot m \mod N$
    end if
  end for
  return c

SCA attack strategies using measured signals are categorized by physical information types. A crypto system provides computation time, power consumption, and EM radiation to the attacker, and each measurement is used for timing attack, power analysis attack, and EM analysis attack, respectively.

Timing attack, which is introduced by Kocher in 1996, reveals secret information using calculation time differences of a crypto algorithm [3]. Because public key algorithms including RSA [4] and ECC [5] have different calculation time depending on message and key values, timing attack is performed for asymmetric key cipher analysis rather than symmetric key cipher analysis in general case.

For example, when the RSA algorithm based on modular exponentiation is implemented, a cipher text c is computed as $c = m^d \mod N$, $N = p \cdot q$ where p, q are the large prime numbers, m is a plaintext, and d is a private key of RSA (see Algorithm 1). Computation of m^d has different calculation time for each m and d value. The time

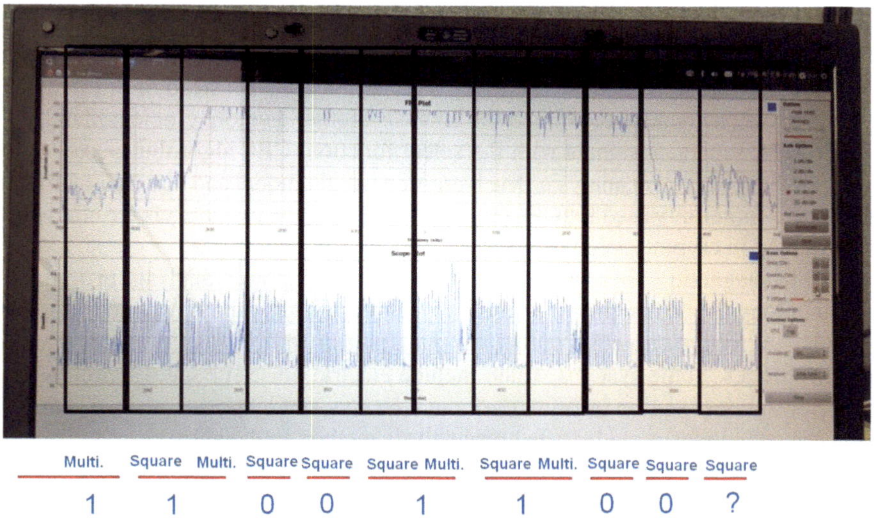

Fig. 3 EM radiation from the smart-phone

difference makes the attacker break the crypto system. Figure 3 shows the electromagnetic radiation from the smart-phone during computing Algorithm 1. Timing attack is not only performed to public key crypto system, but also performed to symmetric key crypto system. Timing attack against the block cipher AES was studied in 2005 [6], but the attack uses cache memory access time rather than calculation time. Another example of practical timing attack is that a remote SSL server using RSA was attacked in 2003 [7].

The power consumption sometimes generates effective information about computation of cryptographic algorithm in the device. Power analysis attack deciphers crypto system using the measured power consumption signals. The basic premise of power analysis is that a power consumption signal while a crypto algorithm is computing has different shape in contrast to power signals in non-crypto operation cycle. Power analysis attack can be classified as simple power analysis, differential power analysis, and correlation power analysis. Simple power analysis (SPA) attack uses a power trace (or few traces) and directly interprets the signal without other special analysis method in contrast to other power analysis methods. Attackers discern differences between exponent operation and multiplication calculation in a power trace, for instance, when SPA is performed to break RSA algorithm. In 1998, differential power analysis (DPA) attack was introduced by Kocher et al. [2]. The basic strategy of DPA attack is statistical signal processing, and the method is one of most powerful SCA algorithms. While SPA uses just one or few signals to analyze crypto system, DPA requires a large number of signals to get statistics of traces. The DPA attack can use various statistics to analyze power traces, but difference of mean proposed by Kocher is popularly used. Difference of mean method groups power consumption signals into two parts using the selection function. Means of each group are calculated, then the maximum difference of means indicates the secret key. Correlation power analysis (CPA) attack calculates correlation coefficient between power traces and hypothetical values to reveal secret values instead of difference of means between signal sets [8], and the analysis method is more efficient than DPA. While DPA separates signals with the select function, CPA algorithm makes leakage models for efficient analysis. The hamming weight model and hamming distance model are commonly used for CPA leakage model, and the two models are efficiently represented in cryptographic devices.

Electromagnetic (EM) analysis is almost same as power analysis in terms of methodology [9]. While power analysis reveals crypto key with power consumption signals, EM analysis uses EM radiation signals measured from cryptographic devices by EM probes. Some systems which include electric noises or special circuits that hide power consumption information cannot be attacked by power analysis, but EM analysis sometimes provides effective analysis method in the situations. As the EM analysis is sensitive to interference and environment, it may require some special techniques for noise reduction.

A different type of SCA is realized through the fault injection on power and clock signals of a cryptographic device and the observation of the corresponding outputs. Differential Fault Analysis (DFA) [10] is using this type of attack and analyzing the outputs. S. Mangard et. al had defined passive non-invasive attacks as side channel attacks, and active non-invasive attacks are defined as fault injection attacks using power glitches or clock glitches in their work [11].

2.2 Introduction to SCARF Project

SCARF project is the government funded side-channel analysis related project of Korea [12] which was started from 2009. in this project, universities, companies, and research institutes are included, and the lead organization of SCARF project is ETRI (Korea Electronics and Telecommunications Research Institute). Major research goal of the SCARF is to develop the key leakage analysis system (SCARF system) including the side channel analysis for the smart devices.

The SCARF system is designed to evaluate the key leakage against side channel analysis and fault-injection attacks on crypto modules implemented by hardware or software languages. The SCARF system is shown in Fig. 4 and is composed of a graphical user interface (GUI) software and evaluation boards.

The SCARF software controls the evaluation boards and an oscilloscope based on the SCARF script, which is the list of execution command that controls the SCARF boards. Users can compose their own script to analyze all SCARF boards at their discretion. By means of the script and its interpreter, the SCARF software collects and analyzes the power or electromagnetic data that are consumed when an electric device is activated or when an encryption module starts an encryption operation; or that is obtained by fault-injection such as voltage glitches. It conducts analyses the signals using various analytical methods such as DPA and CPA, after applying the

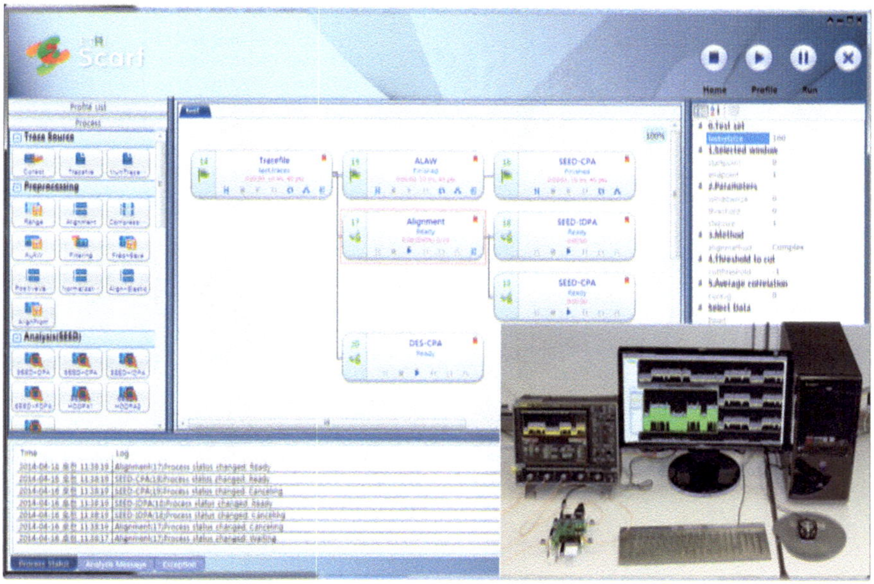

Fig. 4 SCARF system

pre-processing process (e.g. alignment, filtering, compression) to the collected power or electromagnetic waveforms. Unlike side channel analysis, fault injection analysis, which uses the fault result created by injecting a fault such as a voltage glitch or a clock glitch, uses the DFA method for analysis. Users can perform these procedures by only *drag & drop* of a proper *process* and simple editing for properties of the *process*. The *process* of the SCARF software is a GUI tool for waveform collection, pre-processing, encryption algorithm analysis, validation, fault-injection, and fault-injection analysis. The SCARF analysis software as a user-oriented analysis system designated for system development and evaluation, has the following main functions:

- Control of an evaluation board and an oscilloscope
- Visual inspection of collected waveforms
- Pre-processing for improved analysis such as filtering, alignment, and compression of waveforms
- Support for various analysis algorithms
- Support for a user-defined process
- Analysis on SEED, DES, AES, ARIA, RSA-CRT, and ECC algorithms
- A secret key extraction feature by fault injection
- Distributed processing, parallel processing, and profiling.

The type of analysis differs according to the application used for analysis; generally, an analysis technique that is suitable for the encryption algorithm used by the application is required. Therefore, the SCARF analysis software provides

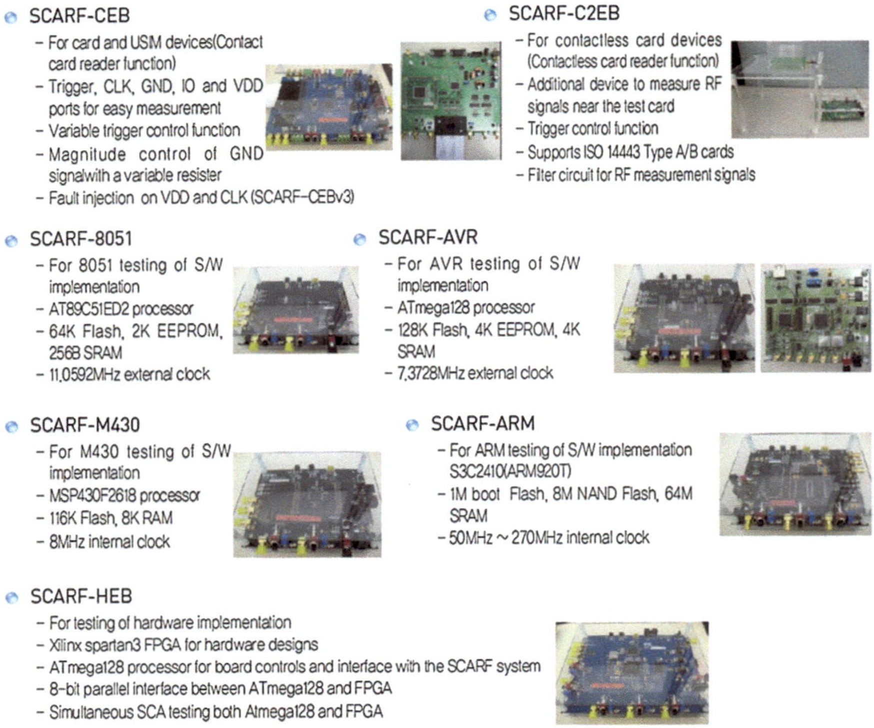

Fig. 5 Main features of SCARF evaluation boards

various type of software, including a method of extracting a secret key by applying various analysis techniques to each algorithm. The software also provides a function for improving the performance of side channel analysis, as well as various types of hardware needed for side channel and fault injection analyses.

SCARF evaluation boards can be classified as card evaluation boards, software evaluation boards and hardware evaluation boards according to testing purpose of them. Main features of the SCARF evaluation boards is shown in Fig. 5. The SCARF-CEB and the SCARF-C2EB is used to analyze the side channel security of card-type electric devices, such as the smart card or SIM card. The former is for a contact card device and the latter is for a contactless card device. The SCARF-CEB ver3 is able also to inject fault on power and clock signals for DFA. The SCARF-8051, the SCARF-AVR, the SCARF-M430 and the SCARF-ARM are used to analyze the side channel security of a software implementation of crypto algorithms on each processor such as AT89C51ED2 8051, ATmega128, MSP430F2618, S3C2410 ARM. The SCARF-AVR ver3 is able also to inject fault on power and clock signals for DFA. The SCARF-HEB is used to analyze the side channel security of a hardware implementation of crypto algorithms, and it is similar to SASEBO board [13] of AIST.

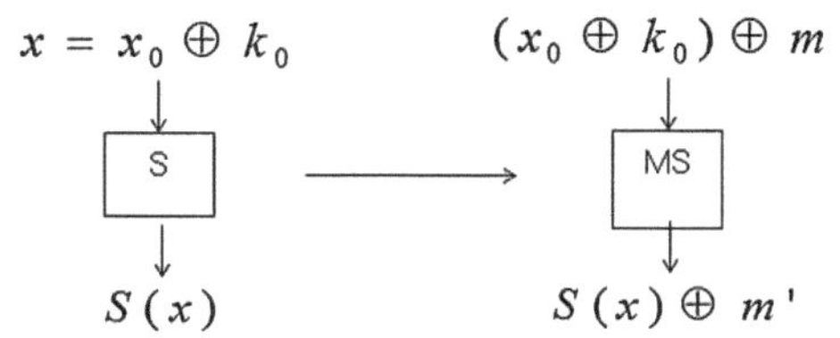

Fig. 6 Concept of masking of an S-box in symmetric-key encryption algorithm

3 Secure Crypto Implementation Against Side Channel Attacks

In this section, we briefly explain basic methods to protect the side channel analysis and give two examples of secure crypto implementations against side channel attacks.

3.1 How to Protect Side Channel Analysis

A typical side channel analysis uses correlations between leakage power and intermediate values during the arithmetic operation of the cryptographic device. The correlation power analysis (CPA) and the differential power analysis (DPA) result from a statistical analysis of two major properties.

The first property is the amplitude dimension. The amplitude is related to power consumption of the cryptographic device. An attacker can find a secret key using the power analysis because the amplitude is not independent of hamming weight of the intermediate values during the arithmetic operation. Therefore, a method to decrease correlations of the amplitude dimension can be a good solution against the power analysis. This method is called the *masking*. In order to remove correlations between the power consumption and the intermediate value, a random number is applied to the intermediate value. Figure 6 shows the concept of the masking method. A simple example of the masking method is as follows.

- **Phase 1.** choose m, m' at random.
- **Phase 2.** $MS(x \oplus m) := S(x) \oplus m'$, where $S(\ )$ is S-box of a symmetric-key encryption algorithm and $MS(\ )$ is a masked S-box.

In this example, generally, $x = x_0 \oplus k_0$ where x_0 is a byte of the input message and k_0 is a byte of the round key. $S(x)$ cannot be calculated directly in order to disturb the power consumption of the original $S(x)$ operation. Therefore, if an attacker analyzes the power traces of the Phase 2 the attacker cannot find a secret key.

The second property is the time dimension. The statistical significance of the power analysis is meaningful when the expected operations occur at the expected time. Therefore, a method to disarrange the operation order can be also a solution

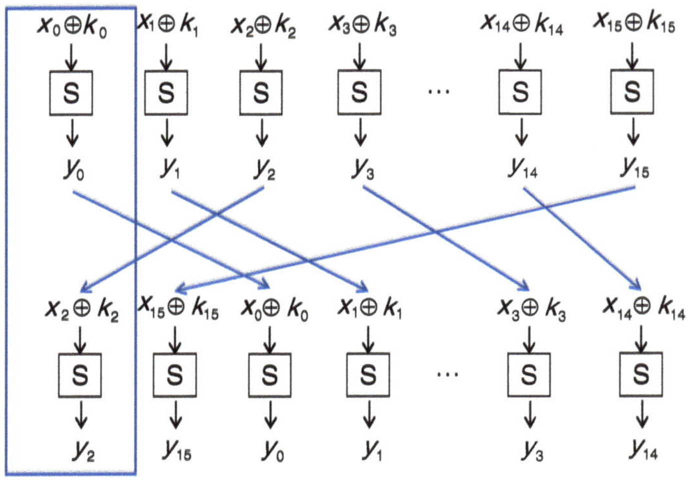

Fig. 7 Concept of shuffling of S-boxes in symmetric-key encryption algorithm

against the power analysis. This method is called the *shuffling*. Figure 7 shows the concept of the shuffling method. The shuffling does not give an attacker the expected operations, so the attacker cannot perform the power analysis just as he/she intended.

3.2 Example 1: Adaptive Random Masking on ARIA

Block cipher ARIA [14, 15], which is based on an involution SPN structure, encrypts and decrypts data in 128-bit blocks. ARIA can have a 128-bit, 192-bit, or 256-bit key size, and a corresponding number of round functions. A round operation in the ARIA procedure has a key addition, substitution layer, and diffusion layer. Round keys are composed by a combination of 512-bit expanded keys (W_0, W_1, W_2, W_3). In addition, the expanded key is made from round functions. The substitution layer of ARIA uses four S-boxes (S_1, S_2, S_1^{-1}, S_2^{-1}), and S_1^{-1} and S_2^{-1} are the inverses of S_1 and S_2.

Recently, Kang et al. proposed an adaptive random masking technique for ARIA and showed some implementation results in terms of software implementation and hardware implementation [16]. The adaptive random masking method basically uses different random numbers for each input block in a round, and the method can defend the SODPA (second-order differential power analysis) without additional protection like S-box shuffling. Since the ARIA algorithm uses four S-boxes at the substitution layer, four masked S-boxes should be defined and, in general, the masked S-box table should be computed and stored at every round. However, the adaptive random masking uses only one masked inversion table, and the masked S-boxes are computed at each round when the algorithm is implemented to software. Furthermore,

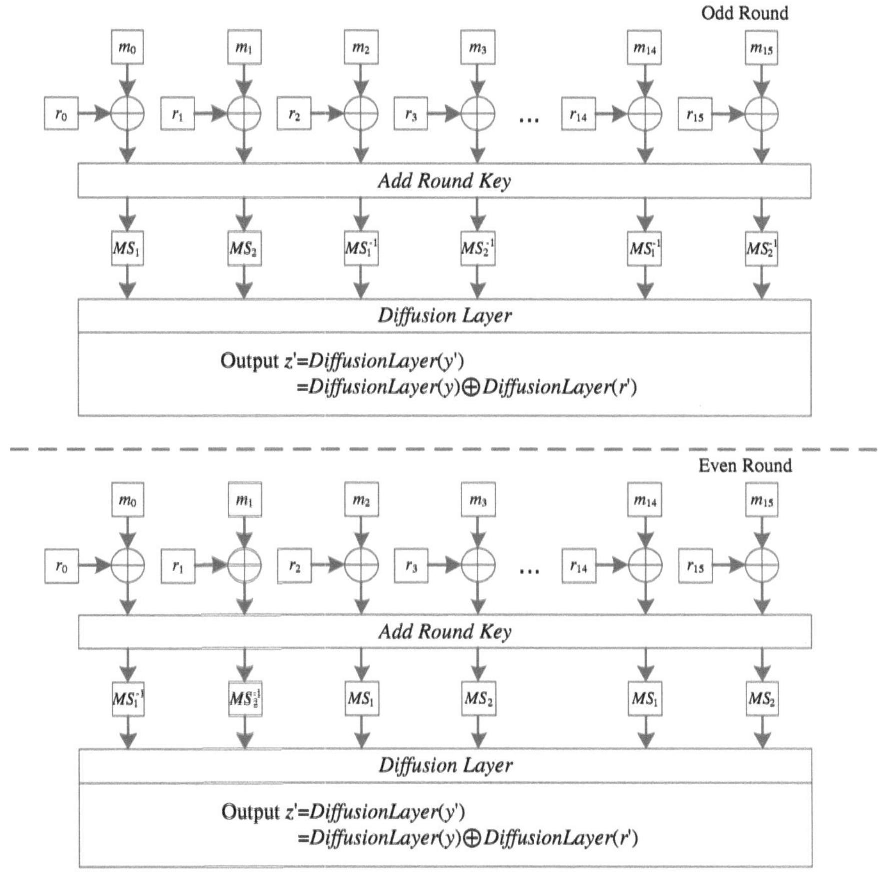

Fig. 8 Adaptive random masking scheme on ARIA

the adaptive random masking uses different random masks for every input block to defend against SODPA. The inversion table and affine transform tables (A, D, A_1, D_1) are precomputed for operation speed in the case of software implementation.

There are sixteen masked S-box calculations in a round, and each calculation uses a different random number r_i , as shown in Fig. 8. The sequence of masked S-boxes in a substitution layer is alternated every round. Since $MS(x_i') = S(x_i) \oplus r_i'$ and $MS^{-1}(x_i') = S^{-1}(x_i) \oplus r_i'$ by the definition of the masked S-boxes, $y' = y \oplus r_i'$, where r_i is an 8-bit input masking random, r_i' is an 8-bit output masking random, x_i is an 8-bit S-box input, x_i' is $x_i \oplus r_i$, and y' is an output of the masked substitution layer. For example, the first block message m_0 in odd round is calculated to $x_0' = m_0 \oplus ek_0 \oplus r_0$ and inserted into the MS_1 box with random masks r_0, r_0', r_1, and r_1', where ek_i is an 8-bit subblock of 128-bit round key, ek, for $i = 0, ..., 15$ and MS_1 is one of masked S-boxes of ARIA. The output of MS_1 box is computed as $y_0' = S_1(x_0) \oplus r_0'$.

Because the adaptive random masking algorithm uses different mask values in a round, the diffusion layer output z' includes random masks as

$$z' = DiffusionLayer(y') = DiffusionLayer(y) \oplus DiffusionLayer(r'). \quad (1)$$

Therefore, the diffusion output with random numbers moves forward the next round of operations. When the n-th round key is denoted as $ek(n)$ and the n-th round random mask r' is defined as $r'(n)$, the $(n+1)$-th round key of the adaptive random masking algorithm is calculated as $DiffusionLayer(r'(n)) \oplus ek(n+1)$.

In terms of software implementation performance, while the ARIA masking algorithm generally needs 1,024 bytes of RAM and 1,024 bytes of ROM to generate making tables, the adaptive random masking method can be implemented with 256 bytes of ROM and 256 bytes of RAM. And, if additional 1,204 bytes of ROM is used for the affine transform tables, the operation speed is improved by approximately 90 %.

In terms of hardware implementation performance, the 8/16-bit architecture module uses only 47.47 μW at 100 kHz 2.5 V, and the overhead of the masking scheme is only a 7.1 % increment in area and a 35.8 % increment in power consumption compared to non-masked ARIA module. It means that the adaptive random masking is appropriate for small devices like RFIDs and smart cards.

3.3 Example 2: Combined Masking on SEED

SEED is a kind of classic Feistel structure block cipher with 16 rounds [17]. The nonlinear operations of SEED are the S-box operations and addition modulo $2^3 2$. In general, these nonlinear parts are considered to construct the masking method. The general masking method requires 512 bytes of RAM for masked S-box (MS) tables having two S-boxes, S_1 and S_2. It may impose a heavy burden on some devices. Furthermore, the use of a general type of MS table after the operation of masked addition in SEED requires one $Secure_{AtoB}$ function call per masked addition. As each round of SEED uses two S-boxes and three modular additions, a masked implementation of SEED requires 512 bytes of RAM corresponding to MS tables and 48 $Secure_{AtoB}$ function calls for 48 addition operations.

Kim et al. proposed a new-style masked S-box which can reduce the amount of operations of the masking addition process as well as the RAM usage [18]. The new-style masked S-boxes are designed to satisfy the following equations, where m and m' are the 8-bit random numbers generated before encryption.

$$MS_1(x) = S_1(x -_8 m) \oplus m', \; MS_2(x) = S_2(x -_8 m) \oplus m'. \quad (2)$$

$x -_8 m$ is defined as $(x - m) \bmod 2^k$, where $x, m \in GF(2^8)$.

The new-style masked S-box table outputs the Boolean masked value from the Arithmetic masked input value. This method reduces the number of $Secure_{AtoB}$ function calls from 48 to 16. in addition, the results of the performance analysis indicate

Table 1 Comparison of size

Implementation	Size	Reference
WB-AES	770 KB	14.5 KB (openSSL 1.0)
WB-DES	4.5 KB	8.3 KB (openSSL 1.0)

that this method reduces the RAM requirements from 512 bytes to 288 bytes and the processing time by 38 % compared with the general masking method. This method can be applicable to some block ciphers that have S-boxes and modular additions as nonlinear components, such as MARS, GOST, and BLOWFISH.

4 White-Box Cryptography

In the white box attack context, an attacker is supposed to have total visibility into execution platforms and software implementations of cryptographic algorithms. This means that a white-box attacker can analyze and tamper with the binary code and the corresponding memory page using disassembler, debugger, and more. Thus, a secret key of cryptographic algorithms stored in plain memory can be unveiled by a white-box attack if they are executed on a device in hostile environment. Software implementations which protect against such white-box attacks are denoted white-box implementations. They are usually deployed in applications where a cryptographic key is involved to protect assets, for example in DRM applications. In the following, we briefly review design principles of white-box implementations and cryptanalysis on them.

4.1 White-Box Implementation

White-box implementations transform a cipher into a network of lookup tables. The secret key is integrated into the lookup tables and protected by random encodings. This is depicted in Fig. 9, where E and G are external encodings, and F is an internal encoding. Since an encoding is followed by its corresponding decoding, the overall functionality of a cryptographic algorithm remains the same.

With this design principle, Chow et al. presented the first white-box implementations on AES in 2002 [19]. Because lightweight logical bitwise operations including XOR operations are performed by table lookups, it comes at quite a substantial price for performance and size. It is reported, for example, that the white-box AES implementation requires approximately 10 times of operations compared to the straightforward AES. In addition, white-box implementations of a cryptographic algorithm require a large amount of memory for lookup tables as shown in Table 1. Thus careful choices should be made where to deploy white-box implementations.

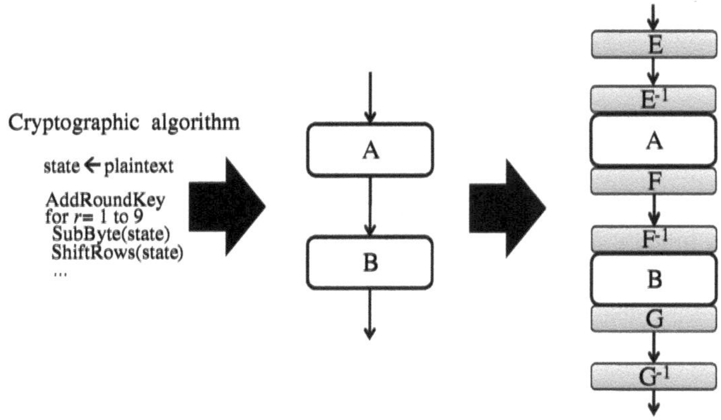

Fig. 9 Encoded lookup table for white-box implementations

Without formal security proofs, its security mainly depends on the white-box diversity and the white-box ambiguity, and it was believed that an attacker is forced to take too many variables into account to succeed in the attack. Unfortunately, several types of cryptanalysis on this approach have been represented thus far.

4.2 White-Box Cryptanalysis

The first white-box AES implementation has been broken by an algebraic cryptanalysis with about 2^{30} time complexity [20]. Without going into depth, the basic idea behind the cryptanalysis is as follows:

1. Let us represent the set of lookup tables of one round of the white-box AES implementation as shown in Fig. 10. P_i and Q_i are internal encodings, T_i are key-integrated S-boxes, and M represents the MixColumns operation.
2. Let y_i be bijective functions between a byte of the input and a byte of the output, and the other bytes be constant but different for each y_i. Define then the functions $f = y_i \circ y_i^{-1}$.
3. Using the set of f, the non-linear part of Q_i can be recovered because f only depends on Q_i and the constants.
4. Because Q_i is the inverse of P_i of the next round, non-linear encodings are canceled out at the boundary of a round. Thus, we can obtain an AES implementation with only linear encodings using mixing bijections by repeating the above steps. It requires about 2^{30} time complexity in total to solve the remaining equations and to derive the secret key.

Michiels et al. generalized this strategy against substitution-linear transformation (SLT) ciphers [21], and Tolhuizen recently improved it to 2^{22} time complexity using

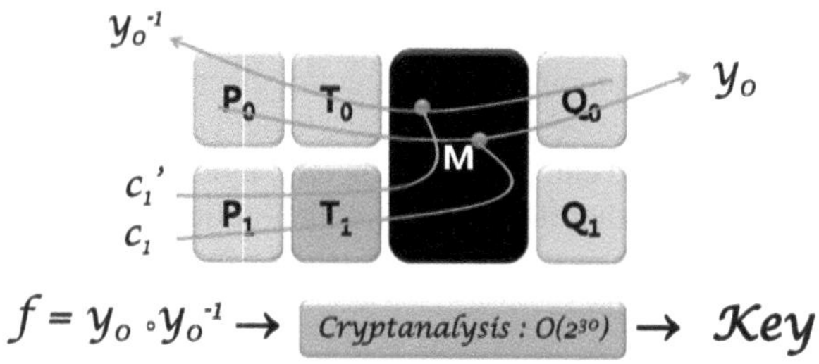

Fig. 10 Simplified cryptanalysis of a round of the white-box AES implementation

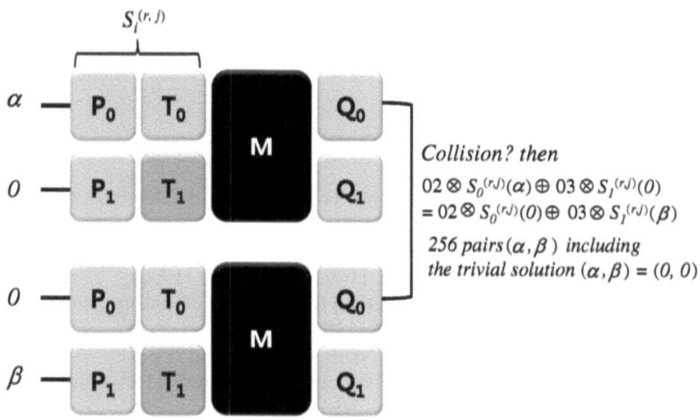

Fig. 11 Simplified collision attack of a round of the white-box AES implementation

a pre-processing step [22]. Among other types of cryptanalysis [23–25] on white-box AES variants, Lepoint and co-workers introduced a collision-based cryptanalysis [26] on white-box AES implementations as shown in Fig. 11. This cryptanalysis finds collisions in order to recover functions S_i and associated key bytes using the equation written in Fig. 11. An attacker in this simplified description constructs a linear system to recover S_0. After that, S_1 can be recovered through an exhaustive search. Once the S_i functions have been recovered, an attacker can easily recover Q_i. The total time complexity of this cryptanalysis is 2^{22}.

5 Conclusions

In this article, we introduced the side channel analysis, secure cyrpto implementation, and white-box cryptography among the various cryptographic engineering research topics. Many basic mathematical backgrounds are required for the crypto engineering.

The cryptographic algorithms, which are based on the mathematical computations, must be interpreted as the computer science and/or electrical engineering language. And then, these are implemented on the various devices (for example, RFID, sensor, smart card, smart-phone, and etc.) with the efficient, lightweight, and/or secure manner. Therefore, we believe that the research collaboration between the mathematicians and engineers is the fastest way to achieve the successful story in the area of the cryptographic engineering.

Acknowledgments This work was supported by the KLA-SCARF project, the ICT R&D program of ETRI (Research on Key Leakage Analysis and Response Technologies).

References

1. Definition of cryptographic engineering by Wikipedia. Available online at http://en.wikipedia.org/wiki/Cryptographic_engineering
2. Kocher, P., Jaffe, J., Jun, B.: Differential power analysis. In: Proceedings of CRYPTO 1999, LNCS 1666, pp. 388–397 (1999)
3. Kocher, P.: Timing attacks on implementations of Diffe-Hellman, RSA, DSS and other systems. In: Proceedings of CRYPTO 1996, LNCS 1109, pp. 104–113 (1996)
4. Rivest, R., Shamir, A., Adleman, L.: A method for obtaining digital signatures and public-key cryptosystems. Commun. ACM **21**(2), 120–126 (1978)
5. Koblitz, N.: Elliptic curve cryptosystems. Math. Comput. **48**(177), 203–209 (1987)
6. Bernstein, D.: Cache-timing attacks on AES. Retrieved 10 Nov 2011. Available online at http://cr.yp.to/antiforgery/cachetiming-20050414.pdf (2011)
7. Brumley, D., Boneh, D.: Remote timing attacks are practical. In: Proceedings of the 12th conference on USENIX Security Symposium, pp. 1–14 (2003)
8. Brier, E., Clavier, C., Olivier, F.: Correlation power analysis with a leakage model. In: Proceedings of CHES 2014, LNCS 3156, pp. 135–152 (2004)
9. Gandolfi, K, Mourtel, C., Oliveier, F.: Electromagnetic analysis: concrete results. In: Proceedings of CHES 2001, LNCS 2162, pp. 255–265 (2001)
10. Biham, E., Shamir, A.: Differential fault analysis of secret key cryptosystems. In: Proceedings of CRYPTO 1997, LNCS 1294, pp. 513–525 (1997)
11. Mangard, S., Oswald, E., Popp, T.: Power analysis attacks: revealing the secrets of smart cards, Springer (2007)
12. SCARF project. Available online at http://www.k-scarf.or.kr
13. SASEBO and SAKURA project. Available online at http://www.morita-tech.co.jp/SAKURA/en/index.html
14. National Security Research Institute: The ARIA Specification. http://210.104.33.10/ARIA/index-e.html
15. Kwon, D., Kim, J., Park, S., Sung, S., Sohn, Y., Song, J., Yeom, Y., Yoon, E., Lee, S., Lee, J., Chee, S., Han, D., Hong, J.: New block cipher: ARIA. In: Proceedings of ICISC 2003, LNCS 2971, pp. 432–445 (2003)
16. Kang, J., Choi, D., Choi, Y., Han, D.-G.: Secure hardware implementation of ARIA based on adaptive random masking technique. ETRI J. **34**(2), 76–86 (2012)
17. Koera Internet & Security Agency: Block Cipher Algorithm SEED. http://seed.kisa.or.kr/eng/about/about.jsp
18. Kim, H., Cho, Y., Choi, D., Han, D.-G., Hong, S.: Efficient masked implementation for SEED based on combined masking. ETRI J. **33**(2), 267–274 (2011)
19. Chow, S., Eisen, P., Johnson, H., Oorschot, P.C.V.: White-box cryptography and an AES implementation. In: Proceedings of SAC 2002, LNCS 2595, pp. 250–270 (2003)

20. Billet, O., Gilbert, H., Ech-Chatbi, C.: Cryptanalysis of a white box AES implementation. In: Proceedings of SAC 2004, LNCS 3357, pp. 227–240 (2004)
21. Michiels, W., Gorissen, P., Hollmann, H.D.: Cryptanalysis of a generic class of white-box implementations. In: Proceedings of SAC 2009, LNCS 5867, pp. 414–428 (2009)
22. Tolhuizen, L.: Improved cryptanalysis of an AES implementation. In: Proceedings of the 33rd WIC Symposium on Information Theory, (2012)
23. Bringer, J., Chabanne, H., Dottax, E.: White box cryptography: another attempt. In: IACR Cryptology ePrint Archive, Report 2006/468, https://eprint.iacr.org/2006/468.pdf
24. Mulder, Y.D., Roelse, P., Preneel, B.: Cryptanalysis of the Xiao-Lai white-box AES implementation. In: Proceedings of SAC 2004, LNCS 3357, pp. 34–49 (2004)
25. Mulder, Y.D., Wyseur, B., Preneel, B.: Cryptanalysis of a perturbated white-box AES implementation. In: Proceedings of INDOCRYPT 2010, LNCS 6498, pp. 292–310 (2010)
26. Lepoint, T., Rivain, M., Mulder, Y.D., Roelse, P., Preneel, B.: Two attacks on a white-box AES implementation. In: Proceedings of the workshop on selected areas in cryptography (2013)

The Continuing Challenge of Steel: How to Win Mathematicians and Influence Scientists in Other Disciplines

Kaoru Sato

Abstract The steel industry is a growing industry. We have seen constant improvements in the steel quality with the help of analytical sciences and the measurement technology. The challenge of steel continues. Collaborations with scientists in other disciplines are vital for the innovation in steel: The steel industry of Japan is currently working closely with neutron scientists. Mathematics is already embedded in the daily activities of the steel industry. Further communication and synergy with mathematicians will play a crucial role for a creation of the new steel era.

Keywords Steel industry · Analytical sciences · Microstructure control · Electron microscopy · Neutron · University-industry collaboration

1 Introduction

The steel industry is a growing industry. The production of crude steel in 2012 was 1.5 billion tons, which is about double the amount in 2000. Steel production will exceed 2.2–2.9 billion tons by 2050 [1]. This is because there will be a dramatic increase in steel consumption per head in developing countries. What is important is that it is not simply the change in the quantity but the quality of steel. When we look back to the history, the tragedy of the Titanic would have been certainly avoided if the steel of today was used [2]. The transition of high towers from the Eiffel Tower, Tokyo Tower to the Tokyo Sky Tree was materialised thanks to the innovation of steel [3, 4]. Sir Alan Cottrellfs remark always encourages us. Above all, there remains the continuing challenge of steel. For many reasons—an abundance of rich and reducible ores, high intrinsic strength and melting point, ease of alloying and richness

K. Sato (✉)
Steel Research Laboratory, JFE Steel Corporation, 1, Kawasaki-cho, Chuo-ku, Chiba-shi, Chiba Pref. 260-0835, Japan
e-mail: ka-sato@jfe-steel.co.jp

M. Wakayama et al. (eds.), *The Impact of Applications on Mathematics*, Mathematics for Industry 1, DOI: 10.1007/978-4-431-54907-9_8,

of phase transformation behaviour—steel remains by far the best source of cheap, reliable tensile strength, and is unlikely ever to be overtaken in this 'mass-market' role [5]. Steel production requires many sciences. Scientists and engineers from many technical backgrounds work in this industry. For the further innovation of steel, it is crucial, I believe, to work with scientists in many more disciplines. In this context, we have great expectations for collaborating with mathematicians.

2 Advanced Steel and Analytical Sciences

The development of the steel industry has been driven by the development of many sciences and technologies. The leap in the steel refinement was achieved through precise measurement of the temperature of molten steels. The need for precise temperature measurements accelerated the development of quantum theory [6]. The development of both science and technology enhances one another.

We saw a great improvement in diffraction contrast theory and experiments in the field of transmission electron microscopy (TEM) in the 1960s. This directly led to the days of understanding and designing metals and alloys for structural use through direct microstructure observation. From late 1970s, we have seen a development in analytical electron microscopy. Thanks to the development of the field-emission gun (FEG) and the realization of spherical aberration (Cs) correction for magnetic objective lenses it has become possible to conduct sub-Å observation and sub-nm microanalysis. By utilizing these techniques, we have attained high strength hot-rolled steel utilizing particle dispersion hardening [7]. In this steel, we are able to control complex carbides whose size is of the order of single nm, thus achieving a tensile strength of 800 MPa. It was state-of-art analytical electron microscopy that revealed that the secret of this high strength is due to the presence of nm-sized complex carbides (Ti, Mo)C [8]. What is remarkable is that this nano-structured steel is produced massively for automotive use under precise control of the thermo-mechanical control process (Fig. 1). The Cs-corrected STEM has been successfully applied to the direct imaging of native passive films on stainless steel [9]. Thanks to the finely focused electron probe and cross-sectional specimens, we were able to obtain a precise elemental profile as well as grasping the structure of the passive film.

Surface technology/engineering is another important area in steel. Our efforts on improving surface properties such as anti-corrosion, lubrication and a good appearance continue. Scanning Electron Microscopy (SEM) has been widely used for surface imaging. We have pioneered the use of low accelerating voltage SEM, which enables higher surface sensitivity and better signal to noise ratio [10] (Fig. 2). Compared with 10 years ago, we have much better understanding of steel surfaces, which has enabled a designing of better surface-coated and sheet steel [11]. The microstructures of new generation multiphase steels are being characterized using new backscattered electron imaging [12].

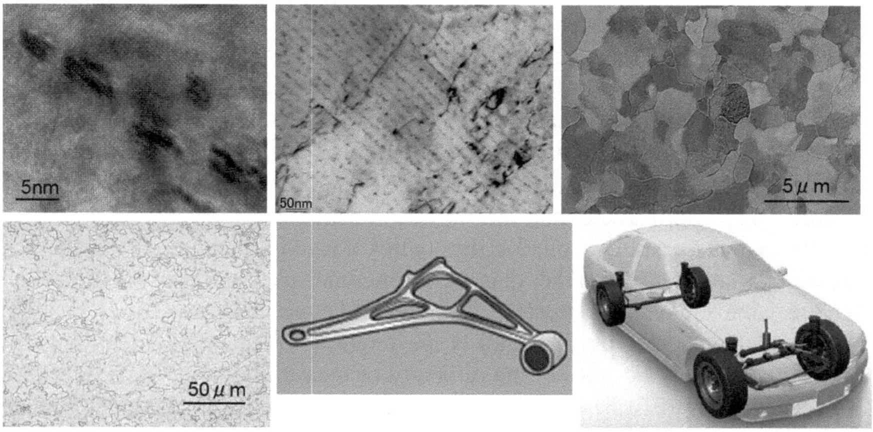

Fig. 1 High strength steel for automotive: from atomic structure to actual car component. Nanometer-sized carbide particles are the secret of high strength. Precise control of precipitates allows us stable production of tons of this advanced steel

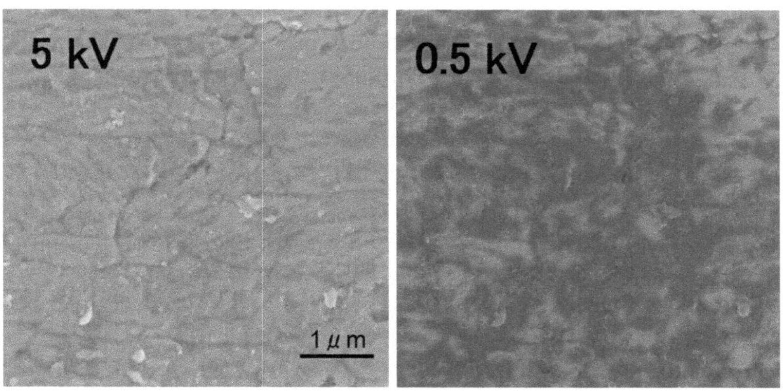

Fig. 2 Scanning electron micrographs of steel surfaces. Image taken at 0.5 kV is more sensitive to surface oxides

3 Mathematics in Steel Industry

Mathematics is embedded in the daily activities in the steel industry. In the field of analytical sciences, for example, we routinely use terms such as statistical error, 3σ, deconvolution of the spectrum, Monte Carlo simulation of the electron beam, maximum entropy method and others. For material designing, thermodynamical calculations, THERMOCAL and FACTSAGE, are being used daily. When simulating how steel products behave when they become, for example, automotive parts, CAE techniques are commonly used [13]. The challenges of constructing reliable modelling for microstructure evolution, fatigue and fracture, corrosion and other dynamic phenomena are being pursued. Control of large scale "chemical reactors", such as blast furnaces or converters requires mathematical modelling and simulation because in these areas experiments are often very difficult to perform. Thus, bridging experimental data obtained for small scale simulators and actual large-scale equipment is essential [14]. For measurements and control and quality design, various mathematical approaches such as regression analysis and multivariate analysis are used routinely [15]. Multivariate analysis for near-infrared spectroscopy i.e. CHEMOMETRICS is successfully applied to the control of acid for pickling. In this application multivariate statistics rather than a conventional spectrum analysis can extract high precision and reliable information on the acid concentrations [16].

Light elements such as B, C and N are important elements in steel, but they are not studied fully because of the difficulty of measuring them. Electron energy loss spectroscopy (EELS) is potentially an excellent technique for light element detection at a nanometer spatial resolution. However, the major problem in detecting small edges in energy loss spectra was channel-to-channel variations in the gain. The standard procedure for removing this noise is to divide by a previously collected "gain spectrum" obtained with the diodes uniformly illuminated. The noise is only reduced by a factor of about 2 as shown in Fig. 3b. Averaging many spectra shifted by one channel relative to each other is an easy method (Fig. 3c), however more than 50 spectra must be acquired to reduce the noise significantly. We developed an iterative averaging method which will eliminate the channel-to-channel variations. The details of this method are described by Boothroyd et al [17]. Only 8–9 spectra and 2–3 iterations are enough to reduce the noise level by a factor of 20–0.005 % (Fig. 3d). This method allows the detection of 0.1 mass% B in Ni_3Al alloy [17] and 0.04 mass% N in a stainless steel [18].

Peak separation or deconvolution is widely used in spectrum analysis. In the characteristic X-ray analysis using energy dispersive X-ray spectroscopy (EDS), separation of X-ray peaks e.g. Ti L-line and N K-line was difficult because of the poor energy resolution of EDS. However, by measuring the energy resolution of the spectrometer you can deconvolute the two peaks. Fig. 4 is such example; the apparent single peak is deconvoluted into N K-peak and Ti L-peak. In this type of analysis you can use either calculated Ti and N peaks or measured Ti and N peaks as references [19]. As discussed in Sect. 2, low-voltage SEM improves sensitivity to surfaces. When this technique is applied, we must use low energy characteristic X-rays for microanalysis. Although the development of a high energy resolution detector is preferred [20], deconvolution software should be utilized fully for better qualitative/quantitative analyses.

As shown in the two examples, mathematical analyses are successfully used in extracting hidden information in experimental data

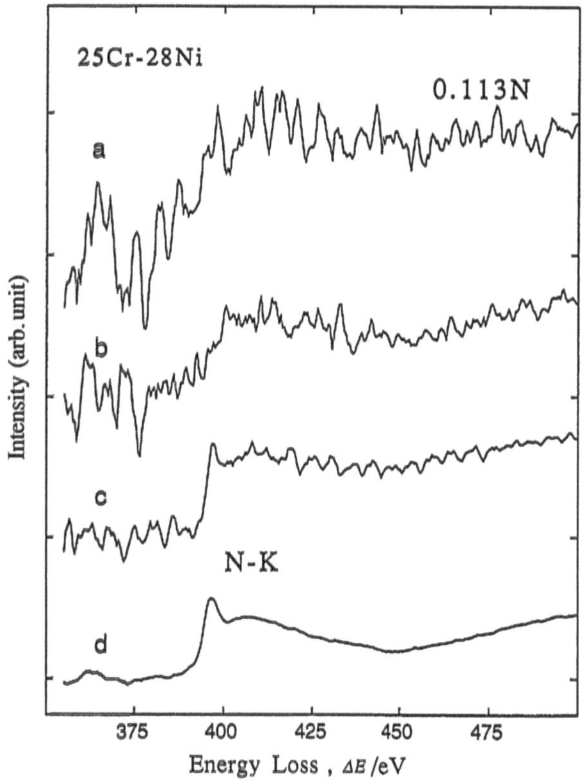

Fig. 3 Electron energy loss spectra of 0.113 %N stainless steel. N-K edge. *a* raw spectrum *b* after divided by gain spectrum *c* average of 9 spectra and *d* after iteration averaging. (Figure 1 of reference [18] with a permission by Japanese Institute of Metals and Materials.)

4 Industry-Academia Collaboration

Traditionally, Japanese industry has had a tendency of doing all research by themselves rather than working with universities. However, increasingly, companies are now working with universities and national institutes.

In Japan, the use of advanced gigantic experimental facilities such as Synchrotron Radiation (SR) high brightness beams and neutron sources in the industry sector have been promoted successfully. Every steel manufacturer utilises SR for their R & D [21]. The Iron and Steel Institute of Japan (ISIJ) has been promoting a research project on the utilization of the neutron for several years [22]. Based on deep and frank discussion among scientists in the steel field and the particle beam field, the steel neutron community is growing. As a member of this community, I enjoy very much working with scientists in other disciplines. Since 2013 we are working closely with scientists of RIKEN who are promoting the development of a compact

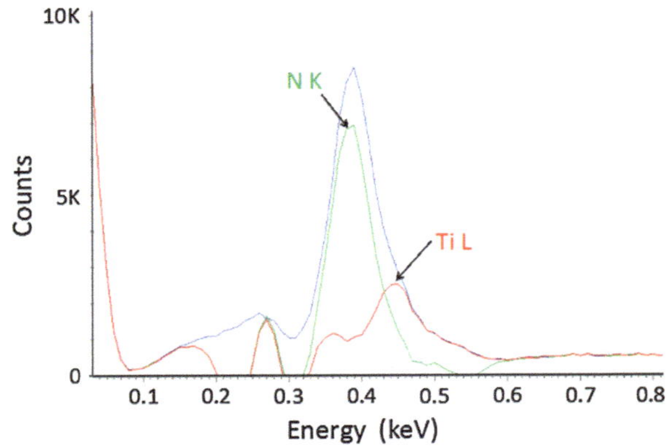

Fig. 4 Deconvolution of energy dispersive X-ray spectrum peak from TiN. A single peak is separated into N K and Ti L peaks

neutron source, RANS (RIKEN Accelerator-driven compact Neutron Source) [23]. The feasibility study group for compact neutron sources was launched in ISIJ. At first we experienced mutual misunderstanding because we used different languages. It is exciting to see the steel engineers understanding physics terminology and the scientists of Riken start using a common language with us. This has been a great lesson; there is no shortcut but to tackle the target research together by sharing the same objectives. Since we started working with mathematicians, our steel scientists have seen many positive changes. Our researchers are now trying to explain basics of steel research to mathematicians. They have started designing several experiments based on mathematicians' suggestions i.e. simpler experiments for understanding the mechanisms rather than the more realistic and complicated experiments that we usually perform. To further promote our synergy, I believe that the exchange of people e.g. internships and accepting postdocs is crucial.

There is vast room for the continuing challenge of steel. Mathematics will play a vital role in processing innovation, design of materials and advanced characterisation. Several examples include high temperature processes, forming and rolling, and plating. For material design, important keywords will be the microstructure evolution, prediction of mechanical properties, fatigue and fracture, corrosion and hydrogen embrittlement. In the category of characterisation/measurements, visualisation, advanced image/spectrum analyses and data mining will see great advancement through collaborating with mathematicians.

5 Concluding Remarks

We live in the steel age. As an analytical scientist, I strongly feel that mass production of steel products today is driven by quick and quantitative process control analytical techniques. Furthermore, high quality steel products are being developed through the precise control of microstructures, precipitates and surface structures. It is advanced analytical techniques, including state-of-the-art electron microscopes, that enable the designing of the nanometer-size particles and the surface oxides of such steel products [24]. Similar innovations are taking place in fields other than analytical sciences. I strongly hope that we can create something totally new, which we cannot even find a word for at this moment. I hope that collaborations with industry will inspire the creation of new mathematics, new sciences for academia. I would like to act as a translator who can win mathematicians and influence scientists in other disciplines as well as to deepen my own speciality. I would like mathematicians to be proactive, too. I highly appreciate the high spirit of the Institute of Mathematics for Industry (IMI) of Kyushu University for their ambition for winning industry people over and influencing engineers in other disciplines.

Acknowledgments The author would like to thank Professor Masato Wakayama, the director of IMI Kyushu University, for giving me this great opportunity to communicate with the distinguished mathematicians.

References

1. http://www.iea.org/media/workshops/2013/ccs/industry/backgroundpaper.pdf 22 Aug 2013
2. Felkins, K., Leighly Jr, H.P., Jankovic, A.: The royal mail ship Titanic: Did a metallurgical failure cause a night to remember? JOM **50**(1), 12 (1998)
3. STEEL CONSTRUCTION TODAY & TOMORROW: Japan Iron and Steel Federation and Iron and Steel Institute of Japan, Nov 2010
4. Sueishi, N., Arakawa, T., Ohmori A., Matsui, T.: JFE technical report, No 14, p. 9 (2009)
5. Pettifor, D.G., Cottrell A.H.: Electron theory in alloy design (maney publishing), chapter 10 future directions p. 284 (1992)
6. Baracca, A.: 1905, annus mirabilis: the roots of the 20th-century revolution in physics and the take-off of the quantum theory. LLULL **28**, 295 (2005)
7. Funakawa, Y., Shiozaki, T., Tomita, K., Yamamoto, T., Maeda, E.: Development of high strength hot-rolled sheet steel consisting of ferrite and nanometer-sized carbides. ISIJ. Int. **44**, 1945 (2004)
8. Sato, K., Nakamichi, H., Yamada, K.: Characterization of nanometer-sized complex carbide in steel using electron microscopy. Kenbikyo (in Japanese) **40**, 183 (2005)
9. Hamada, E., Yamada, K., Nagoshi, M., Makiishi, N., Sato, K., Ishii, T., Fukuda, K., Ishikawa, S., Ujiro, T.: Direct imaging of native passive film on stainless steel by aberration corrected STEM. Corros. Sci. **12**, 3851 (2010)
10. Sato, K., Nagoshi, M., Kawano, T.: Application of low-voltage scanning electron microscopy to the characterization of steel surface. Tetsu to Hagane (in Japanese) **93**, 169 (2007)
11. Sato, K., Sakurai, M., Taira, S, Hamada, E.: Plan-view and cross-sectional characterization of thiourea-treated phosphorus-added steel surface. Electron. Microsc. **53**(5), 553 (2004)
12. Mikmekova, S., Yamada, K.K., Noro, H.: TRIP steel microstructure visualized by slow and very slow electrons. Microscopy **62**(6), 589 (2013)
13. Urabe, M., Ishiwatari, A., Kano, H., Hiramoto, J., Inazumi, T.: Analysis method to identify cause of springback in press forming. IDDRG 2013 conference p. 5 (2013)
14. Natsui, S., Ueda, S., Nogami, H., Kano, J., Inoue, R., Ariyama, T.: Dynamic analysis of gas and solid flows in blast furnace with shaft gas injection by hybrid model of DEM-CFD. ISIJ Int. **51**, 51 (2011)

15. Shigemori, H., Kano, M., Hasebe, S.: Optimum quality design system for steel products through locally weighted regression model. J. Process Control **21**, 293 (2011)
16. Matsushima, T., Inose, M., Aizawa, S., Tahara, K.: CAMP-ISIJ (in Japanese), p. 350 (2013)
17. Boothroyd, C.B., Sato, K., Yamada, K. In: Peachey L.D., Williams D.B., Proceedings of the 12th international congress for electron microscopy, p. 80. San Francisco Press, San Francisco, (1990)
18. Yamada, K., Sato, K., Boothroyd, C.B.: Materials transactions. JIM **33**(6), 571 (1992)
19. McCarthy, J.J., Schamber, F.H.: Least-Squares Fit with Digital Filter. A status report. In: Heinrich, K.F.J., Newbury, D.E., Myklebust, R.L., Fiori, C.E. (eds.) Energy Dispersive X-ray Spectrometry, p. 273. National Bureau of Standards Special Publication 604, Washington (1981)
20. Noro, H., Sato, K., Tanaka, K.: Advantages of transition-edge sensor combined with low voltage. Hyomen Kagaku (in Japanese) **31**, 610 (2010)
21. Nagoshi, M., Kawano, T., Sato, K., Funakawa, M., Shiozaki, T., Kobayashi, K.: Phys. Scr. **T115**, 480 (2005)
22. Yasuhara, H., Sato, K., Toji, Y., Onuma, M., Suzuki, J., Tomota, Y.: Tetsu to Hagane (in Japanese) **96**, 545 (2010)
23. Riken Youtube, Neutrons Transforming Japanese Manufacturing A Record of RANS R&D http://www.youtube.com/watch?v=FXLtM6Lqct0 (10 January 2014)
24. Sato, K. In: Proceedings of the 3rd international symposium on steel science 3 (ISSS 2012), p. 11. (2013)

Implicit Methods for Simulating Low Reynolds Number Free Surface Flows: Improvements on MAC-Type Methods

José A. Cuminato, Cassio M. Oishi and Rafael A. Figueiredo

Abstract This paper is concerned with describing the main improvements introduced to the MAC (Marker-And-Cell) method for the numerical simulation of low Reynolds number free surface flows, namely: a stable implicit treatment of the pressure boundary condition for projection methods, a semi-implicit method based on the Crank–Nicolson (C–N) discretization of the momentum equations, a more accurate method for moving the massless particles representing the free surface and a viscoelastic model based on the Pom-Pom constitutive law, are discussed. Low Reynolds number free surface flows appear in a number of important industrial processes in the oil, food, cosmetic and medical industries and their simulation present a challenge for explicit MAC-type methods due to their parabolic time step constraint. The simulation of moving boundary problems presents a number of difficulties for a numerical method. For the semi-implicit (C–N) MAC method the main difficulty appears in applying the projection method to uncouple velocity and pressure, this is in addition to other difficulties of correctly imposing the boundary conditions on the free surface and the free surface representation itself.

Keywords Navier–Stokes equations · Viscoelastic fluid flows · Free surface · MAC scheme · Implicit strategy · Jet buckling · Extrudate swell · MAC method review · Non-Newtonian fluids

---

J. A. Cuminato (✉) · R. A. Figueiredo
Instituto de Ciências Matemáticas e Computação (ICMC), Universidade de São Paulo, Av. do Trabalhador SãoCarlense 400, São Carlos, SP, Brazil
e-mail: jacumina@icmc.usp.br

R. A. Figueiredo
e-mail: rafigueiredo22@gmail.com

C. M. Oishi
Faculdade de Ciência e Tecnologia (FCT), Universidade Estadual Paulista Julio de Mesquita Filho, Roberto Simonsen, 305, Presidente Prudente, SP, Brazil
e-mail: cassiooishi@gmail.com

M. Wakayama et al. (eds.), *The Impact of Applications on Mathematics*,
Mathematics for Industry 1, DOI: 10.1007/978-4-431-54907-9_9,

1 Introduction

Numerical methods for the simulation of viscoelastic flows with free surface, in general, is a subject that has attracted the attention of many researchers recently. The development of methods to treat free surface flows has been of particular interest as reported by Bonito in [3], that presents the simulation of the stretching of a filament by a finite element method combined with a volume of fluid (VOF) formulation. A smoothed particle hydrodynamics (SPH) method was used by Xu et al. [31] to simulate several experiments of viscoelastic fluids involving free surfaces with application in engineering and proposed a new treatment of rigid boundary conditions for improving computational performance. For the filling process with viscoelastic fluid [32] used a finite volume method combined with level sets to represent the interface. In addition, we can also mention [4], that discusses the numerical analysis and simulation of non-Newtonian flows with complex free surfaces. Related works are also reported by Ciarlet et al. [8], Owens and Phillips [22] that present details on the construction of constitutive models, numerical discretization techniques, numerical methods for solving viscoelastic fluid flows and numerical results comparing different techniques.

In the paper [16] published in 2008 it was presented a review of the improvements introduced in the MAC methodology by the research group at ICMC of University of São Paulo. Since then, several other improvements were introduced into the methodology on top of those reported in the previous paper. It is the main purpose of this paper to describe such improvements and present numerical simulation of practical industrial problems that illustrate the importance of the new features introduced. For a start, we recall that, the MAC method [12] is a finite difference technique for discretizing the Navier–Stokes equations (NSE) on a staggered grid. For the free surface approximation the MAC method employs a front tracking technique, with Lagrangian particles representing the free surface. A projection method is then applied in order to uncouple the solutions for the velocity field from that for the pressure field. Having solved for the velocity and pressure fields the marker particles are moved according to equations $\dot{\mathbf{x}} = \mathbf{u}$.

Originally the MAC method employs an explicit time discretization of the NSE that, frequently requires a very small time step for stability. This is specially serious when the Reynolds number is small. In the simulation of viscoelastic free surface flows this is often the case as this type of flows frequently involves very thick fluids. To overcome the parabolic time step restriction imposed by the explicit time discretization it is necessary to resort, for example, to a semi-implicit discretization of the momentum equations. In so doing it was found that (see [17]) the resulting semi-implicit method did not enjoy unrestricted stability, unless the boundary conditions on the free surface were also discretized implicitly. This in its turn couples again the momentum to the continuity equations, hampering the decoupling of the solvers, and hence increasing the computing time per step.

Another issue that we shall deal with in this paper is the discretization of the dynamical equations for the marker particles. In the original MAC method, as the time step is very small, there is no need for a more accurate high order discretization of these equations. However, when a semi-implicit scheme is employed and a larger time step is allowed, the explicit discretization of the particle dynamics equations needs to be improved in order that mass be conserved accurately. A study of this issue was carried out and will be reported in the main text below.

Oishi et al. [17] proposed a semi-implicit MAC method to simulate free surface flows with low Reynolds numbers. This technique was later extended by Oishi et al. [21] to the simulation of three-dimensional viscoelastic fluid flows for the Oldroyd-B model. The papers [11, 19, 20] applied the same technique for two and three dimension problems. They used a second-order Runge–Kutta method to integrate both; the particle convection equation at the free surface and the constitutive equation of the eXtended Pom-Pom (XPP) fluid model.

Finally regarding the simulation of viscoelastic fluids new constitutive laws have been implemented and tested against experimental results. We shall report the numerical simulation of complex polymer melt flows using the XPP model. Some numerical experiments, simulating benchmarking problems from the literature, will be presented and discussed.

2 The MAC Method

The Navier–Stokes Equations (NSE) for incompressible and isothermal flows can be written in dimensionless conservative form as

$$\frac{\partial \mathbf{u}}{\partial t} + \nabla \cdot (\mathbf{u}\mathbf{u}) = -\nabla p + \zeta \frac{1}{Re} \nabla^2 \mathbf{u} + \theta \nabla \cdot \boldsymbol{\tau} + \frac{1}{Fr^2} \mathbf{g}, \quad \text{in} \quad [0, T] \times \Omega, \tag{1}$$

$$\nabla \cdot \mathbf{u} = 0 \quad \text{in} \quad [0, T] \times \Omega, \tag{2}$$

where t is time, $\mathbf{u}$ is the velocity vector field, p is the pressure and $\mathbf{g}$ is the gravity field. In Eq. (1), $\boldsymbol{\tau}$ is the non-Newtonian extra-stress tensor which is defined by an appropriate constitutive equation. The dimensionless parameters $Re = \frac{\rho L U}{\mu}$ and $Fr = U/\sqrt{gL}$ are the Reynolds and Froude numbers, respectively, where L and U are appropriate length and velocity scales, ρ is density and μ is the viscosity of the fluid. Ω is a domain in $\mathbb{R}^2$ or $\mathbb{R}^3$ and $[0, T]$ is a time interval. In Eq. (1), ζ and θ are chosen according to the fluid model that will be used in the simulation.

In the original version of the MAC method, Harlow and Welch (see [12]) introduced their methodology for Newtonian fluid flows. In this case, the motion equations are simplified by setting $\zeta = 1$ and $\theta = 0$ in Eq. (1). Thus, the equation of motion together with the appropriate boundary conditions are solved by an explicit finite differences discretization on a staggered grid (see Fig. 1), where the velocities are calculated at the cell faces and all other quantities are computed at the cell center. Another fundamental point worth of note about the MAC method is the classification of the grid cells according to their position relative to the fluid. More details on the MAC stencil arrangement and on cell classification can be found in [16].

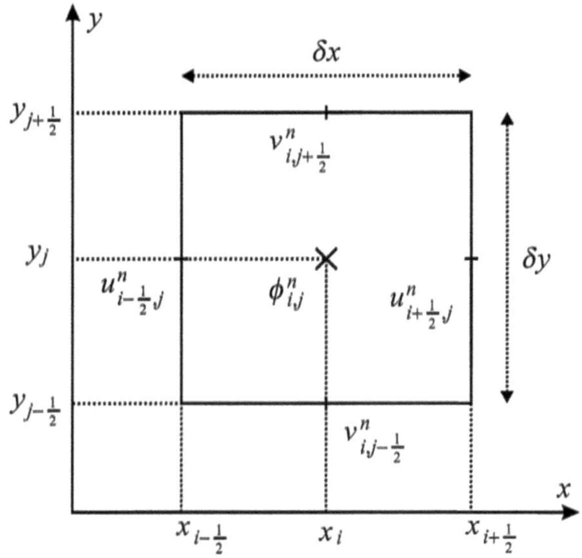

Fig. 1 Staggered grid cell: dependent variables arrangement

For two-dimensional problems, two conditions are imposed on the free surface: normal and tangential stress conditions. For example, for a Newtonian fluid flow, these boundary conditions are given respectively by:

$$\mathbf{n} \cdot \boldsymbol{\sigma} \cdot \mathbf{n}^{\mathrm{T}} = 0, \tag{3}$$

$$\mathbf{m} \cdot \boldsymbol{\sigma} \cdot \mathbf{n}^{\mathrm{T}} = 0. \tag{4}$$

In the equations above $\boldsymbol{\sigma}$ is the total stress tensor given by

$$\boldsymbol{\sigma} = -p\mathbf{I} + \frac{1}{Re}\left[(\nabla\mathbf{u}) + (\nabla\mathbf{u})^{\mathrm{T}}\right], \tag{5}$$

while $\mathbf{n} = (n_x, n_y)$ and $\mathbf{m} = (m_x, m_y)$ are the normal and tangent unit vectors at free surface, respectively. In Cartesian coordinates, Eqs. (3)–(4) become:

$$-p + \frac{2}{Re}\left[\frac{\partial u}{\partial x}n_x^2 + \frac{\partial v}{\partial y}n_y^2 + \left(\frac{\partial u}{\partial y} + \frac{\partial v}{\partial x}\right)n_x n_y\right] = 0, \tag{6}$$

$$2\frac{\partial u}{\partial x}n_x m_x + \frac{\partial v}{\partial y}n_y m_y + \left[\frac{\partial u}{\partial y} + \frac{\partial v}{\partial x}\right](n_y m_x + n_x m_y) = 0. \tag{7}$$

Summarizing this brief introduction to the MAC method, we present below the algorithm for its original version. The mathematical framework supporting the

MAC algorithm is provided by the Helmholtz decomposition theorem, that can be found in [7].

Based on the Helmholtz decomposition and the projection method (see [5]), the MAC algorithm for Newtonian fluid flows can be summarized as:

- **Step 1**—Calculate an intermediate pressure field $\tilde{p}$ on the free surface satisfying the boundary condition (6);
- **Step 2**—Having calculated $\tilde{p}$, an intermediate velocity field $\tilde{\mathbf{u}}$ at $t_{n+1} = t_n + \delta t$, is defined by solving the motion equation:

$$\frac{\partial \tilde{\mathbf{u}}}{\partial t} = \left\{ -\nabla \cdot (\mathbf{uu}) - \nabla \tilde{p} + \frac{1}{Re}\nabla^2 \mathbf{u} + \frac{1}{Fr^2}\mathbf{g} \right\}_{t=t_n}; \tag{8}$$

- **Step 3**—Solve the Poisson equation for the potential ψ,

$$\nabla \psi^{(n+1)} = \nabla \cdot \tilde{\mathbf{u}}, \tag{9}$$

  with homogeneous Neumann type boundary conditions on rigid walls (and inflows) and homogeneous Dirichlet boundary conditions on the free surface (and outflows);
- **Step 4**—Update the velocity field from

$$\mathbf{u}^{(n+1)} = \tilde{\mathbf{u}} - \nabla \psi^{(n+1)}; \tag{10}$$

- **Step 5**—Update the pressure field from

$$p^{(n+1)} = \tilde{p} + \frac{\psi^{(n+1)}}{\delta t}; \tag{11}$$

- **Step 6**—Update the positions of the marker particles by solving the ODE's:

$$\dot{\mathbf{x}} = \mathbf{u}^{(n+1)}. \tag{12}$$

In the original MAC method both the Eqs. (8) and the ODE's (12) are solved explicitly by the forward Euler method. This works quite well when the Reynolds number, in Eq. (8), is neither too large nor too small. For non-Newtonian fluids it is very often the case that the Reynolds number is quite small and the parabolic restriction for the explicit discretization in Eq. (8) sets in, making the time step too small. At first glance, an implicit discretization of the diffusion part in Eq. (8) could be used, as was proposed by Kim and Moin [13], for obtaining stability for confined fluid flows. However, for free surface problems, distinctly of confined problems, it has been shown in [17] and [21] that, when viscous stresses are significant, the usual technique of Crank-Nicolson for discretizing the motion equations is not sufficient to guarantee unrestrictedly stability of the MAC methodology. In the next Section, we shall discuss an implicit strategy to circumvent this difficulty.

3 Improvements on the MAC Method

In this Section, we will present the improvements introduced in the MAC method for low Reynolds number free surface flows. Details will be described for the two-dimensional case since the extension to 3D is straightforward. In addition, we shall present the implicit formulation for the generalized motion Eq. (1) which can be used for Newtonian and non-Newtonian fluid flows.

3.1 Implicit Discretizations: NSE and the Normal Stress Condition at the Free Surface

The first step to obtain a stable method for low Reynolds number problems is to discretize the Navier–Stokes Eqs. (1)–(2) by the Adams–Bashforth/Crank–Nicolson method as

$$\frac{\mathbf{u}^{(n+1)}}{\delta t} - \frac{\zeta}{2Re}\nabla^2\mathbf{u}^{(n+1)} = \frac{\mathbf{u}^{(n)}}{\delta t} + \frac{\zeta}{2Re}\nabla^2\mathbf{u}^{(n)} - \frac{3}{2}\nabla\cdot(\mathbf{uu})^{(n)} + \frac{1}{2}\nabla\cdot(\mathbf{uu})^{(n-1)} - \nabla p^{(n+\frac{1}{2})} + \theta\nabla\cdot\boldsymbol{\tau}^{(n+\frac{1}{2})} + \frac{1}{Fr^2}\mathbf{g}, \tag{13}$$

$$\nabla\cdot\mathbf{u}^{(n+1)} = 0, \tag{14}$$

where the term $\nabla\cdot\boldsymbol{\tau}$ in Eq. (13) is approximated by

$$\nabla\cdot\boldsymbol{\tau}^{(n+\frac{1}{2})} = \frac{1}{2}\Big[\nabla\cdot\boldsymbol{\tau}^{(n)} + \nabla\cdot\overline{\boldsymbol{\tau}}^{(n+1)}\Big]. \tag{15}$$

Details about the calculation of the non-Newtonian extra-stress tensor will be presented in Sect. 3.3.

For modeling viscoelastic fluids, we set $\theta = 1$ in Eq. (13) while ζ is selected according to the viscoelastic model [19].

Thus, following the ideas behind the projection method, a provisional velocity field $\tilde{\mathbf{u}}^{(n+1)}$ is calculated from,

$$\frac{\tilde{\mathbf{u}}^{(n+1)}}{\delta t} - \frac{\zeta}{2Re}\nabla^2\tilde{\mathbf{u}}^{(n+1)} = \frac{\mathbf{u}^{(n)}}{\delta t} + \frac{\zeta}{2Re}\nabla^2\mathbf{u}^{(n)} - \frac{3}{2}\nabla\cdot(\mathbf{uu})^{(n)} + \frac{1}{2}\nabla\cdot(\mathbf{uu})^{(n-1)} - \nabla p^{(n)} + \theta\nabla\cdot\boldsymbol{\tau}^{(n+\frac{1}{2})} + \frac{1}{Fr^2}\mathbf{g}. \tag{16}$$

As a consequence of the use of Eq. (16) in the implicit version, the final pressure field is now obtained from

$$p^{(n+1)} = p^{(n)} + \frac{\psi^{(n+1)}}{\delta t} - \frac{\zeta}{2Re}\nabla^2\psi^{(n+1)}. \quad (17)$$

The implicit methodology described above presents good stability properties (see [23]) for the simulation of confined fluid flows. However, as observed firstly in [17], the presence of a free surface influences adversely stability. In particular, the normal stress condition (3) plays an important role in the construction of a stable scheme for free surface flows. A brief description of the approach in Oishi et al. [17, 21], will be considered here. This formulation combines an implicit discretization of the normal stress condition (3) with the main steps of the projection method.

Initially, Eq. (3) is discretized in time yielding

$$\mathbf{n} \cdot \boldsymbol{\sigma}^{(n+1)} \cdot \mathbf{n}^{\mathrm{T}} = 0, \quad (18)$$

where σ in its more general form is given by:

$$\boldsymbol{\sigma} = -p\mathbf{I} + \zeta\frac{1}{Re}\left[(\nabla\mathbf{u}) + (\nabla\mathbf{u})^{\mathrm{T}}\right] + \theta\boldsymbol{\tau}. \quad (19)$$

Substituting Eq. (19) into Eq. (18), we obtain

$$\mathbf{n} \cdot \left[-p^{(n+1)}\mathbf{I} + \zeta\frac{1}{Re}\left[(\nabla\mathbf{u}^{(n+1)}) + (\nabla\mathbf{u}^{(n+1)})^{\mathrm{T}}\right] + \theta\boldsymbol{\tau}^{(n+1)}\right] \cdot \mathbf{n}^{\mathrm{T}} = 0. \quad (20)$$

In the Newtonian case ($\theta = 0$ and $\zeta = 1$), for a segregated solution, the pressure field needs to be decoupled from the velocity field in Eq. (20). This may be accomplished by the projection method as will be described below. However, in the presence of the non-Newtonian tensor, we firstly use an approximation for $\boldsymbol{\tau}^{(n+1)}$, i.e., $\boldsymbol{\tau}^{(n+1)} \approx \overline{\boldsymbol{\tau}}^{(n+1)}$, where $\overline{\boldsymbol{\tau}}^{(n+1)}$ must be known in this step of the procedure. Thus, applying Eq. (10) and using the approximated value of the non-Newtonian tensor, Eq. (20) can be re-written as:

$$\mathbf{n} \cdot \left[-p^{(n+1)}\mathbf{I} + \zeta\frac{1}{Re}\left[(\nabla(\tilde{\mathbf{u}}^{(n+1)} - \nabla\psi^{(n+1)})) + (\nabla(\tilde{\mathbf{u}}^{(n+1)} - \nabla\psi^{(n+1)}))^{\mathrm{T}}\right] + \theta\overline{\boldsymbol{\tau}}^{(n+1)}\right] \cdot \mathbf{n}^{\mathrm{T}} = 0. \quad (21)$$

Finally, the normal stress condition for the implicit formulation is obtained from substituting the pressure update formula (17) into (21), i.e.,

$$\mathbf{n}\cdot\Bigg[-\left(p^{(n)}+\frac{\psi^{(n+1)}}{\delta t}-\frac{\zeta}{2Re}\nabla^2\psi^{(n+1)}\right)\mathbf{I}$$
$$+\zeta\frac{1}{Re}\left[(\nabla(\tilde{\mathbf{u}}^{(n+1)}-\nabla\psi^{(n+1)}))+(\nabla(\tilde{\mathbf{u}}^{(n+1)}-\nabla\psi^{(n+1)}))^{\mathrm{T}}\right]+\theta\overline{\boldsymbol{\tau}}^{(n+1)}\Bigg]\cdot\mathbf{n}^{\mathrm{T}}=0. \tag{22}$$

The main difference between the implicit formulation and the original explicit MAC version is in the use of Eq. (22) as a boundary condition for ψ at the free surface. This equation, which depends on the normal vector in each cell (denoted here as S-cell) that contains fluid and has one or more faces in contact with empty cells (denoted here as E-cell), will be used to provide the remaining equations for the unknown ψ on the free boundary, so as to complete the linear system arising from the Poisson Eq. (9) applied to the F-cells (cells that do not have any face in contact with empty cells). The resulting linear system will be non-symmetric. This is the main drawback of the implicit formulation. However, as discussed in [20], this system can be effectively solved by employing an appropriate pre-conditioner and CPU times (iterations of the linear solver) reduced to the same levels of those of the symmetric case.

In order to illustrate the resulting equations for ψ on each cell on the free surface, we write Eq. (22), for the two dimensional case, in full as:

$$\frac{\psi^{(n+1)}}{\delta t}-\frac{2\zeta}{Re}\Big[\Big(\frac{\partial^2\psi^{(n+1)}}{\partial y^2}\Big)n_x^2+\Big(\frac{\partial^2\psi^{(n+1)}}{\partial x^2}\Big)n_y^2-2\Big(\frac{\partial^2\psi^{(n+1)}}{\partial x\partial y}\Big)n_xn_y\Big]-\frac{\zeta}{2Re}\nabla^2\psi^{(n+1)}$$
$$=\frac{2\zeta}{Re}\Big[-\Big(\frac{\partial\tilde{v}^{(n+1)}}{\partial y}\Big)n_x^2-\Big(\frac{\partial\tilde{u}^{(n+1)}}{\partial x}\Big)n_y^2+\Big(\frac{\partial\tilde{u}^{(n+1)}}{\partial y}+\frac{\partial\tilde{v}^{(n+1)}}{\partial x}\Big)n_xn_y\Big]$$
$$+\theta[(\overline{\tau}^{xx})^{(n+1)}n_x^2+2(\overline{\tau}^{xy})^{(n+1)}n_xn_y+(\overline{\tau}^{yy})^{(n+1)}n_y^2]-p^{(n)}. \tag{23}$$

Details on the use of these equations can be found in [20] (or in [21]—for the three-dimensional case). In practice it is assumed that the normal vector in (23) can only take three different configurations, namely: $\mathbf{n}=(\pm1,0)$, $\mathbf{n}=(0,\pm1)$, and $\mathbf{n}=(\pm\frac{\sqrt{2}}{2},\pm\frac{\sqrt{2}}{2})$. This is equivalent of assuming that, at each cell, the free surface is approximately either horizontal, vertical or at an angle of 45°.

3.2 Marker Particles Accurate Dynamics

The MAC method is one of the most famous methods for the numerical simulation of free surface flows [6]. The strategy combines the front-tracking method with an Eulerian framework for representing the free surface, so that MAC-type approaches move the interface with good accuracy for Newtonian and non-Newtonian flows [15]. In summary, in this strategy, virtual marker particles are distributed into each cell of an Eulerian grid and the interface is recovered by interpolation with piecewise continuous functions. During the course of one time step the displacement of the marker particles is accomplished by solving the equation

$$\dot{\mathbf{x}} = \mathbf{u}. \tag{24}$$

The methodology described in Sect. 2 is a fully explicit scheme and consequently, the time step tends to be very small for low Reynolds number problems. Thus, the numerical solution of Eq. (24) can accurately be obtained by the forward Euler method [28]. However, when implicit methods are implemented to solve the motion equations, as was discussed in Sect. 3.1, a larger time step may be employed, and special attention should be taken in the solution of Eq. (24). In order to maintain accurate mass conservation and accuracy in the calculation of the velocity field at free surface, a modification in the original MAC method was proposed in [19].

Firstly, the Runge–Kutta scheme RK21 is employed for solving Eq. (24). In this scheme an intermediate particle position $\overline{\mathbf{x}}$ is calculated from:

$$\overline{\mathbf{x}} = \mathbf{x}^{(n)} + \delta t \mathbf{u}\left(\mathbf{x}^{(n)}, t_n\right), \tag{25}$$

and the new updated position is obtained by solving

$$\mathbf{x}^{(n+1)} = \mathbf{x}^{(n)} + \frac{\delta t}{2}\left[\mathbf{u}\left(\mathbf{x}^{(n)}, t_n\right) + \mathbf{u}\left(\overline{\mathbf{x}}, t_{n+1}\right)\right]. \tag{26}$$

It is important to stress that the strategy for obtaining a more accurate scheme overall, should not rely on the use of RK21 for moving the marker particles only, it should also provide more accurate interpolation values for the velocities used by the method. In order to illustrate this strategy, we consider a configuration (see Fig. 2) where there exists one marker particle inside a surface cell with index i, j. The marker particle is represented by a triangle. In this configuration, the normal vector used to calculate the boundary conditions at free surface is approximated by $\mathbf{n} = (\frac{\sqrt{2}}{2}, \frac{\sqrt{2}}{2})$. Using this normal vector and the corresponding tangent vector, the velocities $u_{i+\frac{1}{2},j}$, $u_{i+\frac{1}{2},j-1}$, $v_{i,j+\frac{1}{2}}$ and $v_{i-1,j+\frac{1}{2}}$ depicted in Fig. 2 are obtained from the tangential boundary condition (4) and the continuity Eq. (2), as explained in [19]. Note, for this configuration the closest neighboring cells of the S-cell containing the marker particle are full (F-cell), surface or empty (E-cell) cells. The RK21 scheme needs the value for the velocity at the marker particle $\mathbf{u}_P$ obtained by bilinear interpolation on the four node-velocities closest to the marker particle (in each direction). Some of these velocities may be unavailable, as depicted in Fig. 2 by the filled or blue rectangles, because they belong to an E-cell. These velocities are calculated from:

$$\begin{aligned} u_{i+\frac{1}{2},j+1} &= 2u_{i+\frac{1}{2},j} - u_{i+\frac{1}{2},j-1} \\ v_{i+1,j+\frac{1}{2}} &= 2v_{i,j+\frac{1}{2}} - v_{i-1,j+\frac{1}{2}} \,. \end{aligned} \tag{27}$$

The combination of the first-order extrapolation (27) with the RK21 method results in good mass conservation, even when large time steps are used, as shown by the

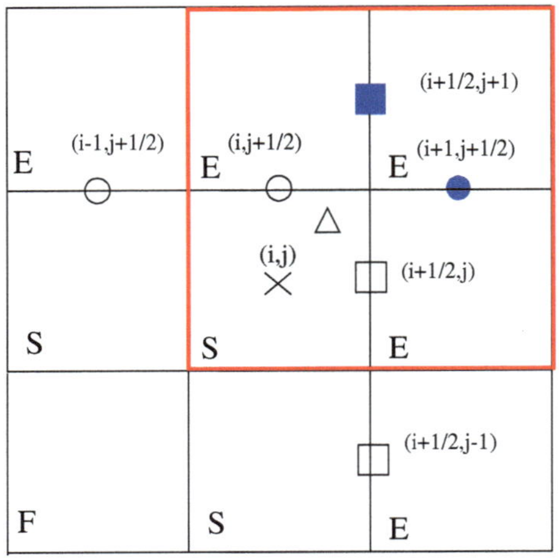

Fig. 2 MAC cell configuration for $\mathbf{n} = (\frac{\sqrt{2}}{2}, \frac{\sqrt{2}}{2})$. *Squares* represent the velocity in x-direction while *circles* represent velocity in y-direction. $\triangle$ represents the location of the marker particle

numerical experiments. Details of a comprehensive mass conservation study can be found in [14].

3.3 Numerical Method for Complex Flows: Viscoelastic Fluids

Another recent development of the MAC methodology for simulating low Reynolds number free surface flows is due to Oishi et al. [20] who consider complex viscoelastic fluids. This approach combines the implicit strategy presented in this work with a differential constitutive equation representing the non-Newtonian tensor for different types of viscoelastic models. In dimensionless form, a general form of constitutive equation for the non-Newtonian tensor $\boldsymbol{\tau}$ is given by:

$$\frac{\partial \boldsymbol{\tau}}{\partial t} + \nabla \cdot (\mathbf{u}\boldsymbol{\tau}) - \left[(\nabla \mathbf{u}) \cdot \boldsymbol{\tau} + \boldsymbol{\tau} \cdot (\nabla \mathbf{u})^{\mathrm{T}}\right] = 2\xi \mathbf{D} - \frac{1}{Wi}\Big\{ f(\lambda, \boldsymbol{\tau})\,\boldsymbol{\tau} + \xi\,[f(\lambda, \boldsymbol{\tau}) - 1]\,\mathbf{I} + \frac{\alpha}{\xi}\boldsymbol{\tau} \cdot \boldsymbol{\tau}\Big\}, \quad (28)$$

$$f(\lambda, \boldsymbol{\tau}) = \frac{2}{\gamma}\left(1 - \frac{1}{\lambda}\right) e^{Q_o(\lambda - 1)} + \frac{1}{\lambda^2}\left[1 - \frac{\alpha}{3\xi^2} tr\,(\boldsymbol{\tau} \cdot \boldsymbol{\tau})\right], \quad (29)$$

$$\lambda = \sqrt{1 + \frac{1}{3\xi}|tr\,(\boldsymbol{\tau})\,|}. \quad (30)$$

The constitutive equation of (28)–(30) is quite general as the XPP, Giesekus and Oldroyd-B models can be derived from it by choosing appropriate parameters, as explained in [19]. The Weissenberg number is defined as $Wi = \frac{\lambda_1 U}{L}$, where λ_1 is the relaxation time of the fluid. The definition of the rheological parameters in Eqs. (28)–(30) and their significance is given in [2].

In order to obtain an accurate method for viscoelastic free surface flows, the constitutive equation (28) for the non-Newtonian extra-stress tensor $\boldsymbol{\tau}$ is discretized using the second-order accurate Runge-Kutta (RK21) method, as follows.

First, Eq. (28) is re-written as

$$\frac{\partial \boldsymbol{\tau}}{\partial t} = \mathbf{F}(\mathbf{u}, \boldsymbol{\tau}), \tag{31}$$

where

$$\begin{aligned}\mathbf{F}(\mathbf{u}, \boldsymbol{\tau}) = &\left[(\nabla \mathbf{u}) \cdot \boldsymbol{\tau} + \boldsymbol{\tau} \cdot (\nabla \mathbf{u})^T\right] + 2\xi \mathbf{D} - [\nabla \cdot (\mathbf{u}\boldsymbol{\tau})] \\ &- \frac{1}{Wi}\left\{f(\lambda, \boldsymbol{\tau})\boldsymbol{\tau} + \xi\left(f(\lambda, \boldsymbol{\tau}) - 1\right)\mathbf{I} + \frac{\alpha}{\xi}(\boldsymbol{\tau} \cdot \boldsymbol{\tau})\right\}.\end{aligned} \tag{32}$$

The next step involves the calculation of an intermediate non-Newtonian extra-stress tensor $\overline{\boldsymbol{\tau}}^{(n+1)}$ by the explicit forward Euler discretization,

$$\frac{\overline{\boldsymbol{\tau}}^{(n+1)} - \boldsymbol{\tau}^{(n)}}{\delta t} = \mathbf{F}\left(\mathbf{u}^{(n)}, \boldsymbol{\tau}^{(n)}\right). \tag{33}$$

The final non-Newtonian extra-stress tensor $\boldsymbol{\tau}^{(n+1)}$ is obtained from solving the equation:

$$\frac{\boldsymbol{\tau}^{(n+1)} - \boldsymbol{\tau}^{(n)}}{\delta t} = \frac{1}{2}\left[\mathbf{F}\left(\mathbf{u}^{(n)}, \boldsymbol{\tau}^{(n)}\right) + \mathbf{F}\left(\mathbf{u}^{(n+1)}, \overline{\boldsymbol{\tau}}^{(n+1)}\right)\right]. \tag{34}$$

3.4 Algorithm for the Improved Implicit MAC-Type Method

Having briefly presented the idea of the new developments, we are now in the position to give the step-by-step outline of the modified MAC algorithm for simulating low Reynolds number problems with complex viscoelastic fluid models. It is assumed that at time $t = t_n$ the solenoidal velocity $\mathbf{u}^{(n)}$, the pressure field $p^{(n)}$, the non-Newtonian extra-stress tensor $\boldsymbol{\tau}^{(n)}$ are known. The solutions $\mathbf{u}^{(n+1)}$, $p^{(n+1)}$ and $\boldsymbol{\tau}^{(n+1)}$ are obtained by the following steps.

- **Step 1**—Calculate $\overline{\boldsymbol{\tau}}^{(n+1)}$ from Eq. (33) and then compute $\boldsymbol{\tau}^{(n+\frac{1}{2})}$ from Eq. (15);

- **Step 2**—Calculate the intermediate velocity $\tilde{\mathbf{u}}^{(n+1)}$ from Eq. (16) using the Adams–Bashforth/Crank–Nicolson scheme. Details about the implementation of the boundary conditions for the intermediate velocity $\tilde{\mathbf{u}}^{(n+1)}$ can be found in [18];
- **Step 3**—Solve the Poisson Eq. (9) simultaneously with the equations for $\psi^{(n+1)}$ obtained from the application of the boundary conditions at the free surface [see Eq. (22)]. The other boundary conditions are the same as those for the original MAC algorithm (see Sect. 2);
- **Step 4**—Update the velocity field from Eq. (10);
- **Step 5**—Update the pressure field from Eq. (17);
- **Step 6**—Calculate $\boldsymbol{\tau}^{(n+1)}$ using Eq. (34);
- **Step 7**—Update the positions of the marker particles by solving the ODE's Eq. (24) using the Runge–Kutta scheme presented in Sect. 3.2 .

4 Numerical Examples

We shall finish this paper by presenting the numerical solution of a couple of examples of free surface problems, chosen because they both are of interest to industry: the time-dependent extrudate swell problem and the jet buckling phenomenon.

4.1 The Time-Dependent Extrudate Swell Problem

The time-dependent extrudate swell problem is a popular benchmark for low Reynolds number free surface flows. This problem consists of a jet of viscous fluid exiting a capillary and due to the normal stress differences, the jet swells and its width expands to a maximum. Numerical results of the extrudate swell problem have been presented in the literature by many researchers (e.g. [1, 9, 25]).

A verification of the numerical scheme proposed in this work for simulating the time-dependent extrudate swell problem was carried out in [20]. In that paper we have described a comparison study about the profile of the free surface with that presented in [25]. A good agreement between results was observed showing the potential of our methodology to simulate this complex fluid flow.

In the present work, we consider a 2D-channel with width L and length $4L$ and an outflow boundary positioned at a distance $6L$ from the channel exit. The Reynolds number adopted is very small, for instance $Re = 0.01$, and a dimensionless mesh spacing $h = 0.025$ is used in all simulations. More details about the computational geometry and boundary conditions used in this test problem can be found in [20].

Firstly, we have simulated the extrudate swell problem for a Newtonian fluid, and results are presented in Fig. 3a. After this, in order to demonstrate that the implicit formulation can deal with complex viscoelastic fluid flows, we used the XPP model with the following parameters: $Wi = 20$, $\zeta = 0.5$, $\gamma = 0.8$, $Q = 8.0$ and $\alpha = 0.01$. The viscoelastic behavior of the XPP model for the time-dependent extrudate swell

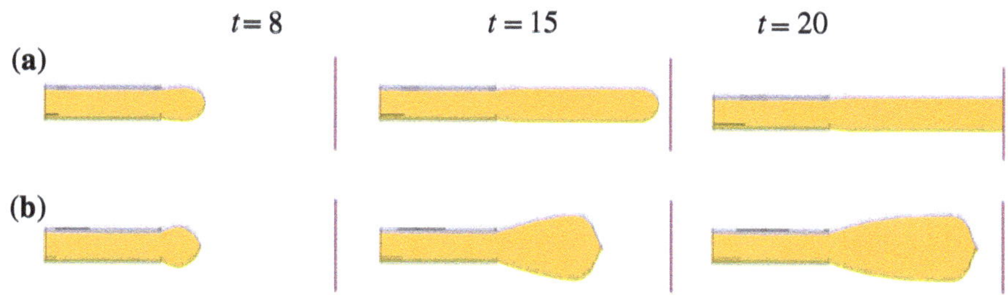

Fig. 3 Numerical simulation for the extrudate swell problem for $Re = 0.01$ of a Newtonian (**a**) and XPP fluid, (**b**) using the implicit formulation. Fluid flow visualization for different dimensionless times

problem can be observed in Fig. 3b. Notice that, the Newtonian jet flows with a slight sign of swelling while the XPP fluid presents clearly the die swell phenomenon. In summary, from Fig. 3, we can observe the memory in the deformation history of the polymer chains of the XPP model while that for Newtonian fluid, in contrast, the jet keeps a constant diameter after the die exit. The characteristic behaviour of the free surface profile of the Newtonian and non-Newtonian fluids, including the amount of swell, agrees with numerical results [25] and Tanner's analysis [26]. Therefore, these results indicate that the implicit MAC-type method can accurately capture the details involved in the simulation of the transient extrudate swell for low Reynolds numbers.

4.2 Jet Buckling Phenomenon

The jet buckling phenomenon occurs when a fluid is injected from a distance H from a plate and builds up as it hits the plate. Due to shear stress, oscillations build up in the fluid column injected. The jet buckling phenomenon is observed in many industrial applications of container filling, in particular for low Re flows. In [10] two important parameters were observed as being important for the buckling in Newtonian fluids: $Re < 1.2$ and $H/D > 7.2$, where D is the diameter of the injector. Recently, Tomé et al. [29] published a comparison study between numerical simulation and experimental data for three-dimensional Newtonian viscous fluid. For viscoelastic fluids, the reader is referred to [11, 19, 24, 27, 30] for numerical investigations of this problem in two and three dimensions.

In this Section we shall present numerical results for the jet buckling using the improved MAC method for the three-dimensional case. The domain geometry is composed of a circular injector of diameter $D = L$, a distance from a plate $H = 10L$, a square plate width $10L$ and a mesh spacing $h = 0.16667$. The scheme employed here was rigorously analysed in [11] by varying different flows conditions in the jet buckling problem, as for example Re and Wi numbers and geometrical parameters.

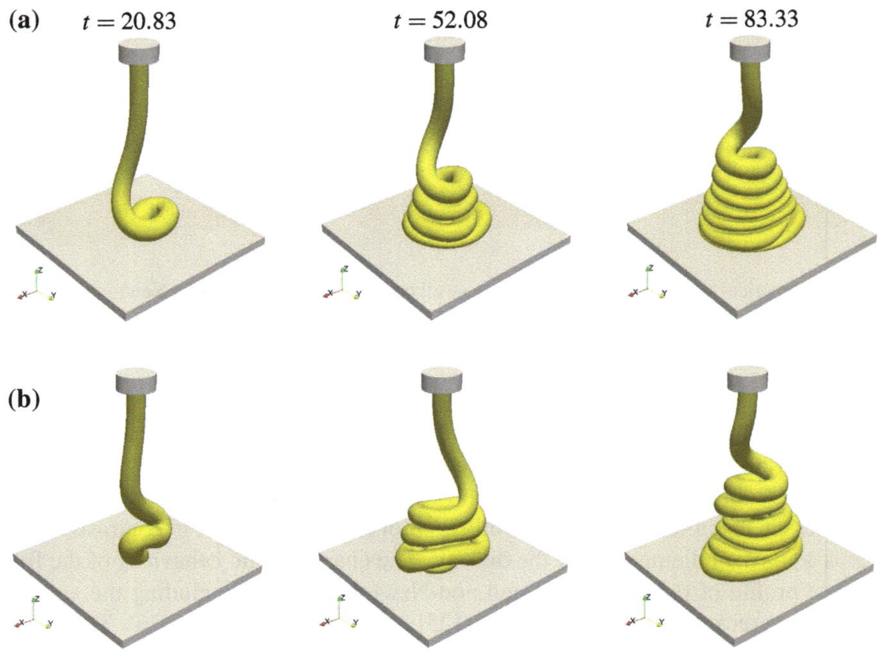

Fig. 4 Influence of the Reynolds number (Re) in the jet buckling problem for a Newtonian fluid: **a** $Re = 0.1$; **b** $Re = 0.06$. Fluid flow visualization for different dimensionless times

In order to demonstrate the advantages of the improvements on MAC method, we performed two tests with low Re, for instance $Re = 0.1$ and 0.06. In the first test it was considered a Newtonian fluid and the results are shown in Fig. 4. A similar experiment is repeated with a polymeric flow characterized by the XPP fluid with the following parameters: $Wi = 5$, $\zeta = 0.4$, $\alpha = 0.1$, $\gamma = 0.7$ and $Q = 4$. Figure 5 displays the numerical results for the viscoelastic fluid flow. From these figures, we can observe the nonlinear dynamic and the instabilities of this free surface fluid flows. It should be noted here that this peculiar behaviour of the fluids in our numerical simulations is in good agreement with Cruickshank's analysis [10]. In summary, the Reynolds number is the most important parameter in this phenomenon and when Re decreases the fluid disperses less due to the viscous forces, independent of the type of the fluid. The occurrence of coiling instabilities, as those depicted in Figs. 4 and 5 at the dimensionless time $t = 83.33$, also was observed in [24]. Thus, even with three-dimensional arbitrary boundaries and complex fluids, the implicit formulation was very efficient and able to simulate low Reynolds number free surface flows.

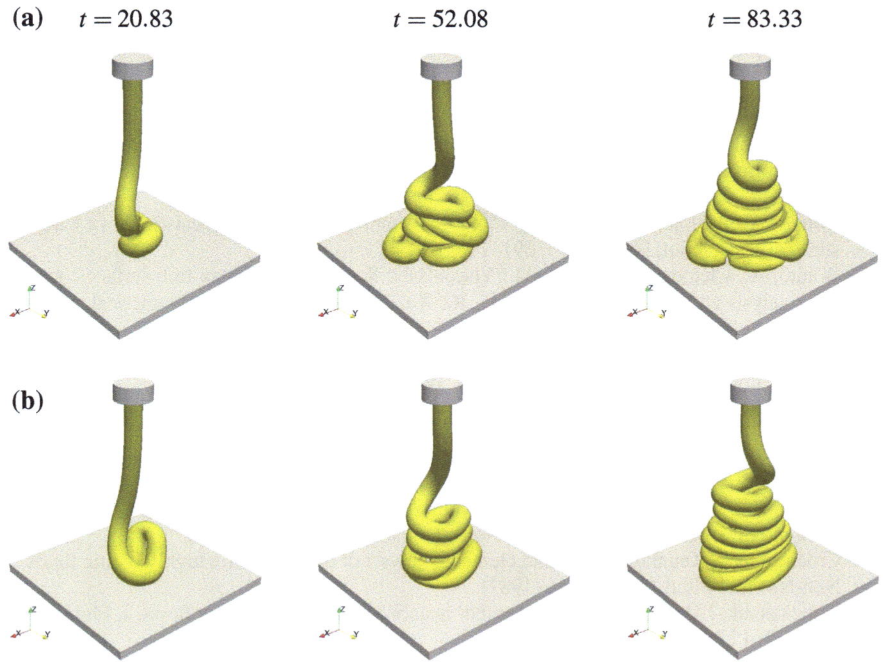

Fig. 5 Influence of the Reynolds number (Re) in the jet buckling problem for a XPP model $(Wi = 5, \zeta = 0.4, \alpha = 0.1, \gamma = 0.7$ and $Q = 4)$: **a** $Re = 0.1$, **b** $Re = 0.06$. Fluid flow visualization for different dimensionless times

5 Conclusion

In this work we have presented a number of improvements to the classical MAC method that enables the use of such methodology for solving viscoelastic low Reynolds number free surface flows. The new features are added to the previous version of the MAC method reported in the paper [16]. This new MAC method is capable of simulating complex problems of academic and industrial interest, as has been illustrated by the numerical examples.

Acknowledgments The authors would like to acknowledge the financial support of FAPESP (projects nos. 2013/07375-0, 2011/09194-7, 2009/15892-9) and CNPq (projects nos. 305447/2010-6 , 473589/2013-3).

References

1. Antonietti, P.F., Fadel, N.A., Verani, M.: Modelling and numerical simulation of the polymeric extrusion process in textile products. Commun. Appl. Ind. Math. **1**, 1–13 (2010)
2. Baltussen, M.G.H.M., Verbeeten, W.M.H., Bogaerds, A.C.B., Hulsen, M.A., Peters, G.W.M.: Anisotropy parameter restrictions for the eXtended Pom-Pom model. J. Non-Newton. Fluid **165**, 1047–1054 (2010)
3. Bonito, A., Picasso, M., Laso, M.: Numerical simulation of 3D viscoelastic flows with free surfaces. J. Comput. Phy. **215**(2), 691–716 (2006)
4. Bonito, A., Clément, P., Picasso, M.: Viscoelastic flows with complex free surfaces: numerical analysis and simulation. Glowinski, R., Xu, J. (eds.) Handbook of Numerical Analysis, Numerical Methods for Non-Newtonian Fluids vol. 16, pp. 305–369 (2011)
5. Brown, D.L., Cortez, R., Minion, M.L.: Accurate projection methods for the incompressible Navier–Stokes equations. J. Comput. Phys. **168**, 464–499 (2001)
6. Caboussat, A.: Numerical simulation of two-phase free surface flows. Arch. Comput. Meth. Eng. **12**, 165–224 (2005)
7. Chorin, A.J., Marsden, J.E.: A Mathematical Introduction to Fluid Mechanics, 3rd edn. Springer, New York (2000)
8. Ciarlet, P.G., Glowinsk, R., Lions, J.L.: Numerical methods for non-newtonian fluids. Handbook of Numerical Analysis, vol. 16, North-Holland, Amsterdam (2011)
9. Crochet, M.J., Keunings, R.: Finite element analysis of die-swell of a highly elastic fluids. J. Non-Newton. Fluid **10**, 339–356 (1982)
10. Cruickshank, J.O.: Low-Reynolds-number instabilities in stagnating jet flows. J. Fluid Mech. **193**, 111–127 (1988)
11. Figueiredo, R.A., Oishi, C.M., Cuminato, J.A., Alves, M.A.: Three-dimensional transient complex free surface flows: numerical simulation of XPP fluid. J. Non-Newton. Fluid **195**, 88–98 (2013)
12. Harlow, F.H., Welch, J.E.: Numerical calculation of time-dependent viscous incompressible flow of fluid with free surface. Phys. Fluids **8**, 2182–2189 (1965)
13. Kim, J., Moin, P.: Application of a fractional-step method to incompressible Navier–Stokes equations. J. Comput. Phys. **59**, 308–323 (1985)
14. Martins, F.P., Oishi, C.M., Sousa, F.S., Cuminato, J.A.: Numerical assessment of mass conservation on a MAC-type method for viscoelastic free surface flows. In: 6th European Congress on Computational Methods in Applied Sciences and Egineering (ECCOMAS 2012), vol. 1, pp. 6545–6562 (2012)
15. McKee, S., Tomé, M.F., Cuminato, J.A., Castelo, A., Ferreira, V.G.: Recent advances in the marker-and-cell method. Arch. Comput. Meth. Eng. **11**, 107–142 (2004)
16. McKee, S., Tomé, M.F., Ferreira, V.G., Cuminato, J.A., Castelo, A., Sousa, F.S., Mangiavacchi, N.: MAC Method. Comput. Fluids **37**, 907–930 (2008)
17. Oishi, C.M., Cuminato, J.A., Ferreira, V.G., Tomé, M.F., Castelo, A., Mangiavacchi, N., McKee, S.: A stable semi-implicit method for free surface flows. J. Appl. Mech. **73**, 940–947 (2006)
18. Oishi, C.M., Cuminato, J.A., Yuan, J.Y., McKee, S.: Stability of numerical schemes on staggered grids. Numer. Linear Algebra Appl. **15**, 945–967 (2008)
19. Oishi, C.M., Martins, F.P., Tomé, M.F., Alves, M.A.: Numerical simulation of drop impact and jet buckling problems using the eXtended Pom-Pom model. J. Non-Newton. Fluid **169**, 91–103 (2012)
20. Oishi, C.M., Martins, F.P., Tomé, M.F., Cuminato, J.A., McKee, S.: Numerical solution of the eXtended Pom-Pom model for viscoelastic free surface flows. J. Non-Newton. Fluid **166**, 165–179 (2011)
21. Oishi, C.M., Tomé, M.F., Cuminato, J.A., McKee, S.: An implicit technique for solving 3d low Reynolds number moving free surface flows. J. Comput. Phys. **227**, 7446–7468 (2008)
22. Owens, R.G., Phillips, T.N.: Computational Rheology. Imperial College Press, London (2002)
23. Quarteroni, A., Saleri, A., Veneziani, A.: Factorization methods for the numerical approximation of Navier–Stokes equations. Comput. Methods Appl. Mech. Eng. **188**, 505–526 (2000)

24. Roberts, S.A., Rao, R.R.: Numerical simulations of mounding and submerging flows of shear-thinning jets impinging in a container. J. Non-Newton. Fluid **166**, 1100–1115 (2011)
25. Russo, G., Phillips, T.N.: Numerical prediction of extrudate swell of branched polymer melts. Rheol. Acta. **49**, 657–676 (2010)
26. Tanner, R.I.: A theory of die-swell revisited. J. Non-Newton. Fluid **129**, 85–87 (2005)
27. Tomé, M.F., Castelo, A., Afonso, A.M., Alves, M.A., Pinho, F.T.: Application of the log-conformation tensor to three-dimensional time-dependent free surface flows. J. Non-Newton. Fluid **175–176**, 44–54 (2012)
28. Tomé, M.F., Castelo, A., Ferreira, V.G., McKee, S.: A finite difference technique for solving the Oldroyd-B model for 3D-unsteady free surface flows. J. Non-Newton. Fluid **154**, 159–192 (2008)
29. Tomé, M.F., Castelo, A., Nóbrega, J.M., Carneiro, O.S., Paulo, G.S., Pereira, F.T.: Numerical and experimental investigations of three-dimensional container filling with Newtonian viscous Fluids. Comput. Fluids **90**, 172–185 (2014)
30. Ville, L., Silva, L., Coupez, T.: Convected level set method for the numerical simulation of fluid buckling. Internat. J. Numer. Methods Fluids **66**, 324–344 (2011)
31. Xu, X., Ouyang, J., Yang, B., Liu, Z.: SPH simulations of three-dimensional non-Newtonian free surface flows. Comput. Methods Appl. Mech. Eng. **256**, 101–116 (2013)
32. Yang, B., Ouyang, J., Wang, F.: Simulation of stress distribution near weld line in the viscoelastic melt mold filling process. J. Appl. Math. (2013)

Robust Naive Bayes Combination of Multiple Classifications

Naonori Ueda, Yusuke Tanaka and Akinori Fujino

Abstract When we face new complex classification tasks, since it is difficult to design a good feature set for observed raw data, we often obtain an unsatisfactorily biased classifier. Namely, the trained classifier can only successfully classify certain classes of samples owing to its poor feature set. To tackle the problem, we propose a robust naive Bayes combination scheme in which we effectively combine classifier predictions that we obtained from different classifiers and/or different feature sets. Since we assume that the multiple classifier predictions are given, any type of classifier and any feature set are available in our scheme. In our combination scheme each prediction is regarded as an independent realization of a categorical random variable (i.e., class label) and a naive Bayes model is trained by using a set of the predictions within a supervised learning framework. The key feature of our scheme is the introduction of a class-specific variable selection mechanism to avoid overfitting to poor classifier predictions. We demonstrate the practical benefit of our simple combination scheme with both synthetic and real data sets, and show that it can achieve much higher classification accuracy than conventional ensemble classifiers.

Keywords Classification · Naive Bayes model · Model combination · Meta-learning · Bayesian learning · Ensemble learning · Real nursing activity recognition

N. Ueda (✉) · A. Fujino
NTT Communication Science Laboratories, 2-4 Hikaridai Seikacho, Sorakugun, Kyoto, Japan
e-mail: ueda.naonori@lab.ntt.co.jp

A. Fujino
e-mail: fujino.akinori@lab.ntt.co.jp

Y. Tanaka
NTT Service Evolution Laboratories, 1-1 Hikarinooka, Yokosuka-shi, Kanagawa, Japan
e-mail: tanaka.yusuke@lab.ntt.co.jp

M. Wakayama et al. (eds.), *The Impact of Applications on Mathematics*,
Mathematics for Industry 1, DOI: 10.1007/978-4-431-54907-9_10,

1 Introduction

Supervised classification problems have been extensively studied in connection with machine learning, and therefore many classifiers have recently become available [11]. However, with classification problems it is much more important to develop a good feature set suitable for classifying instances of specific applications than to find the best general classifier from a practical point of view. When we face new complex classification tasks, since it is difficult to design a good feature set for observed raw data, we often obtain an unsatisfactorily biased classifier. Namely, the trained classifier can only successfully classify certain classes of samples because of its poor feature set. When we have little domain knowledge about a new task to be solved, it is time-consuming and perhaps difficult to design an optimal feature set and/or select a classifier that can discriminate all classes effectively when there are many classes.

We have resorted to an ensemble classification approach that combines the predictions of multiple classifiers in such a case [6]. However, the conventional representative ensemble methods such as simple majority voting, Bagging, and Boosting [2–4, 7] depend strongly on the feature set and/or base classifier, and therefore are limited in terms of improving the classification performance. Unlike these ensemble schemes, our approach involves constructing a classifier and/or feature independent combined classifier (meta-classifier) simply based on multiple classification results, and therefore our combination scheme is applicable to any type of classifiers and any feature set.

In our scheme we regard a set of multiple classifier predictions (class labels) for a sample as a set of realizations of categorical random variables. Intuitively, a set of J predictions for a sample to be classified can be viewed as a J-dimensional categorical feature vector. Since each training sample has a true class label, a class conditional naive Bayes model is trained by using a set of variables within a usual supervised learning framework. Since the predictions obtained by the classifiers in the ensemble can include many misclassifications in our setting, it is not appropriate to model all the variables (i.e., all classifier predictions) equally. Intuitively, when a classifier in the ensemble outputs largely incorrect and distinctive class lables over the samples in a certain class, the naive Bayes classifier unexpectedly overfits to the training data and this results in a poor generalization ability.

To solve the overfitting problem, we introduce a latent variable that can identify whether or not the predictions of a classifier in the ensemble for a set of samples in a certain class are effective in classifying the class. When a classifier in the ensemble is ineffective for a certain class, we exclude the variable corresponding to the classifer in the ensemble. Note that a classifier providing many incorrect predictions in a certain class is not always ineffective; i.e. if almost all the predictions of the classifier are incorrect but the prediction values are almost the same, the classifier's predictions could be effective as regards discriminating the class from the others (see Fig. 1). In our ensemble scheme this intuitively reasonable and simple idea is modeled within a Bayesian framework. In our model, all random variables other than the latent variable are marginalized out, so the derived inference algorithm is very easy to implement.

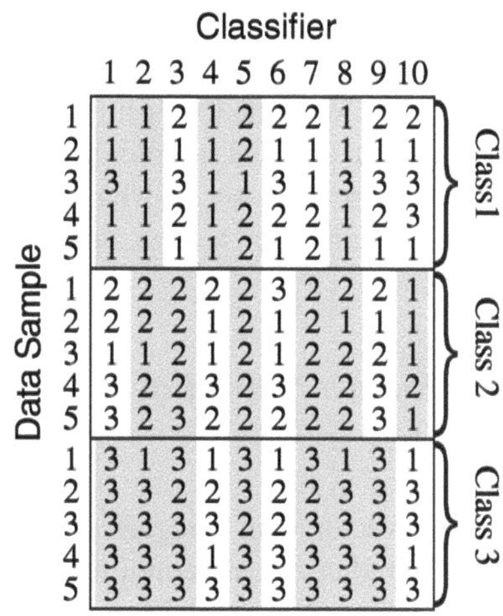

Fig. 1 An example of classfier prediction matrices; $\mathbf{C} = \{\mathbf{C}_1, \mathbf{C}_2, \mathbf{C}_3\}$

We present empirical evaluation results obtained using synthetic data to show that the model works well. We show that our method also performs well when combining multiple classifiers using UCI data sets. More importantly, as a real case study of the combination of poor classifiers, we apply our combination scheme to high-level human activity recognition using accelerometers. The data are real nursing activities collected in a hospital. The conventional single classifiers produced poor classification performance for this data in our preliminary experiments, and therefore it is worth showing the benefit of our combination scheme in relation to this challenging problem.

2 Proposed Model: Robust Naive Bayes Combination

2.1 Problem Setting

We assume there are J classifiers for a K-class classification problem, and multiple predictions for a set of training samples (labeled data) provided by J classifiers have been obtained in many different ways. Many different classifiers, different parameter choices, and different feature representations for observed raw data can all be used to produce a diverse set of predictions. Let $\mathbf{C}_k$ denote a *prediction matrix* with a size of N_k by J for class ω_k samples. The (i, j)th element, $c_{k,i,j}$, corresponds to the jth classifier's prediction for the ith training sample in class ω_k. Clearly, $c_{k,i,j} \in \{1, \ldots, K\}$. Here, N_k is the number of training samples in class ω_k. Figure 1 shows $\mathbf{C} = \{\mathbf{C}_1, \mathbf{C}_2, \mathbf{C}_3\}$.

In the test phase each of the test samples is classified by the same multiple classifiers in the ensemble as used for the training samples. Let $\mathbf{c}_{*,m} = (c_{*,m,1}, \ldots,$

$c_{*,m,J})$ denote a set of J predictions for the mth test sample. Then, each of M test samples is classified by using the meta-classifier obtained in the learning phase. So, our goal is to create the best meta-classifier from a fixed $\mathbf{C}$ so that not only training samples but also test samples are correctly classified as much as possible. The details will be presented later. In what follows, i, j, k and m represent the IDs of the training sample, a classifier, a class, and a test sample, respectively.

2.2 Robust Naive Bayes Combination Model

As shown in Fig. 1, $\mathbf{c}_{k,i} = (c_{k,i,1}, \ldots, c_{k,i,J})$ can be regarded as a J-dimensional feature vector with categorical feature values for the ith sample in ω_k. Assuming that J predictions are conditionally independent given k, we can simply consider a *naive Bayes* (NB) model for ω_k samples, namely $P(\mathbf{c}_{k,i}|\omega_k) = \prod_{j=1}^{J} P(c_{k,i,j}|\omega_k)$. Here, $P(c_{k,i,j}|\omega_k)$ is the multinomial distribution over K discrete symbols because $c_{k,i,j}$ takes a value of 1, ..., K. However, as mentioned in Sect. 1, since predictions are not always correct, such modeling in which all the classifier predictions are equally modeled is not valid. As mentioned in Sect. 1, some classifiers may only be effective for some particular class(es). Considering this, we introduce a binary latent variable $r_{k,j}$ that determines whether or not the jth classifier is effective for class ω_k. As shown in Fig. 1, when almost all the prediction values of the jth classifier for class ω_k samples are the same, the classifier will be informative in terms of discriminating class ω_k and therefore $r_{k,j}$ should be 1. This indicates that the values of $c_{k,i,j}$, $i = 1, \ldots, N_k$ should be almost the same for all $i = 1, \ldots, N_k$ when $r_{k,j} = 1$. On the other hand, the values $c_{k,i,j}$, $i = 1, \ldots, N_k$ do not have to depend on k and j when $r_{k,j} = 0$. The shaded parts in Fig. 1 corresponds to $r_{k,j} = 1$.

The Beta-Bernoulli distributions are assumed to be the priors for the latent variable $r_{k,j}$. Then, the proposed generative process is as follows:

$$\begin{aligned} &\lambda \sim \text{Beta}(a, b), \quad \phi \sim \text{Dirichlet}(\alpha) \\ &r_{k,j}|\lambda \sim \text{Bernoulli}(\lambda), \forall k, j, \quad \theta_{k,j} \sim \text{Dirichlet}(\beta_{k,j}), \forall k, j, \\ &c_{k,i,j}|r_{k,j}, \theta_{k,j}, \phi \sim \text{Multinomial}(\theta_{k,j}), \forall k, j, i, \quad \text{if } r_{k,j} = 1, \\ &\qquad\qquad\qquad\qquad \text{Multinomial}(\phi), \forall k, j, i, \quad \text{if } r_{k,j} = 0. \end{aligned} \tag{1}$$

Here, $\theta_{k,j} = \{\theta_{k,j,l}\}_{l=1}^{K}$ and $\phi = \{\phi_l\}_{l=1}^{K}$. $\alpha = \{\alpha_l\}_{l=1}^{K}$ and $\beta_{k,j} = \{\beta_{k,j,l}\}_{l=1}^{K}$ are hyperparameters of the Dirichlet distributions. $\theta_{k,j,l}$ denotes the probability that the jth classifier outputs class ω_l for a sample of class ω_k when $r_{k,j} = 1$. ϕ_l is the probability that a classifier outputs class ω_l for a sample when $r_{k,j} = 0$. As mentioned above, $r_{k,j} = 0$ indicates that the jth classifier is ineffective for class ω_k. Therefore, ϕ_l does not depend on j and k. Intuitively, $\theta_{k,j}$ should be peaky, while ϕ should be flat.

On the assumption that $\{\mathbf{c}_{k,i}\}_{i=1}^{K}$ are i.i.d. samples, the likelihood of an observed prediction matrix is given by

$$P(\mathrm{C}|\mathrm{R},\Theta,\phi)=\prod_{k=1}^{K}\prod_{i=1}^{N_k}\prod_{j=1}^{J}\prod_{l=1}^{K}\{(\theta_{k,j,l})^{r_{k,j}}(\phi_l)^{1-r_{k,j}}\}^{\delta(c_{k,i,j},l)}$$
$$=\prod_{k=1}^{K}\prod_{j=1}^{J}\prod_{l=1}^{K}\{(\theta_{k,j,l})^{r_{k,j}}(\phi_l)^{1-r_{k,j}}\}^{n_{k,j,l}}. \tag{2}$$

Here, $n_{k,j,l}$ denotes the number of ω_k samples that were predicted as ω_l by the jth classifier. That is, $n_{k,j,l}=\sum_{i=1}^{N_k}\delta(c_{k,i,j},l)$, and $\delta(x,y)=1$ if $x=y$, and 0 otherwise. $\Theta=\{\theta_{k,j}\}$. Moreover, with the help of the conjugacy of the priors, Θ, ϕ and λ can all be marginalized out as

$$P(\mathrm{C}|\mathrm{R};\alpha,\beta)=\int P(\mathrm{C}|\mathrm{R},\phi,\Theta)p(\phi;\alpha)p(\Theta;\beta)d\phi d\Theta$$
$$=\left(\frac{\Gamma(\alpha_\bullet)}{\prod_l\Gamma(\alpha_l)}\frac{\prod_l\Gamma(\sum_{k,j}\delta(r_{k,j},0)n_{k,j,l}+\alpha_l)}{\Gamma(\sum_{k,j}\delta(r_{k,j},0)N_k+\alpha_\bullet)}\right)$$
$$\times\left(\prod_{k=1}^{K}\prod_{j=1}^{J}\frac{\Gamma(\beta_{k,j,\bullet})}{\prod_l\Gamma(\beta_{k,j,l})}\frac{\prod_l\Gamma(r_{k,j}n_{k,j,l}+\beta_{k,j,l})}{\Gamma(r_{k,j}N_k+\beta_{j,\bullet})}\right). \tag{3}$$

Here, we set $\mathrm{R}=\{r_{k,j}\}$, $\alpha_\bullet=\sum_{l=1}^{K}\alpha_l$ and $\beta_{k,j,\bullet}=\sum_{l=1}^{K}\beta_{k,j,l}$. In a similar way, we have

$$P(\mathrm{R};a,b)=\frac{\Gamma(\sum_{k,j}r_{k,j}+a)\Gamma(\sum_k\sum_j(1-r_{k,j})+b)\Gamma(a+b)}{\Gamma(KJ+a+b)\Gamma(a)\Gamma(b)}. \tag{4}$$

Here, $\Gamma(x)$ denotes the gamma function. From Eqs. (3) and (4), we can compute the collapsed posterior distribution as $P(\mathrm{R}|\mathrm{C};\alpha,\beta,a,b)\propto P(\mathrm{C}|\mathrm{R};\alpha,\beta)P(\mathrm{R};a,b)$.

2.3 Estimating Latent Variables

From $P(\mathrm{R}|\mathrm{C};\alpha,\beta,a,b)$, $\mathrm{R}=\{r_{k,j}\}$ can be inferred by Gibbs sampling. Let $r_{\setminus(k,j)}$ denote all $\{r_{s,t}\}$ other than when $(s,t)\neq(k,j)$. Then, $P(r_{k,j}=1|r_{\setminus(k,j)},\mathrm{C})+P(r_{k,j}=0|r_{\setminus(k,j)},\mathrm{C})=1$, so our goal here is simply to compute the ratio $v=P(r_{k,j}=1|r_{\setminus(k,j)},\mathrm{C})/P(r_{k,j}=0|r_{\setminus(k,j)},\mathrm{C})$ because when v is computed, we can obtain

$$P(r_{k,j}=1|r_{\setminus(k,j)},\mathrm{C})=\frac{v}{1+v},\qquad P(r_{k,j}=0|r_{\setminus(k,j)},\mathrm{C})=\frac{1}{1+v}. \tag{5}$$

The v value can be easily computed by using (3) and (4). The details of the derivation will be provided in **Appendix**.

2.4 Classification of Unknown Class Samples

Let $\mathbf{c}_{*,m} = (c_{*,m,1}, \ldots, c_{*,m,J})$ denote a set of J predictions for the mth test sample. Once J classifiers have been trained on the training samples, c_m can be obtained by applying J classifiers to the mth test data sample. The optimal class of $\mathbf{c}_{*,m}$ can be obtained by finding k^* that maximizes the posterior predictive class distribution for $\mathbf{c}_{*,m}$, i.e. $P(\omega_k|\mathbf{c}_{*,m}, \mathrm{C}) \propto P(\mathbf{c}_{*,m}|\omega_k, \mathrm{C})P(\omega_k)$. $P(\omega_k)$ is a class prior probability and is usually set at a uniform distribution.

Then $P(\mathbf{c}_{*,m}|\omega_k, \mathrm{C})$ can be obtained using the following Monte Carlo approximation:

$$P(\mathbf{c}_{*,m}|\omega_k, \mathrm{C}) = \prod_{j=1}^{J} \sum_{r_{k,j}} P(c_{*,m,j}|r_{k,j}, \mathrm{C})P(r_{k,j}|\mathrm{C}) \simeq \prod_{j=1}^{J} \frac{1}{T-t_0} \sum_{t=t_0+1}^{T} P(c_{*,m,j}|r_{k,j}(t), \mathrm{C}), \quad (6)$$

where $r_{k,j}(t)$ is the value of the tth Gibbs sampling via $P(r_{k,j}|r_{\backslash(k,j)}, \mathrm{C})$. Note that t_0 is the burn-in time and the values of $r_{k,j}(t)$, $t = 1, \ldots, t_0$ are discarded from the sampling.

Moreover, the RHS of Eq. (6) can be calculated depending on the value of $r_{k,j}$ as

$$P(c_{*,m,j}|r_{k,j} = 1, \mathrm{C}) = \int P(c_{*,m,j}|\theta_{k,j})p(\theta_{k,j}|\mathrm{C})d\theta_{k,j} = \frac{\prod_{l=1}^{K}(n_{k,j,l} + \beta_{k,j,l})^{\delta(c_{*,m,j},l)}}{N^{(k)} + \beta_{k,j,\bullet}}, \quad (7)$$

$$P(c_{*,m,j}|r_{k,j} = 0, \mathrm{C}) = \int P(c_{*,m,j}|\phi)p(\phi|\mathrm{C})d\phi = \frac{\prod_{l=1}^{K}(n_{\bullet,j,l} + \alpha_l)^{\delta(c_{*,m,j},l)}}{N + \alpha_\bullet}. \quad (8)$$

Here, $n_{\bullet,j,l} = \sum_{k=1}^{K} n_{k,j,l}$. Finally, we assign class ω_{k^*} to the mth test sample as

$$k^* = \mathrm{argmax}_k \{P(\omega_k) \prod_{j=1}^{J} \sum_{t=t_0+1}^{T} \left(\frac{\prod_{l=1}^{K}(n_{k,j,l} + \beta_{k,j,l})^{\delta(c_{*,m,j},l)}}{N_k + \beta_{k,j,\bullet}} \right)^{\delta(r_{k,j}(t),1)} \times \left(\frac{\prod_{l=1}^{K}(n_{\bullet,j,l} + \alpha_l)^{\delta(c_{*,m,j},l)}}{N + \alpha_\bullet} \right)^{\delta(r_{k,j}(t),0)} \}. \quad (9)$$

2.5 Intuitive Explanation

We can see that the class decision rule changes depending on the value of $r_{k,j}$ for each of the classifiers. If $c_{*,m,j} = u$ and hyperparameters are ignored, then the element of the sum in Eq. (9) can be written as $(n_{k,j,u}/N_k)^{\delta(r_{k,j},1)}(n_{\bullet,j,l}/N)^{\delta(r_{k,j},0)}$. Here, $n_{k,j,u}/N_k$ is the ratio of the number of times that the jth classifier judged the class ω_k sample as class ω_u. Therefore, when the jth classifier is effective for class ω_k (i.e. $r_{k,j} = 1$), the decision rule prefers class ω_k with a larger $n_{k,j,u}/N_k$ to class ω_s with a smaller $n_{s,j,u}/N_s$ as the class of the mth test sample. This is intuitively reasonable. Note that we do not necessarily require u to be the true class, since the meta-classifier can learn how each of the classifiers (mis)classifies from the observed multiple predictions. This is one reason why biased classifiers are allowed in our ensemble scheme.

It is important to explain why our meta-classifier can avoid overfitting. When $P(\omega_k|\mathbf{c}_{*,m}, \mathrm{C}) > P(\omega_{\forall s \neq k}|\mathbf{c}_{*,m}, \mathrm{C})$ holds, a test sample $\mathbf{c}_{*,m} \in \omega_k$, is correctly classified as class ω_k As shown in (6), since $P(\mathbf{c}_{*,m}|\cdot)$ consists of j independent factors, we look at each j independently. Let $\omega_{k'}$ be the most similar class to ω_k in the sense that $P(\omega_{k' \neq k}|\mathbf{c}_{*,m}, \mathrm{C}) \geq P(\omega_{\forall s \neq k}|\mathbf{c}_{*,m}, \mathrm{C})$. Overfitting will occur when both $\theta_{k,j}$ and $\theta_{k',j}$ are flat. This is because flat distribution indicates that the predictions (class lables) are diverse and therefore their realizations are often different between training and test data. This means that the jth classifier's predictions (class labels) of both ω_k and $\omega_{k'}$ samples are unreliable. In this case since these parameter values are unstable between training and test data, the magnitude relation of class posteriors can change between training and test data. To address the problem of the overfitting, our model automatically sets $r_{k,j} = 0$ and $r_{k',j)} = 0$, and class-independent (flat-distributed) parameter ϕ is commonly used between ω_k and $\omega_{k'}$. This indicates that when $r_{k,j} = 0$ and $r_{k',j} = 0$, the jth classifier cannot discriminate between ω_k and $\omega_{k'}$ samples since the 2nd term of (9) contributes to $P(c_{*,m,j}|\omega_k, \mathrm{C})$ and $P(c_{*,m,j}|\omega_{k'}, \mathrm{C})$ equally. However, we do not have to give up discriminating them because another classifier(s) in the ensemble can be effective fot the classification.

3 Related Work

Our problem setting is essentially different from usual ensemble classification methods in the sense that our ensemble scheme does not restrict the classifiers that are chosen for the ensemble or the way in which the feature representations are used for training the classifiers. A closely related study, i.e. the Bayesian Classifier Combination (BCC) has been presented [14] to combine the predictions of multiple classifiers by extending the pre-existing observer modeling proposed in [5]. Although they present a basic model (IBCC) and several extensions, we simply focus on IBCC and EBCC models since the classification accuracies of the other extensions are compatible with those of IBCC and EBCC. As graphical models are shown in Fig. 2,

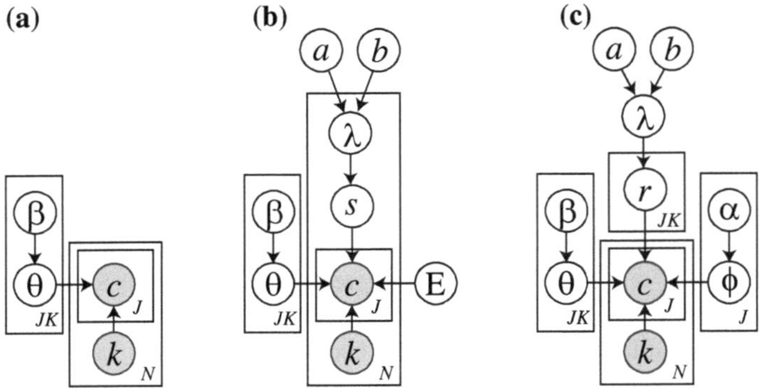

Fig. 2 Graphical models of IBCC, EBCC, and RNBC. **a** IBCC, **b** EBCC, **c** RNBC (Proposed)

the IBCC corresponds to just the naive Bayes model in which the jth classifier outputs a prediction according to a class-conditional multinomial paramter θ_k. Note that although our RNBC looks similar to the EBCC since both introduce latent variables (s and r), their roles are essentially different. In EBCC the latent variable identifies whether or not a sample is easy to classify. Specifically, when $s_n = 1$ ("easy to classify") for the nth sample, the sample c_n is generated by using a predetermined fixed constant multinomial distribution parameter (E). When $s_n = 0$, c_n is generated by using another parameter θ that is estimated by using training data. Since all classifiers predictions in the ensemble are equally used in IBCC and EBCC models, they still suffer from the overfitting problem, as shown in experimental results.

Multiple predictions for each sample are treated as a feature vector of the sample, and therefore $r_{k,j}$ can be regarded as a feature selection indicator latent variable. Therefore it is related to conventional feature selection methods (e.g. [16]). However, usual feature selection is class-independent, while in our model a class-specific feature selection is performed. Note that similar feature selection method to our method have been employed in *unsupervised* problems [9] and typically in subset clustering [8, 10, 12, 17]. The use of this latent variable in supervised classification settings is original based on our survey of the literature.

4 Experiments

4.1 Evaluation Using Synthetic Data

We performed an experimental 20-class classification task where $J = 50$. Each $c_{k,i,j} \in \{1, ..., 20\}$ value included in a synthetic dataset was randomly sampled on a probabilistic distribution defined by using a fixed hyperparameter set of our generative model. We constructed ten different evaluation sets for each dataset, and

Table 1 Classification accuracies (%) on synthetic datasets

	RNBC	IBCC	EBCC	MV	NB	LR+Lasso	SVM(L)	SVM(P)
D1	**99.3** (0.3)	90.9 (1.4)	92.0 (1.6)	24.3 (1.5)	91.1 (1.0)	94.4 (0.7)	91.4 (0.7)	82.7 (0.9)
D2	**100** (0.0)	99.3 (0.6)	99.7 (0.5)	75.4 (0.3)	99.7 (0.2)	99.8 (0.1)	99.6 (0.2)	96.9 (0.6)
D3	**99.9** (0.1)	99.2 (0.2)	99.5 (0.2)	35.2 (1.6)	97.6 (0.7)	97.1 (0.7)	98.0 (0.4)	93.9 (0.6)
D4	**99.9** (0.1)	97.5 (1.2)	97.3 (0.8)	51.7 (1.3)	96.2 (0.5)	99.8 (0.3)	98.7 (0.4)	94.3 (0.6)

examined the percentage, R_I, of the number of (j, k) pairs whose predictions were incorrect but effective, against the total number of (j, k) pairs. We also examined the percentage, R_C, of the number of (j, k) pairs whose predictions were correct for many data samples. The average (R_c, R_i) values were $(5.4, 7.1)$, $(11.7, 2.0)$, $(4.4, 4.9)$, and $(8.5, 1.2)$ for the four datasets, D1, D2, D3, and D4, respectively. In each experiment employing an evaluation set, we used 66 and 34 different data samples per class as training and test samples, respectively.

Table 1 shows the average classification accuracies and standard deviation of ten experiments obtained by employing eight classifiers, the proposed model (RNBC), two BCC models (IBCC and EBCC), naive Bayes (NB), majority voting (MV), logistic regression with L1-Lasso regularization (LR-Lasso) [18], and support vector machines with a linear kernel (SVM(L)) and with a polynomial kernel (SVM(P)) as meta-classifiers. Note that NB corresponds to a variant of RNBC when $r_{k,j} = 1$ for all j and k, as mentioned in Sect. 2. Each number in parenthesis in the table denotes the standard deviation of the ten experiments. As for the BCC models, because the performance of the other variants (DBCC and EDBCC) was fairly similar to those of IBCC and EBCC in the experiments [14], we did not employ them. L1-Lasso is a method for obtaining discriminative classifiers that provide accurate class boundaries of training data from as few features as possible. In this sense L1-Lasso is applicable to our problem, and we included it in the experiment as a *stacked generalization* method [20]. SMVs are also regarded as stacking methods. As for LR and SVM, we transformed each discrete value to an augmented binary vector according to [13], because it is inappropriate for LR and SVM to be applied directly to such categorical values. We set the hyperparameter values of the proposed method and the other methods using the 5-fold cross-validation of training samples.

From the MV results we can see that most predictions were biased. All the methods except the proposed method, RNBC, performed worst for D1 of the four datasets. D1 had the largest R_I of the four datasets, and these methods did not work well for constructing meta-classifiers when there were many incorrect predictions. In contrast, the classification accuracies obtained with RNBC were close to 100 % for all the datasets, as shown in Table 1.

All the methods except RNBC performed better for D2 than for D4. Although D4 had the smallest R_I of the four datasets, the percentage of (j, k) pairs whose predictions were ineffective, calculated as $100 - (R_C + R_I)$, was larger for D4 than for D2. Meta-classifiers based on MV, IBCC, EBCC, and NB combine the predictions equally. On the other hand, RNBC learns which predictions should be employed for classification, by estimating $r_{k,j}$ in a supervised way. The experimental

Table 2 Classification accuracies (%) obtained with single classifiers

	k-NN		SVM(L)		SVM(P)		SVM(R)		C4.5	LR	LDA
	Min.	Max.	Min.	Max.	Min.	Max.	Min.	Max.			
DNA	70.2	73.9	81.3	81.4	86.1	89.0	60.2	66.1	87.9	84.5	90.3
Digit	98.6	98.7	65.0	71.1	97.6	98.8	74.4	98.6	88.2	89.2	90.8
Satellite	90.4	91.0	53.1	54.2	55.8	62.2	73.6	90.5	85.0	86.7	78.6

results also confirmed that RNBC could mitigate the overfitting to ineffective base-classifiers predictions than SVMs and LR+Lasso, which combined the predictions with weights tuned in discriminative ways. The BCC approach requires the sampling of class labels for all test samples as well as unknown model parameters in Gibbs sampling, and therefore it took much more computational time than RNBC. Actually, IBCC and EBCC required a minimum of about 10,000 steps, while RNBC required a maximum of about 100 steps. Note that the results for IBCC and EBCC in Table 1 correspond to the results obtained with 50,000 steps. This indicates that RNBC is much more computational efficient than IBCC and EBCC.

4.2 Real Data

4.2.1 UCI Data

We compared the methods by using the UCI data sets [15] that have been utilized in machine learning experiments. As base classifiers, we employed k-NN, SVM, C4.5, LR, and Linear Discriminant Analysis (LDA). Table 2 shows the classification accuracies obtained with these single classifiers for the three UCI datasets. Table 2 also shows the minimum and maximum values of the average classification accuracies obtained with k-NN and SVMs when using different values for their hyperparameters. The k value of k-NN was selected from three candidates $\{1, 3, 5\}$. The margin parameter value of the SVMs was selected from three candidates $\{1, 100, 10000\}$. The kernel parameter value of SVM(P) was selected from two candidates $\{2, 3\}$. The kernel parameter value of SVM(R) was selected from three candidates $\{0.01, 0.1, 2\}$. Table 2 also shows the average classification accuracies obtained with C4.5, LR, and LDA.

Tables 3 and 4 show the classification accuracies of the meta-classifiers. Here, Table 3 corresponds to the results in which the predictions of all the classifiers were combined, while Table 4 shows the results in which the predictions of the best-tuned k-NN and SVM were combined with the predictions of C4.5, LR, and LDA. In the former (latter) case, J becomes 24 (5). As shown in Tables 3 and 4, BCC, NB and RNBC provided better results than any of the base classifiers. RNBC slightly outperformed BCC and NB. This is valid because almost all $r_{k,j}$ values were 1 in RNBC. Since the UCI data sets provided relatively *easy tasks*, the predictions of the base classifiers were *less biased* and therefore RNBC, BCC, and NB performed similarly. It would be interesting to see whether RNBC obtained better results when

Table 3 Classification accuracies (%) obtained with meta-classifiers ($J = 24$)

	RNBC	IBCC	EBCC	MV	NB	LR + Lasso	SVM(L)	SVM(P)
DNA	**93.0** (0.5)	90.5 (1.2)	91.5 (0.9)	88.2 (1.0)	91.3 (1.2)	90.0 (1.3)	87.3 (1.9)	89.1 (1.2)
Digit	**99.1** (0.2)	**99.1** (0.1)	**99.1** (0.1)	98.0 (0.2)	**99.1** (0.2)	**99.1** (0.2)	99.0 (0.2)	**99.1** (0.1)
Satellite	**92.7** (0.5)	91.9 (0.4)	92.4 (0.6)	90.0 (1.3)	**92.7** (0.5)	91.1 (0.5)	86.3 (0.7)	91.5 (0.5)

Table 4 Classification accuracies (%) obtained with meta-classifiers based on combining best-tuned classifiers ($J = 7$)

	RNBC	IBCC	EBCC	MV	NB	LR + Lasso	SVM(L)	SVM(P)
DNA	92.2 (0.9)	**92.4** (0.6)	**92.4** (0.5)	90.0 (1.3)	91.7 (0.9)	90.8 (0.9)	88.8 (1.7)	89.6 (0.5)
Digit	**99.1** (0.2)	99.0 (0.2)	99.0 (0.2)	98.8 (0.1)	98.9 (0.2)	98.9 (0.1)	98.9 (0.2)	99.0 (0.2)
Satellite	**91.4** (0.4)	91.3 (0.3)	**91.4** (0.4)	91.1 (0.7)	91.2 (0.8)	91.1 (0.3)	90.7 (0.3)	91.1 (0.4)

$J = 24$ than when $J = 7$, although the base classifiers were best tuned when $J = 7$. It can be thought that RNBC tried to use the predictions effectively even when their predictions were incorrect and therefore RNBC could utilize all the classifier predictions effectively to discriminate certain classes.

4.2.2 High-level Real Nursing Activity Recognition Data

As a *real* case study of a combination of biased classifiers, we applied our model to high-level human activity recognition using accelerometers. Four small three-axis accelerometers were attached to nurses and they performed real nursing activities in a hospital. Although the experiments are still on-going, we used the current data (i.e. 22 activity classes, 1,097 instances in total). The activity classes are shown in Table 5. We notice that these activity classes are determined without researcher supervison or observation, and therefore classification is really difficult. Each instance was segmented per action by a certain segmention method in a pre-processing step. Namely, our recogniton task is to classify each segmented activity instance as one of 22 classes. We created five data sets of training and test samples. The average number of training (test) samples per category (activity class) was 39.9 (9.9). We also created a 14-class problem (classes with ◯) that is a subset of the original 22 classes.

Features were calculated on certain time-step windows of acceleration data with a 50 % time overlap between consecutive windows. We used a mean, a standard deviation, a frequency-domain energy, and a frequency-domain entropy as a set of features within a window. These features has been employed in previous studies of activity recognition from acceleration data [1]. By concatenating consecutive feature vectors, each of which was obtained by the sliding window, we have time-series data consisting of 48-dimensional (4 features $\times$ 3 axes $\times$ 4 sensors) vectors per activity instance. A hidden markov model (HMM) can be used as a classifier. However, since the appropriate window size depends on the activity class, it is difficult to choose the best size for all classes. Therefore, we changed the window size (time step) as

Table 5 Nursing activities

	Anamnese (standing)	○	Gatch up	○	Move bed
	Portable X-ray (prone)	○	Assist with portable toilet	○	Record work (PC)
○	Record work (manual)	○	Measure blood pressure (dorsal)		Measure blood sugar
○	Sample blood (dorsal)		Start intravenous injection	○	Assist with wheelchair
○	Wash hands		Attach ECG	○	Remove ECG
○	Measure ECG		Change posture	○	Measure weight (sitting)
○	Find artery		Examine edema (sitting)		Assist walk
○	Set clock				

Table 6 Classification accuracies (%) of base classifiers on a high-level activity recognition dataset

Classes	HMM	C4.5	RF
22	42.4 (1.7)	24.3 (2.9)	32.2 (3.4)
14	54.2 (2.5)	34.1 (1.8)	45.2 (1.0)

2, 4, 6,..., 100 and created 50 kinds of feature representations. Then, we trained an HMM on each of them, and found that the best classification accuracy for test data was around 4 %, as shown in Table 3, when the window size was 48 time steps. To improve the accuracy, we tried to learn a meta-classifier from these multiple predictions obtained by 50 HMMs.

We also show the results of C4.5 and Random Forest (RF) [4] that provided good results for activity recognition [1]. Note that since these classifiers are for static data, we created a set of static samples simply by regarding a four-dimensional feature vector obtained on a window as one sample, as described in [1]. Namely, unlike in the feature representation for HMM, many independent samples per action instance are generated. The best performances of C4.5 and RF are shown in Table 6. Since the sample size was small compared with the number of classes and the action classes themselves were selected in an uncontrolled setting without reseacher supervision, this task was too difficult for conventional classifiers to provide a reasonable result. In both cases (22 and 14 classes), the C4.5 and RF results were worse than the HMM with the best window size, and therefore we can see that a time-series feature representation is appropriate for this task.

Note that the main purpose of the experiment is to show that our model can effectively generate a meta-classifier when the multiple predictions are poorly biased. Table 7 shows the results of several ensemble methods. All meta-classifiers provided better classification accuracies than the single best HMM. This indicates the usefulness of the ensemble scheme. Among them the proposed method (RNBC) obtained the highest accuracies in both cases (22 and 14 classes). From the results of MV we can confirm that the predictions obtained by base classifiers (50 HMMs) were biased. BCC methods (IBCC and EBCC) outperformed the stacking methods (i.e. LR and SVMs). The results of IBCC were close to those of NB. This is intuitively reasonable because IBCC can be considered a transductive version of EBCC with

Table 7 Classification accuracies (%) of meta-classifiers on a high-level activity recognition dataset

Classes	RNBC	IBCC	EBCC	MV	NB	LR + Lasso	SVM(L)	SVM(P)
22	**62.8** (3.3)	57.6 (3.1)	58.2 (2.8)	42.3 (1.9)	56.8 (2.9)	51.0 (2.9)	46.6 (3.8)	43.0 (3.2)
14	**75.6** (1.6)	72.5 (1.6)	72.8 (1.4)	57.7 (1.2)	71.5 (1.4)	71.8 (1.5)	70.2 (1.5)	69.8 (1.5)

dependency modeling that slightly outperformed IBCC, but its performance was inferior to that of RNBC. BCC needed about 30,000 steps to obtain highest accuracy in Gibbs sampling, while RNBC required about only 100 steps.

5 Conclusion

We have developed a new model for learning a meta-classifier from the biased multiple predictions of the classifiers in an ensemble. We confirmed that the proposed method outperformed the conventional methods in our experiments using synthetic and real data sets. Our supervised model learns the relationship between the true class and the multiple predictions of the classifiers in the ensemble. The key feature of our ensemble schme is the introduction of a latent variable that can identify whether or not a classifier is effective for each of the classes. Namely, since our model learns the relationship between the true class and the effective classifier predictions for each class, it could achieve better generalization performance than the conventional methods. We think that this simple and practically useful method surely can contribute to construct a robust classification system with a high generalization ability. Interesting research direction would be to extend the idea to multiple kernel learning [15] and crowd sourcing application [19].

Acknowledgments This research is supported by FIRST program. The authors would like to appreciate the cooperation for experiment by staff of Saiseikai Kumamoto Hospital, Japan.

Appendix

Derivations of Eqs. (3) and (4) are as follows. Equation (3) can be derived as follows:

$$\begin{aligned} P(\mathrm{C}|\mathrm{R};\alpha,\beta) &= \int P(\mathrm{C}|\mathrm{R},\phi,\Theta)p(\phi;\alpha)p(\Theta;\beta)d\phi d\Theta \\ &= \left(\prod_{k=1}^{K}\prod_{j=1}^{J}\frac{\Gamma(\sum_l \beta_{k,j,l})}{\prod_{l'}\Gamma(\beta_{k,j,l'})}\int\prod_{l=1}^{K}(\theta_{k,j,l})^{r_{k,j}n_{k,j,l}+\beta_{k,j,l}-1}d\theta_{k,j,l}\right) \end{aligned}$$

$$\times \frac{\Gamma(\sum_{l'} \alpha_{l'})}{\prod_{l'} \Gamma(\alpha_{l'})} \int \prod_{l=1}^{K} \phi_l^{\sum_k \sum_j (1-r_{k,j}) n_{k,j,l} + \alpha_l - 1} d\phi_l$$

$$= \left(\frac{\Gamma(\alpha_\bullet)}{\prod_l \Gamma(\alpha_l)} \frac{\prod_l \Gamma(\sum_{k,j} \delta(r_{k,j}, 0) n_{k,j,l} + \alpha_l)}{\Gamma(\sum_{k,j} \delta(r_{k,j}, 0) N_k + \alpha_\bullet)} \right)$$

$$\times \left(\prod_{k=1}^{K} \prod_{j=1}^{J} \frac{\Gamma(\beta_{k,j,\bullet})}{\prod_l \Gamma(\beta_{k,j,l})} \frac{\prod_l \Gamma(r_{k,j} n_{k,j,l} + \beta_{k,j,l})}{\Gamma(r_{k,j} N_k + \beta_{j,\bullet})} \right).$$

Here, $B(x, y)$ is the beta function. We used the definition $B(x, y) = \Gamma(x)\Gamma(y)/\Gamma(x + y)$. In a similar manner, Eq. (4) can be derived as follows:

$$P(\mathrm{R}; a, b) = \prod_{k=1}^{K} \prod_{j=1}^{J} \int P(r_{k,j}; \lambda) p(\lambda; a, b) d\lambda$$

$$= \int \lambda^{\sum_k \sum_j r_{k,j} + a - 1} (1 - \lambda)^{\sum_k \sum_j (1 - r_{k,j}) + b - 1} d\lambda / B(a, b)$$

$$= \frac{B\left(\sum_k \sum_j r_{k,j} + a, \sum_k \sum_j (1 - r_{k,j}) + b\right)}{B(a, b)}$$

$$= \frac{\Gamma(\sum_k \sum_j r_{k,j} + a) \Gamma(\sum_k \sum_j (1 - r_{k,j}) + b)}{\Gamma(KJ + a + b)} \cdot \frac{\Gamma(a, b)}{\Gamma(a)\Gamma(b)}$$

$$= \frac{\Gamma(\sum_k \sum_j r_{k,j} + a) \Gamma(\sum_k \sum_j (1 - r_{k,j}) + b) \Gamma(a + b)}{\Gamma(KJ + a + b) \Gamma(a) \Gamma(b)}.$$

Here, we used another definition of the beta function: $B(s, t) = \int x^{s-1}(1-x)^{t-1} dx$.

References

1. Bao, L., Intille, S.: Activity recognition from user-annotated acceleration data. In: Proceedings of International Conference on Pervasive Computing, Pervasive 2004, pp. 1–17. Springer, (2004)
2. Blake, C.L., Merz, C.J.: UCI Repository of Machine Learning Databases. University of California, Department of Information and Computer Science, Irvine (1998)
3. Breiman, L.: Bagging predictors. Mach. Learn. **24**, 123–140 (1996)
4. Breiman, L.: Random forests. Mach. Learn. **45**, 5–32 (2001)
5. Dawid, A., Skene, A.: Maximum likelihood estimation of observer error-rates using the EM algorithm. J. Appl. Stat. **28**, 20–28 (1979)
6. Dietterich, T.G.: Ensemble methods in machine learning. In: Proceedings of the First International Workshop on Multiple Classifier Systems, pp. 1–15. Springer, London (2000)
7. Freund, Y., Schapire, R.E.: Experiments with a new boosting algorithm. In: Proceedings of International Conference on Machine Learning ICML96, pp. 148–156 (1996)
8. Fu, Q., Banerjee, A.: Bayesian overlapping subspace clustering. In: Proceedings of International Conference on Data Mining, ICDM2009 (2009)

9. Guan, Y., Dy, J.G., Jordan, M.I.: A unified probabilistic model for global and local unsupervised feature selection. In: Proceedings of International Conference on Machine Learning ICML2011 (2011)
10. Guyon, I., Elisseeff, A.: An introduction to variable and feature selection. J. Mach. Learn. Res. **3**, 1157–1182 (2003)
11. Hastie, T., Tibshirani, T., Friedman, J.H.: The elements of statistical learning: data mining, inference, and prediction (2009)
12. Hoff, P.D.: Subset clustering of binary sequences, with an application to genomic abnormality data. Biometrics **61**(4), 1027–1036 (2005)
13. Hsu, C., Chang, C., Lin, C.: A practical guide to support vector classification. http://www.csie.ntu.edu.tw/cjlin (2010)
14. Kim, H.C., Ghahramani, Z.: Bayesian classifier combination. In: Proceedings of International Conference on Artificial Intelligence and Statistcs, AISTATS2012. http://www.aistats.org/papers.php (2012)
15. Lanckriet, G.R.G., Cristianini, N., Bartlett, P., Ghaoui, L.E., Jordan, M.I.: Learning the kernel matrix with semidefinite programming. J. Mach. Learn. Res. **5**, 27–72 (2004)
16. Peng, H., Long, F., Ding, C.: Feature selection based on mutual information criteria of max-dependency, max-relevance, and min-redundancy, IEEE Trans. Pattern Anal. Mach. Intell. **27**(8), 1226–1238 (2005)
17. Shan, H., Banerjee, A.: Bayesian co-clustering. In: Proceedings of IEEE International Conference on Data Mining (ICDM), pp. 530–539 (2008)
18. Tibshirani, R.: Regression shrinkage and selection via the Lasso. J. Roy. Stat. Soc. **58**(1), 267–288 (1996)
19. Whitehil, J., Ruvolo, P., Wu, T., Bergsma, L., Movellan, J.: Whose vote should count more: optimal integration of labels from labelers of unknown expertise. Advances in Neural Information Processing Systems, NIPS2009 (2009)
20. Wolpert, D.H.: Stacked generalization. Neural Netw. **5**, 241–259 (1992)

Developing Mathematicians for Industry Research Teams

Murray A. Cameron

Abstract Interdisciplinary teams are of increasing importance and mathematicians can be valuable contributors to them. Mathematicians can be leaders, not just reactive problem solvers in these teams. PhD training for mathematicians should be augmented so that graduates are best able to contribute in this environment. The nature of some new PhD programs is described together with detail of the program of the Industry Doctoral Training Centre of the Australian Technology Network of Universities.

Keywords Industrial mathematics · Doctoral training · Research teams · Education · Workplace

1 Introduction

Measuring, understanding, predicting and controlling complex processes are major challenges for industrial research. These challenges are met with teams of people—problem owners and researchers with knowledge of relevant disciplines. Often those teams need innovative, quantitative skills of a high order, such as can be provided by a mathematician with a PhD. However, to work successfully in that environment a mathematical scientist needs skills not provided in most PhD programs.

Career options for PhD graduates in the mathematical sciences have predominantly been in academia and so in most countries the PhD program has assumed the student will progress to a career of teaching and research in a university. As a result, PhD programs aim to develop appropriate attributes in their graduates.

M. A. Cameron (✉)
Industry Doctoral Training Centre, Australian Technology Network of Universities,
Sydney, Australia
e-mail: murray.cameron@uts.edu.au

M. Wakayama et al. (eds.), *The Impact of Applications on Mathematics*,
Mathematics for Industry 1, DOI: 10.1007/978-4-431-54907-9_11,

In the so-called applied areas of the mathematical sciences such as applied mathematics, computational mathematics, operations research, and statistics, the educational focus for PhD students has been on research to develop methods for previously formulated problems rather than on the more realistic case of taking a problem from another discipline, formulating it as a mathematical problem and solving it. The student concentrates on developing deep knowledge in a specific area of mathematics in order to solve specific research questions. In Australia (as in many other countries) the PhD usually has no formal coursework component. A student may undertake courses, but this is usually informal (not involving the sitting of a formal examination) and is aimed at strengthening knowledge directly related to the PhD research.

In recent times there have been two reasons to review PhD programs and the attributes of their graduates:

- Broad questioning of whether a PhD program is an effective and appropriate way to develop highly capable people
- The growing numbers of interdisciplinary research teams in industry and in scientific research are increasing the demand for appropriately trained mathematicians.

In this paper, I first consider the aims of the 'traditional' PhD program and some of the concerns raised about it recently. I then consider the capabilities required for mathematicians working on industrial projects and other interdisciplinary teams and how those capabilities are developed. I then describe a few of the international initiatives to modify the training of mathematicians. Finally I describe the education program being developed in a collaboration among five Australian Universities and some of the issues that have arisen.

2 Is a PhD a Useful Experience?

If you search the websites of various universities and disciplines, a common thread emerges of the aims of the PhD. They seem to be that a graduate will:

- Demonstrate competence to carry out independent and original academic research.
- Present the results of research to a standard equivalent to that of a peer-reviewed academic publication.
- Demonstrate ability to present and defend academic work in front of peers.
- Engage within the full community of scholars (e.g. networking, dissemination of knowledge, conferences, demonstrating impact and value of research).

These aims are directed towards a career in a university pursuing research in an existing discipline. While these are valuable aims, there are gaps. They fail to mention some capabilities that are important in work environments other than the academic research environment. In fact these capabilities are often useful in an academic environment as well. Because of the gaps, there is questioning of the usefulness of PhD study and suggestions for changing the PhD. For example, an article in 2010 in The Economist [2] listed a number of problems with PhD programs:

- The production of PhDs has far outstripped demand for university lecturers.(In Europe, at least, it seems the majority of PhD graduates are employed by business and industry.)
- Business leaders complain about shortages of high-level skills, suggesting PhDs are not teaching the right things.
- Many PhDs find it difficult to transfer their skills into business and industry

There have been many discussions of the perceived issues around PhD programs and possible approaches to address them. For example a recent discussion paper prepared for a consortium of Australian universities [10] suggested:

- Employability of PhD graduates
- Industry concern about attributes of PhD graduates
- Government concern about inadequate supply of graduates
- Cost & effectiveness of the training provided.

are all issues that need to be considered in developing improved programs for PhD students. There is clear overlap in these findings, and they are relevant to the education of mathematicians for careers that may be in industry.

3 What Does an Industrial Mathematical Scientist Do?

The material in this section is based on personal experience and observation as well as from reading a variety of sources. A more thorough exposition, including the results of a survey on mathematicians in industry and their managers, is presented in a report of the Society of Industrial and Applied Mathematics (SIAM) published in 1998 [11].

3.1 The Role of a Mathematician in Industry

A research mathematician appointed to an organisation will generally be assigned to work on a particular project or in a particular area. The task of the mathematician consists of a series of steps which may be iterated.

Identify an area where there is a problem and broadly describe the problem
Formulate the problem into a mathematical problem and select the assumptions and approximations that will be made in order to make progress
Solve the mathematical problem. Mathematical research often focuses on this step and, in particular, new ways to solve problems that are simpler, faster or able to be used more generally

Explain the applicability of the solution to the initial problem, describe its general features and the relevance and importance of the approximations and assumptions made to reach the solution.

3.2 Capabilities of a Successful Industrial Mathematician

To fulfil this role, the mathematician needs to be good at asking questions so as to identify the key questions to be addressed and the reasonable assumptions that can be made in formulating the problem mathematically.

More broadly, a researcher (including a mathematician) moves in an environment where she or he needs to be in a position to hear about problems. The researcher needs to establish their relevance (so that they are included in the right discussions of problems), credibility (so that the researcher's views are listened to and given due weight) and value (so that the potential benefits of the researcher's proposals are properly recognised and measured against the potential cost). The researcher needs to be able to influence, persuade and negotiate. Achieving this requires

- a commitment to the value of communication
- a thorough understanding of the underlying mathematics, and,
- good communication skills.

Cameron [7], speaking to a statistical audience as an employer of mathematical scientists in a research organisation, argued that the prime requirement is to be a scientist with deep statistical skills rather than to be a statistician looking for statistical problems among the science.

Ultimately, a research team, company or other research enterprise invests in mathematicians so that they might go beyond solving problems and create new technology that provides products or strategic advantage which competitors cannot match. For example, in molecular biology many companies and research groups are collecting and analysing similar sorts of data. If a mathematician can develop a methodology to discover information more reliably than others then that is a significant advantage leading to more discoveries, products or research publications.

3.3 What Industry Values

Industry values quantitative skills of a high order, but is often unconcerned about whether those skills have been developed in mathematics or in another discipline, such as physics, geophysics, engineering or computer science. Ultimately, what is valued is the ability to frame and solve problems and to explain the solution, its implications and its limitations. Except in special cases, those problems will arise in fields other than the mathematical sciences, and so mathematicians need to learn to talk with people in other fields, to work in interdisciplinary teams, formulate and solve mathematical problems relevant to the problems of the team and to explain the relevance and importance of their solution.

For example, in [11] the authors present a list of skills for which preparation was "less than good" for PhD graduates, as rated by the graduates. (The percentages are the percentages of respondents who rated the issue as less than good.)

1. Working well with colleagues (67 %)
2. Communicating at different levels (58 %)
3. Having broad scientific knowledge (47 %)
4. Effectively using computer software and systems (45 %)
5. Dealing with a wide variety of problems (35 %)

The report also lists the responses of managers, who added 'real world problem solving skills' to the list. (I believe that 'real world problem solving skills' implies finding an adequate solution when it is needed rather than waiting until an optimal solution is found.)

The capabilities are summarised in [12]

> This report explores the implications of this interdisciplinary environment for the skills and traits considered essential by employers in industry and government. Our interviewees emphasized communication skills, the ability to work effectively in a team, enthusiasm, self-direction, the ability to complete projects, and a sense of the business.

A research team, and a company, expect solutions to problems and hope for new approaches and technologies that will give longer-term strategic advantage over competitors. *Part of the mathematician's role is to try to identify trends that will become important and that will provide the strategic advantage.* There is the opportunity (and the obligation) for the mathematician to think broadly, persuade others and to be a research leader beyond the mathematical sciences.

The actual mathematics that industry values will depend on the industry and the particular problems being considered by the company at the time. It is likely that the problems, and the relevant mathematics, will change over time and what seemed important will become a commodity routinely available in standard software used by non-specialists. On the other hand, some pieces of mathematics which seemed abstract and not relevant to industry may become highly relevant.

There are other capabilities that will be an advantage for the mathematician to possess. It is important to have a good knowledge of the disciplines most relevant to collaborators or the organisation with which the mathematician is working. Because the mathematician will be likely to come upon problems requiring mathematics outside his or her specialisation, it is important to have a broader mathematical background. It is also helpful to understand the processes of research in particular areas, whether the questions being researched are central or peripheral in the field and what constitutes a valuable advance. Finally it is helpful to develop a network of colleagues, inside and outside the research team, with whom to discuss problems, progress and possible paths to follow.

In a general paper it is not possible to be very precise about the capabilities that mathematicians need. However some organisations are specific. For example, the

Australian Bureau of Statistics (2012) has published "Statistical Skills for Official Statisticians" [3].

4 Changing the Education of Mathematical Scientists

There have been several approaches to broadening the training of mathematical scientists. Informal and formal additions to a PhD program have been tried, as well as post PhD training. For example G.E.P. Box used to have "Monday Night Beer Sessions" [5] in which researchers from many disciplines came, presented their problem and received ideas from the assembled Faculty and PhD students. In Britain, "Maths in Industry Study Groups" began in 1968 and in Australia the concept was copied in 1984 and has occurred annually since. Similar activities have occurred in Europe and North America. Agnew and Keener [1] described a case-study course which taught aspects of problem formulation to senior undergraduate students. It is sobering to read [11] and compare the thinking with [8]—reports written about 15 years apart. While there are innovations and modifications, the same core message is apparent.

The "Committee for Mathematics in 2025" report [8] said in the first "Conclusion" of their Summary (p3):

> ...the value of the mathematical sciences to the overall science and engineering enterprise and to the nation would be heightened if the number of mathematical scientists who share the following characteristics could be increased:
>
> - They are knowledgeable across a broad range of the discipline, beyond their own area(s) of expertise;
> - They communicate well with researchers in other disciplines;
> - They understand the role of the mathematical sciences in the wider world of science, engineering, medicine, defense, and business; and
> - They have some experience with computation.

They went on to say:

> The culture within the mathematical sciences should evolve to encourage development of the characteristics listed in the Conclusion above.

4.1 Recent Programs

Recent approaches (in mathematics and more broadly) have considered either providing training in the development of skills that are relevant to a career in industry or providing a simple path for students to go from a PhD to industry. As well, in mathematics there has been a focus on broadening the mathematical base of students in the manner advocated in [8] and quoted above.

4.1.1 Courses, Formal and Informal

There has been a stark contrast between a North American PhD (which has traditionally had a strong and rigorous component of advanced coursework as a precursor to the research component) and a PhD in countries that have followed the British model of a 'research-only' PhD where students have had no formal requirement to undertake courses, but have often participated in reading groups or 'sat in' on courses, usually without any formal assessment.

Following the concerns about PhD programs described briefly above, some changes have been made in the UK to some PhD programs. The Engineering and Physical Sciences Research Council (EPSRC) has funded the creation of a number of Centres for Doctoral Training (including several in mathematics, statistics and operations research). Students in these Centres typically undertake a PhD over 4 years with the first year devoted to advanced technical courses, developing various research skills and developing a proposal for the research they plan to do for the PhD. As part of the creation of the Centres for Doctoral Training, the EPSRC also funded a number of collaborations among universities called "Doctoral Training Coursework Centres". The courses presented vary from one week intensive residential courses to traditional semester-long courses delivered via internet. The assessment of students also varies, from short and informal to traditional formal examination.

Although most of the Centres for Doctoral Training have industry links, at the present time none of the Mathematical Sciences Centres is deemed to be an Industry Doctoral Centre and the non-technical training for students appears to be general research skills rather than skills for working in industry.

I have mentioned the non-technical capabilities needed to contribute in an industry context. The biggest perceived gaps, according to the SIAM report [11] were "working well with colleagues" and "communicating at different levels". Many courses address this by including assignments and projects undertaken by groups and the presentation of results to the class. Some also explain the principles of successful communication in groups. Beyond the basic elements of working in groups, important aspects of the work of an industrial mathematician include (a) talking with researchers to elicit the aspects of a problem which lead to the key assumptions of the mathematical formulation, and (b) explaining to specialists in *possibly unrelated* disciplines the key aspects of the mathematical approach, solution and results. There are books on aspects of this related to statistics [4, 6] and many of the principles apply in mathematics more broadly. Gibbons and MacGillivray [9] highlight the value of training and working as a tutor in understanding the differences among the backgrounds and learning and working styles of co-workers. This is important in explaining mathematical and statistical approaches and results.

4.2 Internships and Industrial Experience

The SIAM report of 2012, [12], emphasises the value of industrial experience to a new graduate—often provided through internships—when a person (student or recent graduate) spends time working on a project in an organisation. The report on pp. 33–34 cites internship programs by Mitacs (Canada), Matheon (Germany) and the Computational and Applied Mathematics Department at Rice University. There are of course others, for example in Europe there is a link between the University of Kaiserslautern and the Fraunhofer Institute for Industrial Mathematics.

5 The ATN Industry Doctoral Training Centre

The Australian Technology Network of Universities (ATN) is a collaboration of 5 technology universities, one in each of Brisbane, Sydney, Melbourne, Adelaide and Perth. In the past few years they have developed the Industry Doctoral Training Centre (Mathematics & Statistics). (IDTC) The aims of the IDTC are to:

1. Engage in a new form of industry focussed training in Mathematics and Statistics PhDs with a view to embedding this approach in the Australian research training system.
2. Increase the skills, capabilities and job relevance of PhD graduates in Mathematics and Statistics.
3. Form strategic partnerships in research and research training between the ATN and industry and government through the disciplines of Mathematics and Statistics and with an industry-centred, problem-solving focus.
4. Increase Australia's research capacity and the relevance to industry and society of research in the fields of Mathematics and Statistics.
5. To deliver research outcomes of benefit to the Australian community by applying the research expertise residing in the ATN in Mathematics and Statistics in an industry R&D context.

The IDTC program was developed in 2011 and it took its first students in 2012 under the Foundation Director, Professor Lee White. The key features of the IDTC program are:

1. The research project of each student is provided by an industry partner and the partner provides funding for the student as well as nominating a staff member as an industry supervisor. The industry partner owns the intellectual property generated.
2. The student may be an existing employee of the industry partner (and continue to be employed provided their principal task is the research project for their PhD). Employee or not, the student is expected to spend a significant amount of time at the workplace of the partner—in essence this is an extended internship.

3. Each student is expected to complete their PhD within four years (the standard for other students is 3 years) and in the first 3 years the student will complete a number of modules. There is some flexibility in what constitutes a module—appropriate modules will be negotiated by the student and the academic supervisor, but it is assumed that the student will complete at least 6 technical modules—each at least the equivalent of a 20 lecture course—and 4 on-line capability development modules to improve the "employability" skills of the student. These cover teamwork, communication, leadership and commercialisation.
4. The student will also attend an induction course (covering research methodology and ethics) and actively participate in two Mathematics in Industry Study Groups or be involved in an equivalent effort in "formulating problems and finding solutions".

An important aspect is to develop a cohort spirit among the students (despite their geographic separation) and so the students report on their progress once a year at a conference of all the students.

5.1 Some Early Experiences

The first students are only just entering their third year in the program, so it is too early to claim success, but some observations of broader relevance can be made.

- There is less flexibility in a project with an industry objective rather than an academic one and the risks in this need to be acknowledged with possible responses identified at the start.
- It is likely that an academic supervisor and an industry supervisor will have quite different perspectives and drivers. This needs to be recognised and a process for resolving them (for the benefit of the student) should be identified before problems arise.
- The cohort experience can be developed through formal activities (courses, conferences) and by giving students the responsibility (and a small budget) for organising informal activities.
- There is a tendency for all (students, supervisors, other academics) to assume the IDTC Program is the same as a traditional PhD with a few annoying interruptions that should be minimised, rather than an opportunity for much broader personal development. The "culture change" identified by the Committee for Mathematics in 2025 is needed. That means careful explanation of what is different and why, regular repetition of the key points and modification of the processes for reviewing student progress to recognise the differences in the program.
- There are additional challenges (and opportunities) for the academic supervisor in taking on a PhD student with an industry project. The challenges are particularly prominent in the early stages and so some extra acknowledgement for the supervisor can be valuable.

- Identifying and delivering appropriate modules for a relatively small cohort of students with diverse backgrounds and needs is not straight-forward. The solution requires care, pragmatism and opportunism!
- Employability skills training could be a cursory coverage or an extended program of personal development and management training. Finding the right balance requires professional insight.

6 Conclusion

We should hope that the opportunity for mathematical scientists to have satisfying careers in industry will increase, not shrink, and that implies improvement of existing PhD training and the development of new approaches. Experience over the last 40 years has shown successes, but they have been piecemeal and the time has come for larger efforts. There are challenges, not least finding the resources to provide courses and monitored experiences to relatively small numbers of students. Collaboration among institutions, such as that supporting the IDTC is an obvious approach.

Acknowledgments The ATN Industry Doctoral Training Centre was set up by a number of people including Ms Vicki Thomson, Prof Attila Brungs and the Foundation Director, Prof Lee White. I commend their vision and efforts and thank them for their support to continue the development of the IDTC.

References

1. Agnew, J.L., Keener, M.: A case-study course in applied mathematics using regional industries. Am. Math. Mon. **87**(1), 55–59 (1980). Article Stable. http://www.jstor.org/stable/2320385
2. Anonymous: The disposable academic. The Economist. http://www.economist.com/node/17723223 (2010)
3. Australian Bureau of Statistics: Statistical skills for official statisticians. Booklet and web PDF (2012)
4. Boen, J.R., Zahn, D.: The human side of statistical consulting. Lifetime Learning Publications, Belmont (1982)
5. Box, G.: An Accidental Statistician: The Life and Memories of George E.P. Box. Wiley, New Jersey (2013)
6. Cabrera, J., McDougall, A.: Statistical Consulting. Springer, Berlin (2002)
7. Cameron, M.: Training statisticians for a research organisation. In: Bulletin of 57th Session of International Statistical Institute (ISI), vol. 57 (2009)
8. Committee for Mathematical Sciences in 2025: The mathematical sciences in 2025. Technical report, National Research Council (2013)
9. Gibbons, K., MacGillivray, H.: Education for a workplace statistician. In: MacGillivray, H.L., Martin, M., Phillips, B. (eds.) Topics from Australian Conference on Teaching Statistics: OZCOTS 2008–2012. Springer Science + Business Media, LLC, New York (2014)
10. Group of Eight: The changing phd. Technical report, Group of Eight. http://www.go8.edu.au/university-staff/go8-policy-and-analysis/2013/the-changing-phd (2013)
11. SIAM: The SIAM report on mathematics in industry. Technical report, Society for Industrial and Applied Mathematics (1998)
12. SIAM: Mathematics in industry. Technical report, Society for Industrial and Applied Mathematics (2012)

Cryptanalysis of Pairing-Based Cryptosystems Over Small Characteristic Fields

Takuya Hayashi

Abstract There are many useful cryptographic schemes which use bilinear pairings. In particular, η_T pairing over small characteristic fields, such as $GF(2^n)$ and $GF(3^n)$, is one of the most efficient algorithms from the implementation point of view. The security of pairing-based cryptosystems using η_T pairing over $GF(2^n)$ (resp. $GF(3^n)$) relies on the hardness of the discrete logarithm problem over $GF(2^{4n})$ (resp. $GF(3^{6n})$). However, new index calculus methods proposed by Joux and Barbulescu et al. allow us to solve these problems in quasi-polynomial time. Recent experimental results show that these methods are quite practical, implying that the η_T pairing over $GF(2^n)$ and $GF(3^n)$ is unsuitable for pairing-based cryptosystems. In this paper, we survey the recent progress on index calculus methods and related experimental results.

Keywords Cryptanalysis · Discrete logarithm problem · Pairing-based cryptosystem

1 Introduction

There are many useful cryptographic schemes, such as ID-based encryption [13], keyword searchable encryption [12], attribute-based encryption [32], functional encryption [31], that use bilinear pairings. It is important to estimate the security of such pairing-based cryptosystems. Such cryptosystems can be broken, if the discrete logarithm problem (DLP) can be solved.

One of the most efficient algorithms for implementing the pairing is the η_T pairing [6] defined over a supersingular elliptic curve on finite fields of small

T. Hayashi (✉)
Institute of Mathematics for Industry, Kyushu University, 744, Motooka, Nishi-ku, Fukuoka 819-0395, Japan
e-mail: t-hayashi@imi.kyushu-u.ac.jp

M. Wakayama et al. (eds.), *The Impact of Applications on Mathematics*, Mathematics for Industry 1, DOI: 10.1007/978-4-431-54907-9_12,

characteristics, e.g., binary fields $GF(2^n)$ and ternary fields $GF(3^n)$, where n is a positive integer. Since the embedding degree of the curve is 4 for $GF(2^n)$ (resp. 6 for $GF(3^n)$), the η_T pairing can reduce a DLP over the curve on $GF(2^n)$ (resp. $GF(3^n)$) to a DLP over $GF(2^{4n})$ (resp. $GF(3^{6n})$). Therefore, pairing-based cryptosystems using the η_T pairing on $GF(2^n)$ (resp. $GF(3^n)$) are insecure if the DLP over $GF(2^{4n})$ (resp. $GF(3^{6n})$) is solvable in practical time.

In this paper, we survey recent results of the DLP over small characteristic fields related to the pairing-based cryptosystems using the η_T paring. After the development of the function field sieve [3, 27], the complexity for solving the DLP over small characteristic fields $GF(q^n)$ was sub-exponential time $L_{q^n}(1/3)$, where

$$L_{q^n}(\alpha) = \exp(O(\log(q^n)^\alpha \log\log(q^n)^{1-\alpha})),$$

and parameters for pairing-based cryptosystems using η_T pairing were determined using this complexity. But recently, Joux showed that the complexity can be reduced to $L_{q^n}(1/4 + o(1))$ [26], and Barbulescu et al. [5] showed it can be reduced to quasi-polynomial time $2^{O((\log\log q^n)^2)}$ and finally broke the sub-exponential time barrier. This great progress strongly influences to the difficulty of the DLP over small characteristic fields, and the security of pairing-based cryptosystems over small characteristic fields, both in theory and in practice.

This paper is organized as follows. In Sect. 2, we will introduce pairing-based cryptosystems and the relationship between its security and the discrete logarithm problem. Then we will explain the key ideas of new index calculus methods [5, 26] in Sect. 3, and recent experimental results will be introduced in Sect. 4. Finally in Sect. 5, we will conclude and show some future works on this area.

2 Pairing-Based Cryptosystems and Discrete Logarithm Problem

In this section, we briefly explain the security of pairing-based cryptosystems.

Before beginning the discussion, we define the discrete logarithm problem (DLP). Let g be a generator of a finite cyclic group $G = <g>$. For a given $h \in G$, the DLP in G is the problem to find an integer ℓ such that $h = g^\ell$. Generally, ℓ is described as $\log_g h$. In this paper, the problem is called the DLP over $GF(q^n)$, when the group G is a subgroup of the multiplicative group $GF(q^n)^\times$. Also, when G is a subgroup of $E(GF(q^n))$, a group of $GF(q^n)$-rational points on an elliptic curve E, the problem is called the elliptic curve discrete logarithm problem (ECDLP) on $E(GF(q^n))$.

A lot of efficient cryptographic protocols using a bilinear pairing have been proposed (for example [12, 13, 31, 32]), and high-speed implementations for the η_T pairing have been reported (for example [4, 7, 9–11, 21, 29]). We discuss the security of pairing-based cryptosystems with the η_T paring over $GF(2^n)$ (resp. $GF(3^n)$) for an integer n. The security of pairing-based cryptosystems with the η_T paring depends on the difficulty of solving the ECDLP over the supersingular elliptic curves. Additionally, MOV reduction [30] reduces this ECDLP to the DLP over $GF(2^{4n})$ (resp. $GF(3^{6n})$) since the embedding degree of the η_T pairing is 4 (resp. 6).

In particular, the η_T pairing is a bilinear map such that $\eta_T : G_1 \times G_1 \to G_2$, where G_1 is an additive subgroup of a supersingular elliptic curve over $GF(2^n)$ (resp. $GF(3^n)$), G_2 is a cyclic subgroup of $GF(2^{4n})^{\times}$ (resp. $GF(3^{6n})^{\times}$), and the cardinalities of G_1, G_2 are the same prime number P. The security of pairing-based cryptosystems with the η_T pairing depends on the difficulty of not only the ECDLP over G_1 but also the DLP over G_2 by MOV reduction. To explain this fact, we take ID-based encryption constructed on pairing-based cryptosystems as an example. The ID-based encryption has a master key $s_{key} \in Z_P$. Each user ID is deterministically transformed into a point $Q_{ID} \in G_1$, and the secret key S_{ID} is defined by $[s_{key}]Q_{ID}$. Therefore, solving the ECDLP over G_1, namely $S_{ID} = [s_{key}]Q_{ID}$, we obtain the master key $s_{key} = \log_{Q_{ID}} S_{ID}$. Additionally, for an arbitrary point $R \in G_1$, we compute $\eta_T(S_{ID}, R)$, $\eta_T(Q_{ID}, R) \in G_2$, and then have $\eta_T(S_{ID}, R) = \eta_T([s_{key}]Q_{ID}, R) = \eta_T(Q_{ID}, R)^{s_{key}} \in G_2$. This implies that $s_{key} = \log_{\eta_T(Q_{ID},R)} \eta_T(S_{ID}, R)$ is also available by solving the DLP over G_2, which is a subgroup of $GF(2^{4n})^{\times}$ (resp. $GF(3^{6n})^{\times}$). Hereafter, we deal with $GF(q^n)$, where $q = p^k$ for a prime number p and a positive integer k, instead of $GF(2^{4n})$ and $GF(3^{6n})$ for simplicity.

3 New Index Calculus Algorithms for Solving DLP Over Small Characteristic Fields

Before introducing new index calculus algorithms, we explain the scheme of the index calculus approach briefly. Here, suppose that we wish to solve the DLP over $GF(q^n)$, whose elements are represented as polynomials in $GF(q)[x]$ of degree smaller than n. The index calculus approach contains two phases:

1. *Finding logarithms of small degree polynomials*: Let S be a set of polynomials of degree not larger than B, where B is a small integer. Find linear equations of logarithms of polynomials in S

$$\sum_{p_i \in S} a_i \log_g p_i \equiv 0 \pmod{q^n - 1}. \tag{1}$$

   When slightly more than $\#S$ linear equations are obtained, we can compute these logarithms by solving a linear system constructed by these linear equations. The way to find these linear equations is depends on the algorithm and the implementation.
2. *Descending from target element to small degree polynomials*: Let h be a target element, we wish to compute its logarithm, and let $\deg h \approx n - 1$. Since logarithms of polynomials of degree not larger than B are already obtained in

previous phase, the logarithm can be computed if a B-smooth polynomial $g^e h$ for a randomly chosen integer e is found, where B-smooth means the polynomial has no irreducible factor of degree not larger than B. However it is hard to find if n is much larger than B, so instead of this, the descent approach is usually used. For an element Q, which is firstly equal to h, find an expression for $\log_g Q$ in the logarithms of polynomials of degree smaller than $\deg Q$. When we obtain the expression, $\log_g Q$ is represented by a linear combination of logarithms of smaller degree polynomials. If there are any polynomials of degree larger than B, pick Q from these polynomials, and compute the expression recursively. The degree of Q will decrease and eventually reach B, then we can compute logarithms of each Q, and finally obtain $\log_g h$.

In the previous index calculus methods, such as function field sieve [3, 27], both phase need sub-exponential time, but phase 1 is actually dominant and needs much more computation than phase 2.

In 2013, Joux invented new index calculus approach for solving DLP over small characteristic fields, which has the complexity $L_{q^n}(1/4 + o(1))$ [26] (independently, similar approach is also invented by Göloğlu et al. [17]). Then Barbulescu et al. [5] improved the approach and reduced the complexity to quasi-polynomial time $2^{O((\log \log q^n)^2)}$. These algorithms made great progress both in theory and in practice. In this section, we introduce key ideas of these algorithms. To utilize these ideas, we consider the DLP over $GF(q^{2n})$ instead of that over $GF(q^n)$.

3.1 Polynomial Time Algorithm for Finding Logarithms of Small Degree Polynomials

In [26], Joux invented a new index calculus method which has a polynomial time algorithm for finding logarithms of small degree polynomials. Here we introduce this polynomial time algorithm.

Let $h_1 x^q - h_0$ has an irreducible factor $f \in GF(q^2)[x]$ of degree n, where $h_0, h_1 \in GF(q^2)[x]$ are small degree polynomials. Then $GF(q^{2n})$ can be represented as $GF(q^2)[x]/(f)$. Note that, in this representation, $x^q \equiv h_0/h_1 \pmod{f}$.

To produce linear relations between logarithms of degree-1 polynomials, Joux introduced the well-known systematic equation

$$y^q - y = \prod_{\alpha \in GF(q)} (y - \alpha). \tag{2}$$

Substituting y for $(ax + b)/(cx + d)$ where $a, b, c, d \in GF(q^2)$, and multiplying by $(cx + d)^{q+1}$, (2) yields

$$(a^q h_0 + b^q h_1)(cx + d) - (ax + b)(c^q h_0 + d^q h_1)$$
$$\equiv h_1(cx + d) \prod_{\alpha \in GF(q)} ((a - \alpha c)x + (b - \alpha d)) \pmod{f}, \tag{3}$$

where the left hand side is a polynomial of degree at most $D = \max(\deg h_0, \deg h_1) + 1$ and the right hand side is factored into linear polynomials and h_1[1]. Once the left hand side is factored into linear polynomials, we can obtain a linear relation of logarithms of linear polynomials and the logarithm of h_1, by taking logarithm on both sides. When we obtain enough (more than q^2) linear relations, we can compute logarithms of linear polynomials and the logarithm of h_1 by solving a linear system.

Under the heuristic that the left hand side factors into linear polynomials with a probability close to that for a random polynomial of the same degree, we need to perform $D!$ trials. Since D can heuristically be chosen as a constant, we expect to find enough relations by considering $O(q^2)$ non-duplicate candidate Eq. (3) generated by (a, b, c, d). By selecting (a, b, c, d) from P_q, a set of distinct representatives of the left cosets of $\mathrm{PGL}_2(GF(q))$ in $\mathrm{PGL}_2(GF(q^2))$, we can avoid the duplicates. The cardinality of P_q is $q^3 + q$, thus, we expect to obtain enough non-duplicate candidate equations and so enough linear relations.

The complexity to produce linear relations is $O(q^2)$ smoothness tests. Since the linear system constructed by the linear relations has dimension $O(q^2)$ with $O(q)$ non-zero elements per row, the cost for solving the system is $O(q^5)$ arithmetic operations using sparse matrix techniques which has $O(d^2)$ complexity for dimension d.

For finding logarithms of quadratic polynomials and higher degree polynomials, the approach can be extended by substituting y for $\sum(a_i x^i)/\sum(b_i x^i)$ in (3), instead of $(ax + b)/(cx + d)$.

3.2 *Quasi-Polynomial Time Algorithm for Descent Phase*

Barbulescu et al. [5] proposed a quasi-polynomial time algorithm for the DLP over small characteristic fields, which uses Joux's approach for finding logarithms of small degree polynomials. The key idea is a quasi-polynomial time descent algorithm, therefore we introduce it here.

Let $Q \in GF(q^2)[x]$ and let m be $\lceil \deg Q/2 \rceil$. We wish to find an expression for $\log_g Q$ in the logarithms of polynomials of degree at most m. Like Joux's polynomial time algorithm for finding logarithms of small degree polynomials, This approach uses the systematic Eq. (2). For $(a, b, c, d) \in P_q$, substituting y for $(aQ + b)/(cQ + d)$ and multiplying $(cQ + d)^{q+1}$, (2) yields

[1] When $a, b, c, d \in GF(q)$, (3) yields a trivial equation. This is the reason why we need to embed the original DLP to the DLP over $GF(q^{2n})$.

$$(a^q \bar{Q}(h_0/h_1) + b^q)(cQ + d) - (aQ + b)(c^q \bar{Q}(h_0/h_1) + d^q)$$
$$\equiv (cQ + d) \prod_{\alpha \in GF(q)} ((a - \alpha c)Q + (b - \alpha d)) \pmod{f}, \quad (4)$$

where $\bar{Q}$ is obtained by q-th powering each coefficient of Q. The denominators of the left hand side can be eliminated by multiplying by h_1^D. Note that the polynomial of the left hand side has degree at most $D = (\max(\deg h_0, \deg h_1) + 1) \deg Q$. If this polynomial is m-smooth, then (4) yields a linear relation of the logarithms of polynomials of degree at most m and logarithms of translates of Q. After collecting slightly more than q^2 relations, we search a linear combination of these relations that eliminates all translates of Q except for Q. To achieve this, consider a row vectors with coordinates indexed by elements $\beta \in GF(q^2)$. For each relation, we define a vector v whose entry v_β is 1 if $Q - \beta$ appeared in the right hand side of (4), otherwise 0. If the matrix has full rank, by linear algebra on the matrix, we can obtain an expression for $\log_g Q$ in the logarithms of polynomials of degree at most m.

In this approach, it is required to obtain at least q^2 linear relations. To ensure this,

$$P_{q^2}(m, D)(q^3 + q) \gg q^2 \quad (5)$$

should hold, where $P_{q^2}(m, D)$ describes the probability that uniformly random degree D polynomials are m-smooth. The condition would hold if q is enough large, but otherwise, m should increase such that the condition is satisfied. The condition would hold if q is enough large, but otherwise, m should be changed larger to hold it.

The complexity for producing linear relations is polynomial time in q, between $O(q^2)$ and $O(q^3)$. Since the matrix has dimension $O(q^2)$ with $O(q)$ non-zero elements per row, the linear algebra computation of the matrix can be done in $O(q^5)$ arithmetic operations using sparse matrix techniques. For each descent this approach produces $O(q^2 n)$ nodes in the descent tree, and the depth of the tree is in $O(\log n)$ because the degree of Q will decrease by half for each descent. Hence, the number of nodes of the descent tree is $(q^2 n)^{O(\log n)}$. Since any polynomial in q and n is absorbed in the O notation in the exponent, the complexity of the whole descent phase is $\max(q, n)^{O(\log n)}$. If the characteristic p of the field is bounded by $(\log q^n)^{O(1)}$, the time complexity is $2^{O((\log \log q^n)^2)}$, which is quasi-polynomial.

4 Recent Experimental Progress

Before the algorithms described in the previous section was developed, the world record for solving the DLP over small characteristic, specifically the DLP over $GF(3^{6 \cdot 97})$ whose cardinality is 923-bit, was held by Hayashi et al. [22]. The record was achieved using the function field sieve [3, 27] and the computation took about 900,000 CPU hours.

After the aforementioned developments, the record was updated drastically. On February 19th, 2013, Granger et al. succeeded in solving the DLP over $GF(2^{1971}) \cong GF((2^{27})^{73})$ using a new index calculus method [15]. This computation took 3,132 CPU hours, this is much faster compared to the previous record. One month later, on March 22th, Joux [24] updated the record to $GF(2^{4080}) \cong GF((2^{16})^{255})$ using a new $L_{q^n}(1/4 + o(1))$ algorithm. The computation used Kummer extension to define the field for efficient computation. Again, Granger et al. updated the record to $GF(2^{6120}) \cong GF((2^{24})^{255})$ on April 11th [16, 18]. This also used Kummer extension improvement. On May 21th, Joux [25] again replaced the record by $GF(2^{6168}) \cong GF((2^{24})^{257})$ which is used twisted Kummer extension instead of Kummer extension. This computation took 467 CPU hours. The current world record is $GF(2^{9234}) \cong GF((2^{18})^{513})$, whose cardinality is 9,234-bit size, achieved by Granger et al. [19] on January 31th, 2014. The computation required 397,422 CPU hours.

The above experiments are performed for demonstration of the new index calculus algorithms, and these fields are actually not important in cryptography because no cryptosystems uses them. But Adj et al. [2] showed this issue is serious for pairing-based cryptosystems over small characteristic fields.

The previous security estimation for pairing-based cryptosystems using η_T pairing over ternary fields was made by Shinohara et al. [33], based on the function field sieve. The estimation is precise and it shows that the complexity for solving the DLP over $GF(3^{6\cdot 509})$ is $2^{111.4}$, which means that these cryptosystems were expected to be secure at least 30 years by taking the parameter $n = 509$. However, based on the new index calculus methods, Adj et al. [2] showed that the complexity is reduced to $2^{73.7}$, and the cryptosystems can be broken in a year and hence must not be used. Additionally, Adj et al. [1] experimented to solve the DLP over $GF(3^{6\cdot 137})$ whose cardinality is 1303-bit, by 918 CPU hours computation. Granger et al. [20] also solved the DLP over $GF((2^{12})^{367})$, this field contains $GF(2^{4\cdot 367})$ as a subfield, which means the pairing-based cryptosystems using η_T pairing over binary fields with the parameter $n \leq 367$ are insecure. The above estimations imply that the pairing-based cryptosystems over small characteristic are totally insecure and should be avoided.

5 Concluding Remarks

In this paper, we survey the recent progress of the discrete logarithm problem over small characteristic fields. The new index calculus methods [5, 17, 26] are much faster than the previous index calculus method like the function field sieve [3, 27] both in theory and in practice. The experiments and estimates [1, 2, 15, 16, 18–20, 24, 25] show the efficiency of these methods, and imply the pairing-based cryptosystems over small characteristic are totally insecure.

To keep the paring-based cryptosystems secure, one can use the pairings over large characteristic fields, such as Ate pairing [23]. In such case, the new index calculus cannot apply efficiently. However, the difficulty for solving the DLP over large characteristic extension fields related to pairing-based cryptosystems, such as $GF(p^{12})$ for Barreto-Naehrig curves [8] and $GF(p^{18})$ for Kachisa-Schaefer-Scott curves [28], is not precisely estimated yet, so that further research is needed.

Another important problem for paring-based cryptosystems is the pairing inversion problem, which is to compute the map of inverse of the pairing. The problem must also be intractable to make some cryptographic protocols secure. There are several papers [14, 34, 35] about the problem, however its difficulty is still not well-known.

In conclusion, secure use of the pairing-based cryptosystems requires further studies on the above mentioned points

References

1. Adj, G., Menezes, A., Oliveira, T., Rodríguez-Henríquez, F.: Computing discrete logarithms in $F_{3^{6\cdot137}}$ using magma. IACR Cryptology ePrint Archive, Report 2014/057 (2014)
2. Adj, G., Menezes, A., Oliveira, T., Rodríguez-Henríquez, F.: Weakness of $F_{3^{6\cdot509}}$ for discrete logarithm cryptography. In: Cao Z., Zhang F. (eds.) Proceedings of 6th International Conference on Pairing-based Cryptography (Pairing 2013). Lecture Notes in Computer Science, vol. 8365, pp. 20–44. Springer, Berlin (2013)
3. Adleman, L.M.: The function field sieve. In: Adleman L.M., Huang M.D.A. (eds.) Proceedings of 1st Algorithmic Number Theory Symposium (ANTS-I). Lecture Notes in Computer Science, vol. 877, pp. 108–121. Springer, Berlin (1994)
4. Ahmadi, O., Hankerson, D., Menezes, A.: Software implementation of arithmetic in F_{3^m}. In: Carlet C., Sunar B. (eds.) Proceedings of 1st International Workshop on the Arithmetic of Finite Fields (WAIFI 2007). Lecture Notes in Computer Science, vol. 4547, pp. 85–102. Springer, Berlin (2007)
5. Barbulescu, R., Gaudry, P., Joux, A., Thomé, E.: A quasi-polynomial algorithm for discrete logarithm in finite fields of small characteristic. IACR Cryptology ePrint Archive, Report 2013/400 (2013)
6. Barreto, P.S.L.M., Galbraith, S.D., O'Eigeartaigh, C., Scott, M.: Efficient pairing computation on supersingular abelian varieties. Des., Codes Crypt. **42**(3), 239–271 (2007)
7. Barreto, P.S.L.M., Kim, H.Y., Lynn, B., Scott, M.: Efficient algorithms for pairing-based cryptosystems. In: Yung M. (ed.) Proceedings of Advances in Cryptology: CRYPTO 2002, 22nd Annual International Cryptology Conference. Lecture Notes in Computer Science, vol. 2442, pp. 354–368. Springer, Berlin (2002)
8. Barreto, P.S.L.M., Naehrig, M.: Pairing-friendly elliptic curves of prime order. In: Preneel B., Tavares S.E. (eds.) Proceedings of Selected Areas in Cryptography 2005 (SAC 2005). Lecture Notes in Computer Science, vol. 3897, pp. 319–331. Springer, Berlin (2005)
9. Beuchat, J.L., Brisebarre, N., Detrey, J., Okamoto, E.: Arithmetic operators for pairing-based cryptography. In: Paillier P., Verbauwhede I. (eds.) Proceedings of 9th International Workshop on Cryptographic Hardware and Embedded Systems (CHES 2007). Lecture Notes in Computer Science, vol. 4727, pp. 239–255. Springer, Berlin (2007)
10. Beuchat, J.L., Brisebarre, N., Detrey, J., Okamoto, E., Shirase, M., Takagi, T.: Algorithms and arithmetic operators for computing the η_T pairing in characteristic three. IEEE Trans. Comput. **57**(11), 1454–1468 (2008)
11. Beuchat, J.L., Brisebarre, N., Shirase, M., Takagi, T., Okamoto, E.: A coprocessor for the final exponentiation of the η_T pairing in characteristic three. In: Carlet C., Sunar B. (eds.) Proceedings of 1st International Workshop on the Arithmetic of Finite Fields (WAIFI 2007). Lecture Notes in Computer Science, vol. 4547, pp. 25–39. Springer, Berlin (2007)

12. Boneh, D., Crescenzo, G.D., Ostrovsky, R., Persiano, G.: Public key encryption with keyword search. In: Cachin C., Camenisch J. (eds.) Proceedings of Advances in Cryptology: EUROCRYPT 2004, 23rd Annual International Conference on the Theory and Applications of Cryptographic Techniques. Lecture Notes in Computer Science, vol. 3027, pp. 506–522. Springer, Berlin (2004)
13. Boneh, D., Franklin, M.K.: Identity-based encryption from the Weil pairing. In: Kilian J. (ed.) Proceedings of Advances in Cryptology: CRYPTO 2001, 21st Annual International Cryptology Conference. Lecture Notes in Computer Science, vol. 2139, pp. 213–229. Springer, Berlin (2001)
14. Galbraith, S.D., Hess, F., Vercauteren, F.: Aspects of pairing inversion. IEEE Trans. Inf. Theory **54**(12), 5719–5728 (2008)
15. Göloglu, F., Granger, R., McGuire, G., Zumbrägel, J.: Discrete logarithms in $GF(2^{1971})$. Number Theory Mailng List (2013). http://listserv.nodak.edu/cgi-bin/wa.exe?A2=NMBRTHRY;f7755cbe.1302
16. Göloglu, F., Granger, R., McGuire, G., Zumbrägel, J.: Discrete logarithms in $GF(2^{6120})$. Number Theory Mailng List (2013). http://listserv.nodak.edu/cgi-bin/wa.exe?A2=NMBRTHRY;fe9605d9.1304
17. Göloglu, F., Granger, R., McGuire, G., Zumbrägel, J.: On the function field sieve and the impact of higher splitting probabilities—application to discrete logarithms in $F_{2^{1971}}$ and $F_{2^{3164}}$. In: Canetti R., Garay J.A. (eds.) Proceedings of Advances in Cryptology:- CRYPTO 2013, 33rd Annual International Cryptology Conference. Lecture Notes in Computer Science, vol. 8043, pp. 109–128. Springer, Berlin (2013)
18. Göloglu, F., Granger, R., McGuire, G., Zumbrägel, J.: Solving a 6120-bit DLP on a desktop computer. IACR Cryptology ePrint Archive, Report 2013/306 (2013)
19. Granger, R., Kleinjung, T., Zumbrägel, J.: Discrete logarithms in $GF(2^{9234})$. Number Theory Mailng List (2014). http://listserv.nodak.edu/cgi-bin/wa.exe?A2=NMBRTHRY;49bb494e.1305
20. Granger, R., Kleinjung, T., Zumbrägel, J.: Discrete logarithms in the jacobian of genus 2 supersingular curve over $GF(2^{367})$. Number Theory Mailng List (2014). http://listserv.nodak.edu/cgi-bin/wa.exe?A2=NMBRTHRY;23651c2.1401
21. Granger, R., Page, D., Stam, M.: Hardware and software normal basis arithmetic for pairing-based cryptography in characteristic three. IEEE Trans. Comput. **54**(7), 852–860 (2005)
22. Hayashi, T., Shimoyama, T., Shinohara, N., Takagi, T.: Breaking pairing-based cryptosystems using η_T pairing over $GF(3^{97})$. In: Wang X., Sako K. (eds.) Proceedings of 18th Annual International Conference on the Theory and Application of Cryptology and Information Security (ASIACRYPT 2012). Lecture Notes in Computer Science, vol. 7658, pp. 43–60. Springer, Berlin (2012)
23. Hess, F., Smart, N.P., Vercauteren, F.: The eta pairing revisited. IEEE Trans. Inf. Theory **52**(10), 4595–4602 (2006)
24. Joux, A.: Discrete logarithms in $GF(2^{4080})$. Number Theory Mailng List (2013). http://listserv.nodak.edu/cgi-bin/wa.exe?A2=NMBRTHRY;71e65785.1303
25. Joux, A.: Discrete logarithms in $GF(2^{6168})[= GF((2^{257})^{24})]$. Number Theory Mailng List (2013). http://listserv.nodak.edu/cgi-bin/wa.exe?A2=NMBRTHRY;49bb494e.1305
26. Joux, A.: A new index calculus algorithm with complexity $L(1/4 + o(1))$ in very small characteristic. IACR Cryptology ePrint Archive, Report 2013/095 (2013)
27. Joux, A., Lercier, R.: The function field sieve in the medium prime case. In: Vaudenay S. (ed.) Proceedings of Advances in Cryptology: EUROCRYPT 2006, 25th Annual International Conference on the Theory and Applications of Cryptographic Techniques. Lecture Notes in Computer Science, vol. 4004, pp. 254–270. Springer, Berlin (2006)
28. Kachisa, E.J., Schaefer, E.F., Scott, M.: Constructing Brezing-Weng pairing-friendly elliptic curves using elements in the cyclotomic field. In: Galbraith S.D., Paterson K.G. (eds.) Proceedings of 2nd International Conference on Pairing-based Cryptography (Pairing 2008). Lecture Notes in Computer Science, vol. 5209, pp. 126–135. Springer, Berlin (2008)

29. Kawahara, Y., Aoki, K., Takagi, T.: Faster implementation of η_T pairing over $GF(3^m)$ using minimum number of logical instructions for $GF(3)$-addition. In: Galbraith S.D., Paterson K.G. (eds.) Proceedings of 2nd International Conference on Pairing-based Cryptography (Pairing 2008). Lecture Notes in Computer Science, vol. 5209, pp. 282–296. Springer, Berlin (2008)
30. Menezes, A., Okamoto, T., Vanstone, S.A.: Reducing elliptic curve logarithms to logarithms in a finite field. IEEE Trans. Inf. Theory **39**(5), 1639–1646 (1993)
31. Okamoto, T., Takashima, K.: Fully secure functional encryption with general relations from the decisional linear assumption. In: Rabin T. (ed.) Proceedings of Advances in Cryptology: CRYPTO 2010, 30th Annual International Cryptology Conference. Lecture Notes in Computer Science, vol. 6223, pp. 191–208. Springer, Berlin (2010)
32. Sahai, A., Waters, B.: Fuzzy identity-based encryption. In: Cramer R. (ed.) Proceedings of Advances in Cryptology: EUROCRYPT 2005, 24th Annual International Conference on the Theory and Applications of Cryptographic Techniques. Lecture Notes in Computer Science, vol. 3494, pp. 457–473. Springer, Berlin (2005)
33. Shinohara, N., Shimoyama, T., Hayashi, T., Takagi, T.: Key length estimation of pairing-based cryptosystems using η_T pairing. In: Ryan M.D., Smyth B., Wang G. (eds.) Proceedings of 8th International Conference on Information Security Practice and Experience (ISPEC 2012). Lecture Notes in Computer Science, vol. 7232, pp. 228–244. Springer, Berlin (2012)
34. Vercauteren, F.: The hidden root problem. In: Galbraith S.D., Paterson K.G. (eds.) Proceedings of 2nd International Conference on Pairing-based Cryptography (Pairing 2008). Lecture Notes in Computer Science, vol. 5209, pp. 89–99. Springer, Berlin (2008)
35. Verheul, E.R.: Evidence that XTR is more secure than supersingular elliptic curve cryptosystems. J. Cryptology **17**(4), 277–296 (2004)

Applied Algebraic Geometry in Model Based Design for Manufacturing

Hirokazu Anai

Abstract In this paper we show the interplay of real algebraic geometry and control system design in manufacturing from the standpoint "how applications affect to algorithm development in real algebraic geometry". One of important perspectives of the interaction is how we overcome the inherent computational complexity for solving practical problems and the key point is making good use of their special structures of the practical problems.

Keywords Real algebraic geometry · Quantifier elimination · Symbolic optimization · Control system design · Manufacturing design

1 Introduction

Nowadays model-based design (MBD) is rapidly introduced in a wide range of industrial fields. MBD is a mathematical and visual method of addressing problems associated with designing complex control, signal processing and communication systems. In fact it is used in many manufacturing applications such as motion control, industrial equipment, aerospace, and automotive. One of the main steps in MBD is "controller analysis and synthesis".

The controller synthesis/design process requires to solve complicated optimization problems more accurately and efficiently. Furthermore, numerous problems in science and engineering can be reduced to that of solving constraints and optimization

H. Anai (✉)
Knowledge Platforms Laboratories, Fujitsu Laboratories Ltd.,
4-1-1 Kamikodanaka, Nakaharaku, Kawasaki, Kanagawa 211-8588, Japan
e-mail: anai@jp.fujitsu.com, h.anai@kyudai.jp

H. Anai
Institute of Mathematics for Industry, Kyushu University,
744 Motooka, Nishi-ku, Fukuoka 819-0395, Japan

M. Wakayama et al. (eds.), *The Impact of Applications on Mathematics*, Mathematics for Industry 1, DOI: 10.1007/978-4-431-54907-9_13,

problems, hence algorithms for solving constraints, such as polynomial systems, are of great importance both in theory and practice. Though conventional approaches to solve optimization problems are based on numerical iterative methods, here we focus on another approach by symbolic and algebraic methods in algebraic geometry over the real numbers (Note that engineers usually seek for information on the real solutions).

Real algebraic geometry deals with the solution set of (possibly quantified) systems of polynomial equations and/or inequalities over the real numbers, thus problems arising from manufacturing process are related to real algebraic geometry problems. Typical questions in real algebraic geometry is to determine the properties of the solution sets such as non-emptiness, dimension and quantifier free description as a semi-algebraic set. Such tasks are carried out by symbolic and algebraic algorithms in real algebraic geometry: cylindrical algebraic decomposition (CAD) or quantifier elimination (QE). Various algorithms and deep complexity results about CAD and QE have been studied during the last several decades [4]. Moreover, practically efficient software systems of QE have been developed and also are applied to many nontrivial application problems, see [11, 13, 21].

In this paper we show recent developments in the interplay of real algebraic geometry and controller design in manufacturing. Moreover we explain how applications affect to algorithm development in real algebraic geometry in the context of manufacturing design.

2 Computational Real Algebraic Geometry and Control Theory

There is a close relation between the history of development of control theory and computational methods which are available at that time (see Fig. 1).

The stability criterion of Routh-Hurwitz, for example, provides a method to check stability only in terms of four fundamental arithmetic operations without finding roots of a characteristic polynomials. This was an epoch-making new technology at the times when it was difficult to find the polynomial roots due to lack of computers. Nichols diagram provides the frequency characteristics of a closed-loop system from that of a open-loop system in view of diagram (without computation). This is also one of the wisdom at the times with no computers. Moreover, control system design method based on Bode diagram is regarded as the method which enables us to obtain the compensators with desired characteristics only by graphical addition and subtraction.

The methods to solve Ricatti equations and various optimal control problems (including nonlinear cases) were research fields at the times of modern control theory. Therefore, once the method to solve Ricatti equations based on the eigenvalues of Hamilton matrices has been established, it became to be one of the main targets of control theory to reduce control problems into that of existence of the solution to Ricatti equations. That is, reducing control problems to those of Ricatti equations was considered to give an answer to the problems. Solution to H_∞ control problems is one of their typical examples.

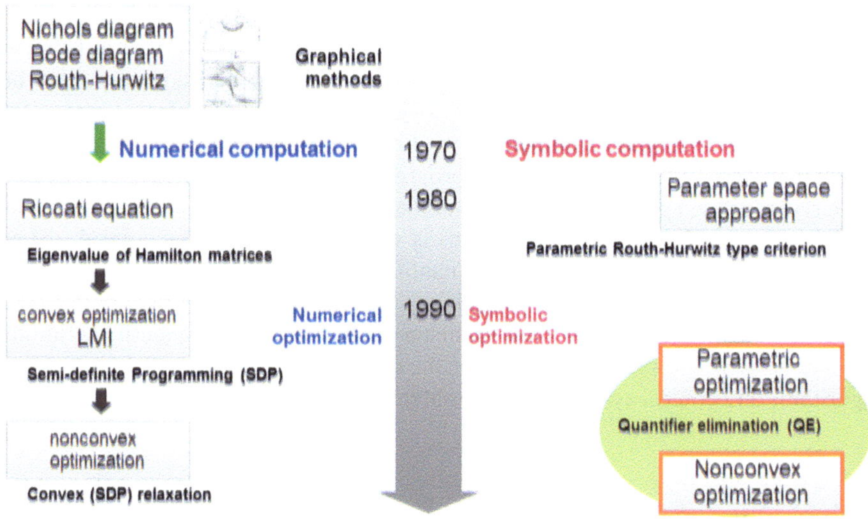

Fig. 1 Brief history of control theory with computational methods

In 1990, control system design based on numerical optimization methods has attracted considerable attention. The typical example is the method to solve robust control system design and analysis problems. Robust control system analysis and design problems are reduced to convex optimization problems described by LMI (Linear Matrix Inequality) and then solve them numerically by using SDP (Semi-Definite Programming), for which efficient numerical methods have been studied vigorously based on interior point methods in the field of mathematical optimization. This method gives solution to the problems which have no analytic solution by finding global solution (though it is a numerical method) and, in particular, breaks new way in robust control design and multi-objective design problems. Recently, The interest in this direction is shifting to the study of non-convex optimization problems derived from control problems in order to solve more practical problems Ricatti equations, LMI convex optimization, and non-convex optimization are control system design methods based on numerical computations. These design methods have become practically effective by virtue of improvement in ability and accuracy of computers and development of efficient algorithms.

While there exists another computational way "symbolic and algebraic computation" that is opposite to numerical computation in some sense and its research field is called "computer algebra". In general symbolic computations take much time and the size of the problems which can be solved by symbolic computation within a reasonable amount of time is limited compared with numerical computations. However, computer algebra methods have a good property that the output is easy to understand for designing parameters in manufacturing because they can deal with design parameters symbolically. Viewed in this light, there were several attempts to apply computer algebra techniques to control theory. The first such trials were at the middle of 1980s. For example, one of initiative works was achieved by Saito [18] and control system design environment using symbolic computation was proposed in [1]. The ideas of such works are fascinating but they are suitable for educational use but far from employing them for practical control system design and analysis because the ability of computers is not sufficient at that time,

After that in 1990s there was great progress in computer algebra. In particular, QE has made much progress. In fact, several effective algorithms and good software for QE were developed since 1990s. This is vital to the research of control system analysis and design methods based on quantifier elimination and many results have been presented so far (see [9, 11, 15, 17]).

In the sequel, we first briefly show the history of development of quantifier elimination and its useful properties, then explain the relation between quantifier elimination algorithms and control system design problems.

3 Real Quantifier Elimination

Many mathematical and engineering problems can be translated to formulas consisting of polynomial equations, inequalities, quantifiers ($\forall, \exists$) and Boolean operators ($\wedge, \vee, \neg, \rightarrow$, etc). Such formulas construct sentences in the so-called first-order theory of the real closed field and are called first-order formulas. Quantifier elimination is a symbolic and algebraic algorithm which deals with first-order formulas. QE outputs an equivalent quantifier free formula for a given first-order formula. For example, QE derives an equivalent quantifier free formula $b^2 - 4c < 0$ for the input formula $\forall x(x^2 + bx + c > 0)$ over the field of real numbers. If all variables in a given first-order formula are quantified, QE returns true/false of the input formula. In this paper we consider real quantifier elimination. See [4] for the detail of QE.

QE is regarded as one of the tools for constraint solving and optimization problems. For example, here we briefly mention how we solve optimization problems by QE.

$$\begin{array}{lll} \text{Objective function:} & f(x_1, \ldots, x_n) & \rightarrow \min \\ \text{Constraints:} & g_1(x_1, \ldots, x_n)\rho_1 0, \; \ldots, & g_l(x_1, \ldots, x_n)\rho_l 0 \end{array} \tag{1}$$

where $\rho_i \in \{=, \neq, <, >, \leq, \geq\}$. The optimization problem (1) is translated to the corresponding QE problem (first-order formula) as follows:

$$\exists x_1 \cdots \exists x_n (k - f(x_1, \ldots, x_n) = 0 \wedge \varphi(x_1, \ldots, x_n)) \tag{2}$$

where $\varphi(x_1, \ldots, x_n) \equiv \bigwedge_i (g_i(x_1, \ldots, x_n)\rho_i 0)$ and k is a newly introduced variable assigned to the objective function $f(x_1, \ldots, x_n)$. Performing QE to the first-order formula (2), we obtain a formula $\psi_1(k)$ which shows the possible range of k, i.e., f. So the minimum of k in $\psi_1(k)$ is the minimum of the objective function f and it is the globally minimum. See [13] for the detail of various kind of optimization techniques accomplished by QE and their industrial applications.

Optimization by QE has following properties: QE enables us to

- obtain not only one feasible solution but also feasible region of solutions,
- deal with non-convex optimization,
- and examine decision problems exactly.

These features (advantages) of QE are useful to resolve many unsolved problems, if we utilize numerical methods only, in engineering and industrial problems.

The history of the algorithms for QE begins with Tarski-Seidenberg decision procedure in 1950s [19, 22]. But this is very intricate and far from feasible. In 1975, Collins presented a more efficient general purpose QE algorithm based on cylindrical algebraic decomposition [5]. The algorithm has improved by Collins and Hong [6] and was implemented as "QEPCAD" by Hong. Weispfenning has presented another QE algorithm by using Comprehensive Gröbner basis and the real root counting for multivariate polynomial systems [25].

A general-purpose QE by CAD has a bad computational complexity. A lot of efforts for improving its efficiency have been done so far [4, 14, 20]. Many control applications of QE appeared since the 1990s, for example see [9, 15]. In parallel, to circumvent the inherent computational complexity of a QE algorithm, several researchers have focused on developing QE algorithms specialized to particular types of input formulas in order to make good use of their specialties. This direction is quite promising in practice since a number of important problems in engineering have been successfully reduced to such particular input formulas and resolved efficiently by using the specialized QE algorithms.

Weispfenning presented a more efficient QE algorithm based on virtual substitution [16, 23]. Though there is some degree restriction of a quantified variable in input formulas for virtual substitution, this approach is actually applied to many application problems [24]. Implementation of the method was done on REDUCE as "REDLOG" by Dolzman and Sturm [8].

Moreover, González-Vega et.al. [10] also presented a special QE algorithm for definite conditions of polynomials. This approach has been tailored for a special class of QE problem called a "sign definite condition (SDC)"

$$\forall x\ (x \geq 0 \rightarrow f(x) > 0) \tag{3}$$

where $f(x) \in \mathbb{R}[x]$. Quite a lot of practical control system design problems can be recast as the SDC (3), see [2, 3, 7, 11, 12]. In the following section we focus the special QE algorithm for the SDC.

4 Illustrative Application

We again point out the difficulty to apply general quantifier elimination algorithms to control design problems in view of efficiency. It is quite promising to utilize special quantifier elimination algorithms for a subclass of input formulas originated from control problems. Here we introduce successful examples of such direction in control system design via a special QE algorithm for SDC.

4.1 A Special QE Algorithm for SDC

A special QE method based on the Sturm-Habicht sequence for the first-order formula $\forall x\ (f(x) > 0)$, where $f(x) \in \mathbb{R}[x]$ with parametric coefficients was first proposed in [10]. The algorithm is desired to be modified for checking a SDC (3) since a quite wide range of the important problems in robust control can be reduced to the SDC [3, 7, 11, 12].

We briefly sketch a special QE algorithm using the Sturm-Habicht sequence for the SDC (see [12] for the detail of the algorithm and its effective implementation). The Sturm-Habicht sequence of a polynomial $f(x) \in \mathbb{R}[x]$ with degree n is defined as the subresultant sequence starting from $f(x)$ and $f'(x)$ modulo some specified sign changes.

Let P, Q be polynomials in $\mathbb{R}[x]$; $P = \sum_{k=0}^{n} a_k x^k$, $Q = \sum_{k=0}^{m} b_k x^k$, where n, m are non-negative integers. For $i = 0, 1, \ldots, \ell = \min(n, m)$ we define the subresultant $Sres_i(P, n, Q, m)$ associated to P, n, Q and m of index i as $\sum_{j=0}^{i} M_j^i(P, Q)x^j$, where $M_j^i(P, Q)$ is the determinant of the matrix composed by the columns $1, 2, \ldots, n+m-2i-1$ and $n+m-i-j$ in the matrix $s_i(P, n, Q, m)$:

$$s_i(P, n, Q, m) := \overbrace{\begin{pmatrix} a_n & \cdots & a_0 & & \\ & \ddots & & \ddots & \\ & & a_n & \cdots & a_0 \\ b_m & \cdots & b_0 & & \\ & \ddots & & \ddots & \\ & & b_m & \cdots & b_0 \end{pmatrix}}^{n+m-i} \begin{matrix} \left.\vphantom{\begin{matrix} a \\ a \\ a \end{matrix}}\right\} m-i \\ \left.\vphantom{\begin{matrix} a \\ a \\ a \end{matrix}}\right\} n-i \end{matrix}, \tag{4}$$

Let $\nu = n + m - 1$ and $\delta_k = (-1)^{\frac{k(k+1)}{2}}$. The Sturm-Habicht sequence associated to P and Q is defined as the list of polynomials $\{SH_j(P, Q)\}_{j=0,\ldots,\nu+1}$ given by $SH_{\nu+1}(P, Q) = P$, $SH_\nu(P, Q) = P'Q$, $SH_j(P, Q) = \delta_{\nu-j} \cdot Sres_j(P, n, P'Q, \nu)$ for $j = 0, 1, \ldots, \nu - 1$, where $P' = \frac{dP}{dx}$. In particular, $\{SH_j(P, 1)\}_{j=0,\ldots,\nu+1}$ is called the Sturm-Habicht sequence of P. We simply denote it by $\{SH_j(P)\}$.

The Sturm-Habicht sequence can be used for real root counting (like the Sturm sequence) and it has better properties in terms of specialization and computational complexity [10]: Let $P(x) \in \mathbb{R}[x]$ and $\{g_0(x), \ldots, g_k(x)\}$ be a set of polynomials obtained from $\{SH_j(P(x))\}$ by deleting the identically zero polynomials. Let $\alpha, \beta \in \mathbb{R} \cup \{-\infty, +\infty\}$ s.t. $\alpha < \beta$. We define $W_{SH}(P; \alpha)$ as the number of sign variations on $\{g_0(\alpha), \ldots, g_k(\alpha)\}$. We note that the way of counting the number of sign variations is different from the case of conventional Sturm sequence, see [10, 12] for how to count the number of sign variation for the Sturm-Habicht sequence.

Then $W_{SH}(P; \alpha, \beta) \equiv W_{SH}(P; \alpha) - W_{SH}(P; \beta)$ gives the number of real roots of P in $[\alpha, \beta]$. We denote the principal j-th Sturm-Habicht coefficient of $SH_j(f)$, i.e., the coefficient of degree j of $SH_j(f)$, by $st_j(f)$ and the constant term of $SH_j(f)$ by $ct_j(f)$ for all j. Then, when the Sturm-Habicht sequence is regular, we have

$$\begin{aligned} W_{SH}(f; 0, +\infty) &\equiv W_{SH}(f; 0) - W_{SH}(f; +\infty) \\ &\equiv V(\{ct_n(f), \ldots, ct_0(f)\}) - V(\{st_n(f), \ldots, st_0(f)\}), \end{aligned} \quad (5)$$

where $V(\{a_i\})$ stands for the number of sign changes over the sequence $\{a_i\}$. The SDC holds if and only if $W_{SH}\ (f; 0, +\infty) = 0$, $st_n(f) > 0$ and $ct_n(f) > 0$ hold. Hence an equivalent condition to the SDC can be obtained by choosing all sign conditions that satisfy $W_{SH}(f; 0, +\infty) = 0$, $st_n(f) > 0$ and $ct_n(f) > 0$. The obtained condition is the form of a union of semi-algebraic sets.

4.2 Multi-objective Fixed-Structure Controller Design

H_∞ control design and μ synthesis are recognized as useful methods for practical control design under plant uncertainty, but the order of controller designed is fairly high in general, which sometimes causes the difficulty of implementation. Hence, the lower order or fixed-order controller such as PID controller is preferable in many control applications. However, the synthesis problems are non-convex for most of the cases, meaning, it is not so easy to get the desirable controller by numerical optimization. A parameter space approach is known to be one of the effective methods for robust control synthesis and multi-objective design using a fixed-order controller. The approach can be utilized to determine the set of certain controller parameters to be designed so that the resultant feedback control system satisfies the given design specifications.

Multi-objective controller design problem considered here is the following: Consider a feedback control system shown in Fig. 2, where $\mathbf{p} = [p_1, p_2, \ldots, p_d]$ is the vector of uncertain real parameters in the plant G and $\mathbf{x} = [x_1, x_2, \ldots, x_e]$ is the vector of real parameters of the controller C. Assume that the controller considered here is of fixed order and all its coefficients of the characteristic polynomial $g(s, \mathbf{x}, \mathbf{p})$ are linear in terms of the plant parameter vector $\mathbf{p}$. We refer to this as the *linear case*. The performance of the control system can often be characterized by a vector $\mathbf{c} = [c_1, \ldots, c_t]$ which are functions of the plant and controller parameters $\mathbf{p}$ and $\mathbf{x}$, *i.e.*, $c_i = c_i(\mathbf{x}, \mathbf{p})$, for $i = 1, \ldots, t$ and the target specifications are usually given as follows:

$$c_i(\mathbf{x}, \mathbf{p}) < \tau_i, \quad i = 1, \ldots, t. \quad (6)$$

The goal of robust control synthesis is to find the region in the controller parameter space which meets the design specifications.

Fortunately, many important design specifications (6) such as H_∞ norm constraints, frequency restricted norms, phase/gain margins, and $\mathbf{D}$-stability constraint

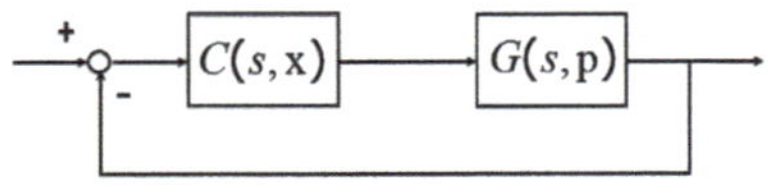

Fig. 2 A standard feedback system

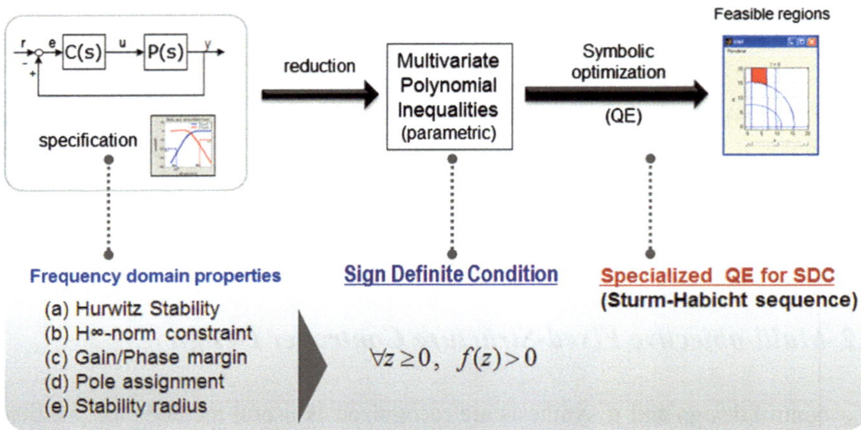

Fig. 3 A parameter space approach by QE

for robustness in control can be reduced to a special class of first-order formula of the form of (3) as seen in [3, 11, 17]. So we can utilize the special QE algorithm shown in Sect. 4.1 for obtaining feasible regions of controller parameters for each specification. The design flow of this parameter space approach to multi-objective controller design accomplished by the special QE algorithm for the SDC is illustrated in Fig. 3.

Application examples: One of the typical examples is the (frequency restricted) H_∞ norm constraint: An H_∞ norm constraint of a strictly proper transfer function $P(s) = n(s)/d(s)$ expressed as

$$\|P(s)\|_\infty := \sup_{\omega} |P(j\omega)| < \gamma,$$

where j is the imaginary unit, is equivalent to

$$\forall \omega (\gamma^2 d(j\omega)d(-j\omega) > n(j\omega)n(-j\omega)).$$

Since we can find a function $f(\omega^2)$ which satisfies $f(\omega^2) = \gamma^2 d(j\omega)d(-j\omega) - n(j\omega)n(-j\omega) > 0$, letting $x = \omega^2$ lead to SDC. Other design specifications in robust control reduced to SDCs are listed in [11, 17] with reduction procedures.

A parameter space approach to multi-objective controller design accomplished by the special QE algorithm for the SDC has been successfully applied several practical controller design problems: The method is utilized to design PI control type AVR

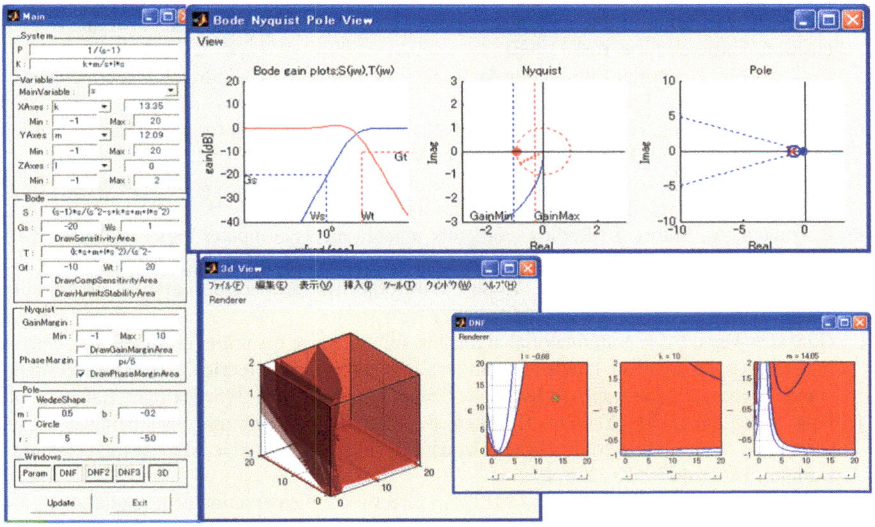

Fig. 4 Screenshot of our MATLAB toolbox: The specifications are inputted via the left vertically long window. The upper horizontally long window contains Bode diagram, Nyquist plot and pole/zero location. The lower right two windows shows feasible regions of controller parameters (here we have 3 parameters) so that the system satisfies given specifications

(Automatic Voltage Regulator) of the excitation control [26] and also to design a controller for a power supply unit [17]. Moreover, the parameter space approach to fixed-structure controller design explained here was implemented as a toolbox on MATLAB (Fig. 4), see [11, 26].

5 Conclusion

In this paper we discussed the interplay of QE algorithms in real algebraic geometry and control system design problems in manufacturing and we see that applications provide us new useful directions of the mathematical algorithm development.

We hope that deep interactions between applications and mathematics would explore a novel field of mathematics.

References

1. Akahori, I., Hara, S.: Computer aided control system analysis and design based on the concept object-orientation (in Japanese). Trans. SICE **24**(5), 506–513 (1988)
2. Anai, H., Hara, S.: Fixed-structure robust controller synthesis based on sign definite condition by a special quantifier elimination. In: Proceedings of American Control Conference, pp. 1312–1316 (2000)
3. Anai, H., Hara, S.: A parameter space approach to fixed-order robust controller synthesis by quantifier elimination. Int. J. Control **79**(11), 1321–1330 (2006)

4. Caviness, B.F., Johnson, J.R. (eds.): Quantifier Elimination and Cylindrical Algebraic Decomposition. Springer, New York (1998)
5. Collins, G.E.: Quantifier Elimination for Real Closed Fields by Cylindrical Algebraic Decomposition, LNCS 32. Springer, Berlin (1975)
6. Collins, G.E., Hong, H.: Partial cylindrical algebraic decomposition for quantifier elimination. J. Symbolic Comput. **12**(3), 299–328 (Sept. 1991)
7. Didier, H., Andrea, G. : Positive Polynomials in Control. Lecture Notes in Control and Information Sciences, vol. 312. Springer, Berlin (2005)
8. Dolzmann, A., Sturm, T.: Redlog: computer algebra meets computer logic. ACM SIGSAM Bull. **31**(2), 2–9 (1997)
9. Dorato, P., Yang, W., Abdallah, C.: Robust multi-objective feedback design by quantifier elimination. J. Symb. Comp. **24**, 153–159 (1997)
10. González-Vega, L.: A combinatorial algorithm solving some quantifier elimination problems. In: Caviness, B., Johnson, J. (eds.) Quantifier Elimination and Cylindrical Algebraic Decomposition, Texts and Monographs in Symbolic Computation, pp. 365–375. Springer, Berlin (1998)
11. Hyodo, N., Hong, M., Yanami, H., Hara, S., Anai, H.: Solving and visualizing nonlinear parametric constraints in control based on quantifier elimination. Appl. Algebra Eng. Commun. Comput. **18**(6), 497–512 (2007)
12. Iwane, H., Higuchi, H., Anai, H.: (2013) An effective implementation of a special quantifier elimination for a sign definite condition by logical formula simplification. In: ASC 2013: Lecture Notes in Computer Science, vol. 8136, pp. 194–208. Springer, Berlin (2013)
13. Iwane, H., Yanami, H., Anai, H.: A symbolic-numeric approach to multi-objective optimization in manufacturing design. Math. Comput. Sci. **5**(3), 315–334 (2011)
14. Iwane, H., Yanami, H., Anai, H., Yokoyama, K.: An effective implementation of symbolic-numeric cylindrical algebraic decomposition for quantifier elimination. Theor. Comput. Sci. **479**, 43–69 (2013)
15. Jirstrand, M.: Nonlinear control system design by quantifier elimination. J. Symb. Comp. **24**(2), 137–152 (1997)
16. Loos, R., Weispfenning, V.: Applying linear quantifier elimination. Comput. J. **36**(5), 450–462 (1993)
17. Matsui, Y., Iwane, H., Anai, H.: Development of Computer Algebra Research and Collaboration with Industry, MI Lecture Note Series, vol 49, pp. 43–52. Kyushu University (2013)
18. Saito, O.: Computer aided control engineering: periphery of control engineering and computer algebraic manipulation (in Japanese). Syst. Control **29**(12), 785–794 (1985)
19. Seidenberg, A.: A new decision method for elementary algebra. Ann. Math. **60**, 365–374 (1954)
20. Strzeboński, A.W.: Cylindrical algebraic decomposition using validated numerics. J. Symbolic Comput. **41**(9), 1021–1038 (2006)
21. Sturm, T.: New domains for applied quantifier elimination. In: Proceedings of the 14th International Workshop on Computer Algebra (CASC) 2006, pp. 295–301 (2006)
22. Tarski, A.: Decision Methods for Elementary Algebra and Geometry. University of California Press, Berkeley (1951)
23. Weispfenning, V.: The complexity of linear problems in fields. J. Symbolic Comput. **5**(1–2), 3–27 (1988)
24. Weispfenning, V.: Simulation and optimization by quantifier elimination. J. Symb. Comput. **24**(2), 189–208 (1997)
25. Weispfenning, V.: A new approach to quantifier elimination for real algebra. In: Caviness, F., Johnson, J.R. (eds.) Quantifier Elimination and Cylindrical Algebraic Decomposition, Texts and Monographs in Symbolic Computation, pp. 376–392. Springer, Berlin (1998)
26. Yoshimura, S., Iki, H., Uriu, Y., Anai, H., Hyodo, N.: Generator excitation control using a parameter space design method. In: 43rd International Universities Power Engineering Conference (UPEC 2008), pp. 1–4 (2008)

The Method of Cyclic Intrepid Projections: Convergence Analysis and Numerical Experiments

Heinz H. Bauschke, Francesco Iorio and Valentin R. Koch

Abstract The convex feasibility problem asks to find a point in the intersection of a collection of nonempty closed convex sets. This problem is of basic importance in mathematics and the physical sciences, and projection (or splitting) methods solve it by employing the projection operators associated with the individual sets to generate a sequence which converges to a solution. Motivated by an application in road design, we present the method of cyclic intrepid projections (CycIP) and provide a rigorous convergence analysis. We also report on very promising numerical experiments in which CycIP is compared to a commerical state-of-the-art optimization solver.

Keywords Convex set · Feasibility problem · Halfspace · Intrepid projection · Linear inequalities · Projection · Road design

AMS 2010 Subject Classification Primary 65K05 · 90C25 · Secondary 90C05

H. H. Bauschke (✉)
Mathematics, University of British Columbia, Kelowna, BC V1V 1V7, Canada
e-mail: heinz.bauschke@ubc.ca

F. Iorio
Autodesk Research, 210 King Street East Suite 600, Toronto, ON M5A 1J7, Canada
e-mail: francesco.iorio@autodesk.com

V. R. Koch
Information Modeling and Platform Products Group (IPG), Autodesk, Inc.,111 Mc Innis Parkway, San Rafael, CA 94903, USA
e-mail: valentin.koch@autodesk.com

M. Wakayama et al. (eds.), *The Impact of Applications on Mathematics*, Mathematics for Industry 1, DOI: 10.1007/978-4-431-54907-9_14,

1 Introduction

Throughout this paper, we assume that

$$X \text{ is a real Hilbert space} \tag{1}$$

with inner product $\langle\cdot,\cdot\rangle$ and induced norm $\|\cdot\|$. (We also write $\|\cdot\|_2$ instead of $\|\cdot\|$ if we wish to emphasize this norm compared to other norms. We assume basic notation and results from convex analysis and fixed point theory; see, e.g., [4, 6, 15, 16, 21].)

Let $(C_i)_{i\in I}$ be a finite family of closed convex subsets of X such that

$$C := \bigcap_{i\in I} C_i \neq \varnothing. \tag{2}$$

We aim to find a point in C given that the individual constraint sets C_i are simple in the sense that their associated projections[1] are easy to compute. To solve the widespread *convex feasibility problem* "find $x \in C$", we employ *projection methods*. These splitting-type methods use the individual projections P_{C_i} in order to generate a sequence that converges to a point in C. For further information, we refer the reader to, e.g., [2, 4, 8, 9, 11, 15, 16, 19].

In previous work on a feasibility problem arising in road design [5], the *method of cyclic intrepid projections (CycIP)* was found to be an excellent overall algorithm. Unfortunately, CycIP was applied heuristically without an underlying convergence result.

The goal of this paper is two-fold. First, we present a checkable condition sufficient for convergence and provide a rigorous convergence proof. In fact, our main result applies to very general feasibility problems satisfying an interiority assumption. Second, we numerically compare CycIP to a commercial LP solver for test problems that are both convex and nonconvex to evaluate competitiveness of CycIP.

The remainder of the paper is organized as follows. In Sect. 2 we provide basic properties of projection operators. Useful results on Fejér monotone sequences are recalled in Sect. 3. Our main convergence results are presented in Sect. 4. In Sect. 5, we review the feasibility problem arising in road design and obtain a rigorous convergence result for CycIP. We report on numerical experiments in Sect. 6 and offer concluding remarks in Sect. 7.

We end this section with notation. The closed ball centered at $y \in X$ of radius r is $B(y;r) := \{z \in X \mid \|y-z\| \leq r\}$. Finally, we write $\mathbb{R}_+$ and $\mathbb{R}_{++}$ for the nonnegative real numbers and strictly positive reals, respectively.

---

[1] Given a nonempty subset S of X and $x \in X$, we write $d_S(x) := \inf_{s\in S} \|x-s\|$ for the *distance* from x to S. If S is also closed and convex, then the infimum defining $d_S(x)$ is attained at a *unique* vector called the *projection* of x onto S and denoted by $P_S(x)$ or $P_S x$.

2 Relaxed and Intrepid Projectors

In this section, we introduce the key operators used in the projection methods studied later.

Fact 1 (relaxed projector) Let C be a nonempty closed convex subset of X, and let $\lambda \in\,]0,2[$. Set $R := (1-\lambda)\operatorname{Id} + \lambda P_C$, let $x \in X$, and let $c \in C$. Then

$$\|x-c\|^2 - \|Rx-c\|^2 \geq \frac{2-\lambda}{\lambda}\|x-Rx\|^2 = (2-\lambda)\lambda d_C^2(x). \tag{3}$$

Proof Combine [2, Lemma 2.4 (iv)] with [4, Proposition 4.8]. □

In fact, the relaxed projector is an example of a so-called *averaged map*; see, e.g., [1, 4, 13] for more on this useful notion.

Definition 2 *(enlargement)* Given a nonempty closed convex subset Z of X, and $\alpha \in \mathbb{R}_+ := \{\xi \in \mathbb{R} \mid \xi \geq 0\}$, we write

$$Z_{[\alpha]} := \{x \in X \mid d_Z(x) \leq \alpha\} = Z + B(0;\alpha) \tag{4}$$

and call $Z_{[\alpha]}$ the α*-enlargement* of Z.

Note that $Z_{[0]} = Z$, that $Z_{[\alpha]}$ is a nonempty closed convex subset of X, and that if $\alpha < \beta$, then $Z_{[\alpha]} \subseteq Z_{[\beta]}$. We mention in passing that the *depth*[2] of each $z \in Z$ (with respect to $Z_{[\alpha]}$), i.e., $d_{X \smallsetminus Z_{[\alpha]}}(z)$, is at least α.

Fact 3 (See, e.g., [4, Proposition 28.10].) Let C be a nonempty closed convex subset of X, and let $\beta \in \mathbb{R}_+$. Set $D := C_{[\beta]}$. Then

$$(\forall x \in X)\quad P_D x = \begin{cases} x, & \text{if } d_C(x) \leq \beta; \\ P_C x + \beta \dfrac{x - P_C x}{d_C(x)}, & \text{otherwise.} \end{cases} \tag{5}$$

Definition 4 *(intrepid projector)* Let Z be a nonempty closed convex subset of X, let $\beta \in \mathbb{R}_+$, and set $C := Z_{[\beta]}$. The corresponding *intrepid projector* $Q := Q_C$ *onto* C *(with respect to* Z *and* β*)* is defined by

$$\begin{aligned} Q\colon X \to X\colon x \mapsto\ & x + \left(1 - \tfrac{P_{[\beta,2\beta]} d_Z(x)}{\beta}\right)(x - P_Z x) \\ =\ & \begin{cases} P_Z x, & \text{if } d_Z(x) \geq 2\beta; \\ x, & \text{if } d_Z(x) \leq \beta; \\ x + \left(1 - \tfrac{d_Z(x)}{\beta}\right)(x - P_Z x), & \text{otherwise.} \end{cases} \end{aligned} \tag{6}$$

---

[2] This function is considered, e.g., in [7, Exercise 8.5].

We refer to these three steps as the *projection step*, the *identity step*, and the *reflection step*, respectively.

Example 5 (intrepid projector onto a hyperslab à la Herman) Suppose that $a \in X \smallsetminus \{0\}$, let $\alpha \in \mathbb{R}$, let $\beta \in \mathbb{R}_+$, and set $Z := \{x \in X \mid \langle a, x\rangle = \alpha\}$. Then Z is a hyperplane and $Z_{[\beta]}$ is a hyperslab. Moreover, the associated intrepid projector onto $Z_{[\beta]}$ is precisely the operator considered by Herman in [18].

Proposition 6 (basic properties of the intrepid projector) *Let Z be a nonempty closed convex subset of X, let $\beta \in \mathbb{R}_+$, set $C := Z_{[\beta]}$, and denote the corresponding intrepid projector onto C (with respect to Z and β) by Q. Now let $\alpha \in [0, \beta]$, and let $y \in Z_{[\alpha]}$, and let $x \in X$. Then*

$$Qx \in [x, P_Z x] \cap C \tag{7}$$

and exactly one of the following holds:

(i) $d_Z(x) \leq \beta$, $x = Qx \in C$, *and* $\|x - y\|^2 - \|Qx - y\|^2 = 0$.
(ii) $d_Z(x) \geq 2\beta$ *and*

$$\begin{aligned}\|x - y\|^2 - \|Qx - y\|^2 &\geq 2(\beta - \alpha)\|x - Qx\| = 2(\beta - \alpha) d_Z(x)\\ &= 2(\beta - \alpha)\big(\beta + d_C(x)\big) \geq 4\beta(\beta - \alpha).\end{aligned} \tag{8}$$

(iii) $\beta < d_Z(x) < 2\beta$ *and*

$$\begin{aligned}\|x - y\|^2 - \|Qx - y\|^2 \geq 2(\beta - \alpha)\|x - Qx\| &= \frac{2(\beta - \alpha)}{\beta} d_Z(x)\big(d_Z(x) - \beta\big)\\ &= \frac{2(\beta - \alpha)}{\beta} d_C(x)\big(\beta + d_C(x)\big).\end{aligned} \tag{9}$$

Consequently, in every case, we have

$$\|x - y\|^2 - \|Qx - y\|^2 \geq 2(\beta - \alpha)\|x - Qx\| \geq 2(\beta - \alpha) d_C(x). \tag{10}$$

Proof Set $\delta := d_Z(x)$, $p := P_Z x$, and write $y = z + \alpha b$, where $z \in Z$, $b \in X$ and $\|b\| \leq 1$. Note that if $\delta > \beta$, then $d_C(x) = \delta - \beta$ using Fact 3.

(i) This follows immediately from the definition of Q.
(ii) Using Cauchy–Schwarz in (11a), and the projection theorem (see, e.g., [4, Theorem 3.14]) in (11b), we obtain

$$\begin{aligned}
\|x-y\|^2-\|Qx-y\|^2 &= \|x-(z+\alpha b)\|^2-\|p-(z+\alpha b)\|^2\\
&= \|x\|^2-\|p\|^2-2\langle x, z+\alpha b\rangle+2\langle p, z+\alpha b\rangle\\
&\geq \|x\|^2-\|p\|^2+2\langle p-x, z\rangle-2\alpha\|p-x\|\|b\| \qquad (11a)\\
&\geq \|x\|^2-\|p\|^2+2\langle p-x, z-p\rangle+2\langle p-x, p\rangle-2\alpha\delta\\
&\geq \|x\|^2-\|p\|^2+2\langle p-x, p\rangle-2\alpha\delta \qquad (11b)\\
&= \|x\|^2+\|p\|^2-2\langle x, p\rangle-2\alpha\delta=\|x-p\|^2-2\alpha\delta\\
&= \delta^2-2\alpha\delta=\delta\big(\delta-2\alpha\big)\geq 2\delta(\beta-\alpha)\geq 4\beta(\beta-\alpha).
\end{aligned}$$

(iii) Set $\eta := (\delta-\beta)/\beta \in \,]0,1[$. Then $Qx=(1-\eta)x+\eta p$ and hence

$$\|x-Qx\|=\eta\|x-p\|=\eta\delta=\frac{\delta-\beta}{\beta}\delta. \qquad (12)$$

Using, e.g., [4, Corollary 2.14] in (13a), and Cauchy–Schwarz and [4, Theorem 3.14] in (13b), we obtain

$$\begin{aligned}
\|x-y\|^2-\|Qx-y\|^2 &= \|x\|^2-\|Qx\|^2-2\langle x, y\rangle+2\langle Qx, y\rangle\\
&= \|x\|^2-\|(1-\eta)x+\eta p\|^2+2\langle(1-\eta)x+\eta p-x, y\rangle\\
&= \|x\|^2-(1-\eta)\|x\|^2-\eta\|p\|^2 \qquad (13a)\\
&\quad +\eta(1-\eta)\|x-p\|^2+2\eta\langle p-x, z+\alpha b\rangle\\
&= \eta\big(\|x\|^2-\|p\|^2\big)+(1-\eta)\eta\|x-p\|^2\\
&\quad +2\eta\big(\langle p-x, z-p\rangle+\langle p-x, p\rangle+\alpha\langle p-x, b\rangle\big)\\
&\geq \eta\big(\|x\|^2-\|p\|^2+(1-\eta)\|x-p\|^2+2\langle p-x, p\rangle\big) \qquad (13b)\\
&\quad -2\alpha\eta\|x-p\|\\
&= \eta\big(\|x-p\|^2+(1-\eta)\|x-p\|^2-2\alpha\|x-p\|\big)\\
&= \delta\eta\big((2-\eta)\delta-2\alpha\big)\\
&= \frac{\delta}{\beta}(\delta-\beta)\Big(\big(2-(\delta-\beta)\beta^{-1}\big)\delta-2\alpha\Big)\\
&= \frac{\delta}{\beta}(\delta-\beta)\big(-\beta^{-1}\delta^2+3\delta-2\alpha\big).
\end{aligned}$$

Now the quadratic $q\colon [\beta, 2\beta] \to \mathbb{R}\colon \xi \mapsto -\beta^{-1}\xi^2+3\xi-2\alpha$ has a maximizer at $\xi=(3/2)\beta$ and it satisfies $q(\beta)=q(2\beta)=2(\beta-\alpha)\geq 0$. It follows that $\min q\big([\beta, 2\beta]\big)=2(\beta-\alpha)$. Therefore, (13) and (12)

$$\|x-y\|^2-\|Qx-y\|^2\geq\frac{\delta}{\beta}(\delta-\beta)2(\beta-\alpha)=2(\beta-\alpha)\|x-Qx\|. \qquad (14)$$

The proof of the "Consequently" part follows easily. □

Proposition 6 implies that the intrepid projector is *quasi nonexpansive*; see, e.g., [3, 8, 22–24] for further results utilizing this notion.

3 Fejér Monotonicity

We now review the definition and basic results on Fejér monotone sequences. These will be useful in establishing our convergence results.

Definition 7 Let $(x_k)_{k\in\mathbb{N}}$ be a sequence in X, and let C be a nonempty closed convex subset of X. Then $(x_k)_{n\in\mathbb{N}}$ is *Fejér monotone with respect to C* if

$$(\forall k \in \mathbb{N})(\forall c \in C) \quad \|x_{k+1} - c\| \leq \|x_k - c\|. \tag{15}$$

Fact 8 Let $(x_k)_{k\in\mathbb{N}}$ be a sequence in X that is Fejér monotone with respect to some nonempty closed convex subset C of X. Then the following hold:

(i) If int $C \neq \varnothing$, then $(x_k)_{k\in\mathbb{N}}$ converges strongly to some point in X.
(ii) If each weak cluster point of $(x_k)_{k\in\mathbb{N}}$ lies in C, then $(x_k)_{k\in\mathbb{N}}$ converges weakly to some point in C.
(iii) If $d_C(x_k) \to 0$, then $(x_k)_{k\in\mathbb{N}}$ converges strongly to some point in C.

Proof See, e.g., [2, 4, Chap. 5], or [12]. □

4 The Method of Cyclic Intrepid Projections

We now assume that each C_i is a closed convex subset of X, with

$$C := \bigcap_{i\in I} C_i \neq \varnothing. \tag{16}$$

The index set is split into two sets, corresponding to enlargements and regular sets:

$$I_0 := \{i \in I \mid C_i = (Z_i)_{[\beta_i]}, \text{ where } \beta_i > 0\} \quad \text{and} \quad I_1 := I \smallsetminus I_0. \tag{17}$$

Assume an index selector map

$$\mathrm{i}\colon \mathbb{N} \to I, \tag{18}$$

where $(\forall i \in I)\ \mathrm{i}^{-1}(i)$ is an infinite subset of $\mathbb{N}$. We say that the *control is quasicyclic with quasiperiod* $M \in \{1, 2, \ldots\}$ if $(\forall k \in \mathbb{N})\ I = \{\mathrm{i}(k), \mathrm{i}(k+1), \ldots, \mathrm{i}(k+M-1)\}$.

Let $(\lambda_i)_{i \in I_1}$ be a family in $]0, 2[$. We define a family of operators

$$(T_i)_{i \in I} \tag{19}$$

from X to X as follows. If $i \in I_0$, then T_i is the intrepid projector onto C_i (with respect to Z_i and β_i); if $i \in I_1$, then T_i is the relaxed projector onto C_i with relaxation parameter λ_i.

Algorithm 9 (method of cyclic intrepid projections) Given a starting point $x_0 \in X$, the *method of intrepid projections* proceeds via

$$(\forall k \in \mathbb{N}) \quad x_{k+1} := T_{\mathrm{i}(k)} x_k. \tag{20}$$

We begin our analysis with a simple yet useful observation.

Lemma 10 *The sequence $(x_k)_{k \in \mathbb{N}}$ generated by Algorithm 9 is Fejér monotone with respect to C.*

Proof Combine Fact 1 with Proposition 6. □

We now deepen our convergence analysis. We start with the purely intrepid case.

Theorem 11 (intrepid projections only) *Suppose that $I_1 = \varnothing$ and that* $\operatorname{int} C \neq \varnothing$. *Then $(x_k)_{k \in \mathbb{N}}$ converges strongly to some point in C.*

Proof Combining Lemma 10 with Fact 8(i), we deduce that $(x_k)_{k \in \mathbb{N}}$ converges strongly to some point $\bar{x} \in X$. Let $i \in I$. Then there exists a subsequence $(x_{n_k})_{k \in \mathbb{N}}$ of $(x_k)_{k \in \mathbb{N}}$ such that $(\forall k \in \mathbb{N})\ \mathrm{i}(n_k) = i$. By Proposition 6, the subsequence $(x_{n_k+1})_{k \in \mathbb{N}} = (T_i x_{n_k})_{k \in \mathbb{N}}$ lies in C_i. Since C_i is closed, we deduce that $\bar{x} \in C_i$. □

The proof of the following result follows that of Herman [18] who considered more restrictive controls. (See also [10].)

Corollary 12 (parallelotope) *Suppose that X is finite-dimensional, that $I_1 = \varnothing$, that each Z_i is a hyperplane, and that* $\operatorname{int} C \neq \varnothing$. *Then $(x_k)_{k\in\mathbb{N}}$ converges to some point in the parallelotope C in finitely many steps.*

Proof By Theorem 11, $(x_k)_{k\in\mathbb{N}}$ converges to some point $\bar{x} \in C$. If $\{x_k \mid k \in \mathbb{N}\} \cap \operatorname{int} C \neq \varnothing$, then $(x_k)_{k\in\mathbb{N}}$ is eventually constant. Assume to the contrary that $(x_l)_{k\in\mathbb{N}}$ is not eventually constant. Then $\bar{x} \notin \{x_k \mid k \in \mathbb{N}\} \cup \operatorname{int} C$. Since each C_i is a hyperslab, $\operatorname{bdry} C_i$ is the union of two disjoint hyperplanes parallel to Z_i. We collect these finitely many hyperplanes in a set H. The finite collection of these hyperplanes containing $\bar{x}$, which we denote by $H(\bar{x})$, is nonempty. Moreover, $(x_k)_{k\in\mathbb{N}}$ cannot have arisen with infinitely many projection steps as these only occur at a minimum distance from the sets. Therefore, infinitely many reflection steps have been executed. Hence there exists $K_1 \in \mathbb{N}$ such that iteration index k onwards, we only execute identity or reflection steps. Now let $\varepsilon > 0$ be sufficiently small such that $B(\bar{x}; \varepsilon)$ makes an empty intersection with every hyperplane drawn from $H \smallsetminus H(\bar{x})$. Since $x_k \to \bar{x}$, there exists $K_2 \in \mathbb{N}$ such that $(\forall k \geq N_2)\ x_k \in B(\bar{x}; \varepsilon)$. Since $\bar{x} \in C$ and $(x_k)_{k\in\mathbb{N}}$ is Fejér monotone with respect to C, it follows that $(\forall k \in \mathbb{N})\ \|x_{k+1} - \bar{x}\| \leq \|x_k - \bar{x}\|$. Hence the aforementioned reflection steps from K_2 onwards must be all with respect to hyperplanes taken from $H(\bar{x})$. Set $K := \max\{K_1, K_2\}$. It follows altogether that $(\forall k \geq K)\ 0 < \|x_{k+1} - \bar{x}\| = \|x_k - \bar{x}\|$. But this is absurd since $x_k \to \bar{x}$. ☐

Remark 13 Both Theorem 11 and Corollary 12 fail if $\operatorname{int} C = \varnothing$: indeed, consider two hyperslabs C_1 and C_2 in $\mathbb{R}^2$ such that C is a line (necessarily parallel to Z_1 and Z_2). If we start the iteration sufficiently close to this line, but not on this line, then $(x_n)_{n\in\mathbb{N}}$ will oscillate between two point outside C.

Furthermore, finite convergence may fail in Corollary 12 without the interiority assumption: indeed, consider a hyperslab in $\mathbb{R}^2$ which is intersected by a line at an angle strictly between 0 and $\pi/4$.

We now present our fundamental convergence result.

Theorem 14 (main result) *Suppose that $\bigcap_{i\in I_1} C_i \cap \bigcap_{i\in I_0} \operatorname{int} C_i \neq \varnothing$ and that the control is quasicyclic. Then the sequence $(x_k)_{k\in\mathbb{N}}$ generated by Algorithm 9 converges weakly to some point in C. The convergence is strong provided one of the following conditions holds:*

(i) *X is finite-dimensional.*
(ii) *I_1 is either empty or a singleton.*

Proof By Lemma 10, $(x_k)_{k\in\mathbb{N}}$ is Fejér monotone with respect to C. Take $y \in \bigcap_{i\in I_1} C_i \cap \bigcap_{i\in I_0} \operatorname{int} C_i$. Writing $\|x_0 - y\|^2 = \sum_{k\in\mathbb{N}} \|x_k - y\|^2 - \|x_{k+1} - y\|^2$, and recalling Fact 1 and Proposition 6, we deduce that $x_k - x_{k+1} \to 0$ and that $d_{C_{\mathrm{i}(n)}}(x_k) \to 0$. The quasicyclicality of the control now yields

$$\max \left\{ d_{C_i}(x_k) \mid i \in I \right\} \to 0. \tag{21}$$

Therefore, every weak cluster point of $(x_k)_{k\in\mathbb{N}}$ lies in C. By Fact 8(ii), there exists $\bar{x} \in X$ such that

$$x_k \rightharpoonup \bar{x} \in C \tag{22}$$

as announced.

Let us turn to strong convergence. Item (i) is obvious since strong and weak convergence coincide in finite-dimensional Hilbert space.

Now consider (ii). If $I_1 = \varnothing$, then strong convergence follows from Theorem 11. Thus assume that I_1 is a singleton. By, e.g., [2, Theorem 5.14],

$$d_C(x_k) \to 0. \tag{23}$$

Hence, using Fact 8(iii), we conclude that $x_k \to \bar{x}$. □

Remark 15 Our sufficient conditions for strong convergence are sharp: indeed, Hundal's example [20] shows that strong convergence may fail if (i) X is infinite-dimensional and (ii) I_1 contains more than one element.

5 CycIP and the Road Design Problem

From now on, we assume that

$$X = \mathbb{R}^n, \tag{24}$$

and that we are given n breakpoints

$$t = (t_1, \dots, t_n) \in X \quad \text{such that } t_1 < \cdots < t_n. \tag{25}$$

The problem is to

$$\text{find } x = (x_1, \dots, x_n) \in X \tag{26}$$

such that all of the following constraints are satisfied:

- **interpolation constraints**: For a subset J of $\{1, \dots, n\}$, we have $x_j = y_j$, where $y \in \mathbb{R}^J$ is given.
- **slope constraints**: each slope $s_j := (x_{j+1} - x_j)/(t_{j+1} - t_j)$ satisfies $|s_j| \le \sigma_j$, where $j \in \{1, \dots, n-1\}$ and $\sigma \in \mathbb{R}^{n-1}_{++}$ is given.

- **curvature constraints**: $\gamma_j \geq s_{j+1} - s_j \geq \delta_j$, for every $j \in \{1, \dots, n-2\}$, and for given γ and δ in $\mathbb{R}^{n-2}$.

This problem is of fundamental interest in road design; see [5] for further details.

By grouping the constraints appropriately, this feasibility problem can be reformulated as the following convex feasibility problem involving six sets:

$$\text{find } x \in C = \bigcap_{i \in I} C_i = C_1 \cap \cdots \cap C_6, \tag{27}$$

where $I := \{1, \dots, 6\}$; see [5, Sect. 2] for details. These sets have additional structure: C_1 is an affine subspace incorporating the interpolation constraints, C_2 and C_3 are both intersections of hyperslabs with normal vectors having disjoint support modeling the slope constraints, and the curvature constraints are similarly incorporated through C_4, C_5, and C_6. All these sets have explicit and easy-to-implement (regular and intrepid) projection formulas. Since only the set C_1 has no interior, we set $T_1 = P_{C_1}$. For every $i \in \{2, \dots, 6\}$, we set Q_{C_i}. (If we set $T_i = P_{C_i}$, we get the classical method of cyclic projections.) This gives rise to the algorithm, which we call the method of **cyclic intrepid projections (CycIP)**.

We thus obtain the following consequence of our main result (Theorem 14):

Corollary 16 (strict feasibility) *Suppose that* $C_1 \cap \bigcap_{2 \leq i \leq 6} \operatorname{int} C_i \neq \varnothing$, *i.e., there exists a* strictly feasible *solution to* (27), *i.e., it satisfies the interpolation constraints, and it satisfies the slope and curvature constraint inequalities* strictly. *Then the sequence generated by CycIP converges to a solution of* (27).

In [5], which contains a comprehensive comparison of various algorithms for solving (27), CycIP was found to be the best overall algorithm. However, due to the interpolation constraint set C_1, which has *empty interior*, the convergence of CycIP is not guaranteed by Theorem 11 or convergence results derived earlier. Corollary 16 is the first *rigorous justification* of CycIP in the setting of road design.

Remark 17 (nonconvex minimum-slope constraints) In [5] we also considered a variant of the slope constraints with an imposed minimal strictly positive slope. This is a setting of significant interest in road design as zero slopes are not favoured because of, e.g., drainage problems. The accordingly modified sets C_2 and C_3 are in that case *nonconvex*; however, explicit formulas for (regular and intrepid) projections are still available. The application of CycIP must then be regarded as a heuristic as there is no accompanying body of convergence results.

In the following section, we will investigate the numerical performance of CycIP and compare it to a linear programming solver.

6 Numerical Results

We generate 87 random test problems[3] as in [5]. The size of each problem, n, satisfies $341 \leq n \leq 2735$. These problems are significantly larger than those of [5] because we wish to compare execution time rather than number of iterations.

Consider the following two measures of infeasibility:

$$(\forall x \in X) \quad d_2(x) := \sqrt{\sum_{i=1}^{6} d_{C_i}^2(x)} \quad \text{and} \quad d_\infty(x) := \max_{i \in I} \|x - P_{C_i} x\|_\infty, \tag{28}$$

where $\|x\|_\infty$ is the max-norm[4] of x. Note that $d_2(x) = d_\infty(x) = 0$ if and only if $x \in C$. Set $\varepsilon := 5 \cdot 10^{-4}$, and let $(x_k)_{k \in \mathbb{N}}$ be a sequence generated by CycIP. We employ either $d_2(x_k) < \varepsilon$ or $d_\infty(x_k) < \varepsilon$ as stopping criterion.

Let $\mathscr{P}$ be the set of test problems, and let $\mathscr{A}$ be the set of algorithms. Let $(x_k^{(a,p)})_{k \in \mathbb{N}}$ be the sequence generated by algorithm $a \in \mathscr{A}$ applied to the problem $p \in \mathscr{P}$. To compare the performance of the algorithms, we use *performance profiles*[5]: for every $a \in \mathscr{A}$ and for every $p \in \mathscr{P}$, we set

$$r_{a,p} := \frac{\tau_{a,p}}{\min\left\{\tau_{a',p} \mid a' \in \mathscr{A}\right\}} \geq 1, \tag{29}$$

where $\tau_{a,p} \in \{1, 2, \ldots, \tau_{\max}\}$ is the time that a requires to solve p and $\tau_{\max}$ is the maximum time allotted for all algorithms. If $r_{a,p} = 1$, then a uses the least amount of time to solve problem p. If $r_{a,p} > 1$, then a requires $r_{a,p}$ times more time for p than the algorithm that uses the least amount of time for p. For each algorithm $a \in \mathscr{A}$, we plot the function

$$\rho_a \colon \mathbb{R}_+ \to [0,1] \colon \kappa \mapsto \frac{\operatorname{card}\left\{p \in \mathscr{P} \mid \log_2(r_{a,p}) \leq \kappa\right\}}{\operatorname{card} \mathscr{P}}, \tag{30}$$

where "card" denotes the cardinality of a set. Thus, $\rho_a(\kappa)$ is the percentage of problems that algorithm a solves within factor 2^κ of the best algorithms. Therefore, an algorithm $a \in \mathscr{A}$ is "fast" if $\rho_a(\kappa)$ is large for κ small; and a is "robust" if $\rho_a(\kappa)$ is large for κ large.

To compare CycIP with a linear programming solver, we model (27) as the constraints of a *Linear Program (LP)*. As objective function, we use $x \mapsto \|x - x_0\|_1$, where $\|x\|_1$ denotes the 1-norm[6] of x. As LP solver, we use Gurobi 5.5.0, a state-

---

[3] In [5], the authors compared CycIP with a Swiss Army Knife. The Wenger Swiss Army Knife version XXL, listed in the Guinness Book of World Records as the world's most multi-functional penknife, contains 87 tools.

[4] Recall that if $x = (\xi_1, \ldots, \xi_n) \in X$, then $\|x\|_\infty = \max\{|\xi_1|, \ldots, |\xi_n|\}$.

[5] For further information on performance profiles, we refer the reader to [14].

[6] Recall that if $x = (\xi_1, \ldots, \xi_n) \in X$, then $\|x\|_1 = |\xi_1| + \cdots + |\xi_n|$.

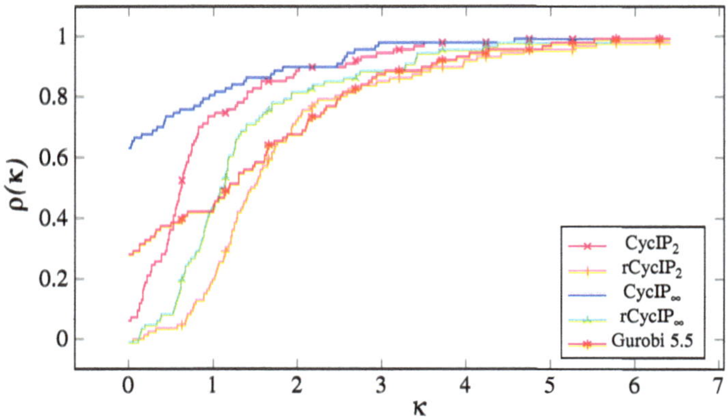

Fig. 1 Performance profile for convex problems

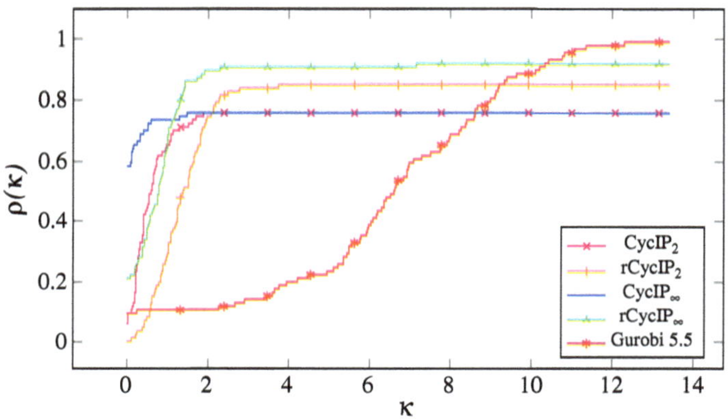

Fig. 2 Performance profile for nonconvex problems

of-the-art mathematical programming solver [17]. CycIP was implemented with the C++ programming language. We run the experiments on a Linux computer with a 2.4 GHz Intel® Xeon® E5620 CPU and 24 GB of RAM. As time measurement, we use *wall-clock time.*[7] We limit the solving time to $\tau_{\max} := 150$ s for each problem and algorithm.

Figure 1 shows the performance profile for the convex case. Here, CycIP uses a cyclic control with period 6 and the randomized **rCycIP** variant has quasicyclic control satisfying $(\forall k \in \mathbb{N})\ \{\mathrm{i}(6k), \mathrm{i}(6k+1), \dots, \mathrm{i}(6k+5)\} = \{1, 2, \dots, 6\}$, i.e., for every $k \in \mathbb{N}$, $(\mathrm{i}(6k), \mathrm{i}(6k+1), \dots, \mathrm{i}(6k+5))$ is a randomly generated permutation of

[7] To allow for a more fair comparison, we included in wall-clock time only the time required for running the solver's software itself (and not the time for loading the problem data or for setting up the solver's parameters).

$(1, 2, \ldots, 6)$. Depending on whether d_2 or d_∞ was used as the infeasibility measure, we write CycIP_2 and CycIP_∞, respectively, and similarly for rCycIP.

For the nonconvex case, shown in Fig. 2, we included a minimum slope constraint as mentioned in Remark 17. For the LP solver, the resulting model becomes a *Mixed Integer Linear Program.*

We infer from the figures that for convex problems, CycIP_∞ solves the test problems quickly and robustly. For nonconvex problems, CycIP_∞ is still fast, but less robust than the slower randomized variant rCycP_∞. Gurobi is the slowest—but also the most robust—algorithm.

7 Conclusion

In this work, we proved that the method of cyclic intrepid projections converges to a feasible solution under quasicyclic control and an interiority assumption. Specialized to a problem arising in road design, this leads to the first rigorous proof of convergence of CycIP. Numerical results show that CycIP is competitive compared to a commercial optimization solver, especially in terms of speed. Randomization strategies increase robustness in case of nonconvex problems for which there is no underlying convergence theory. Future work will focus on obtaining theoretical convergence results and on experimenting with other algorithms to increase robustness in the nonconvex setting.

Acknowledgments The authors thank the referee for very careful reading and constructive comments, Dr. Ramon Lawrence for the opportunity to run the numerical experiments on his server, and Scott Fazackerley and Wade Klaver for technical help. HHB also thanks Dr. Masato Wakayama and the Institute of Mathematics for Industry, Kyushu University, Fukuoka, Japan for their hospitality —some of this research benefited from the extremely stimulating environment during the "Math-for-Industry 2013" forum. HHB was partially supported by the Natural Sciences and Engineering Research Council of Canada (Discovery Grant and Accelerator Supplement) and by the Canada Research Chair Program.

References

1. Baillon, J.B., Bruck, R.E., Reich, S.: On the asymptotic behavior of nonexpansive mappings and semigroups in Banach spaces. Houston J. Math. **4**, 1–9 (1978)
2. Bauschke, H.H., Borwein, J.M.: On projection algorithms for solving convex feasibility problems. SIAM Rev. **38**, 367–426 (1996)
3. Bauschke, H.H., Combettes, P.L.: A weak-to-strong convergence principle for Fejér-monotone methods in Hilbert spaces. Math. Oper. Res. **26**, 248–264 (2001)
4. Bauschke, H.H., Combettes, P.L.: Convex analysis and monotone operator theory in Hilbert spaces. Springer, New York (2011)
5. Bauschke, H.H., Koch, V.R.: Projection methods: swiss army knives for solving feasibility and best approximation problems with halfspaces. Contemporary Mathematics (in press)
6. Borwein, J.M., Vanderwerff, J.D.: Convex functions. Cambridge University Press, Cambridge (2010)

7. Boyd, S., Vandenberghe, L.: Convex optimization. Cambridge University Press, Cambridge (2004)
8. Cegielski, A.: Iterative methods for fixed point problems in Hilbert spaces. Springer, New York (2012)
9. Censor, Y., Zenios, S.A.: Parallel Optimization. Oxford University Press, Oxford (1997)
10. Chen, W., Herman, G.T.: Efficient controls for finitely convergent sequential al gorithms. ACM Trans. Math. Software **37**, 14 (2010)
11. Combettes, P.L.: Hilbertian convex feasibility problems: convergence of projection methods. Appl. Math. Optim. **35**, 311–330 (1997)
12. Combettes, P.L.: Inherently parallel algorithms in feasibility and optimization and their applications. In: Butnariu, D., Censor, Y., Reich, S. (eds.) Quasi-Fejérian analysis of some optimization algorithms, pp. 115–152. Elsevier, New York (2001)
13. Combettes, P.L.: Solving monotone inclusions via compositions of nonexpansive averaged operators. Optimization **53**, 475–504 (2004)
14. Dolan, E.D., Moré, J.J.: Benchmarking optimization software with performance profiles. Math. Program. (Ser. A) **91**, 201–213 (2002)
15. Goebel, K., Kirk, W.A.: Topics in metric fixed point theory. Cambridge University Press, Cambridge (1990)
16. Goebel, K., Reich, S.: Uniform convexity, hyperbolic geometry, and nonexpansive mappings. Marcel Dekker, New York (1984)
17. Gurobi optimization Inc: gurobi optimizer reference manual. http://www.gurobi.com, (2013)
18. Herman, G.T.: A relaxation method for reconstructing objects from noisy x-rays. Math. Program. **8**, 1–19 (1975)
19. Herman, G.T.: Fundamentals of computerized tomography: image reconstruction from projections, 2nd edn. Springer, New York (2009)
20. Hundal, H.: An alternating projection that does not converge in norm. Nonlinear Anal. **57**, 35–61 (2004)
21. Rockafellar, R.T., Wets, R.J.-B.: Variational analysis. Springer-Verlag, New York (1998)
22. Slavakis, K., Yamada, I.: The adaptive projected subgradient method constrained by families of quasi-nonexpansive mappings and its applicaton to online learning. SIAM J. Optim. **23**, 126–152 (2013)
23. Yamada, I., Ogura, N.: Hybrid steepest descent method for variational inequality problem over the fixed point set of certain quasi-nonexpansive mappings. Numer. Funct. Anal. Optim. **25**, 619–655 (2004)
24. Yamada, I., Yukawa, M., Yamagishi, M.: Minimizing the moreau envelope of nonsmooth convex functions over the fixed point set of certain quasi-nonexpansive mappings. In: Bauschke, H.H., Burachik, R.S., Combettes, P.L., Elser, V., Luke, D.R., Wolkowicz, H. (eds.) Fixed-point algorithms for inverse problems in science and engineering, springer optimization and its applications, pp. 345–390. Springer, New York (2011)

Analytical Optimization of Local Quantum Operation and Classical Communication

Go Kato

Abstract This manuscript treats the situation that a quantum state is prepared as an input and is manipulated by local quantum operations and classical communications (LOCC). We analytically optimize the LOCC. Optimizations with respect to LOCC frequently appear in theories of quantum communication and computation. However, this optimization problem is difficult, since we have to fix infinite number of parameters in order to identify the LOCC. In fact, there is no recipe for optimizing LOCC except for the case that the input state is a pure state, which is a special case of a quantum state. In this manuscript, we optimize the LOCC for non-pure input states. As a result, this analytical method enables us to solve a fundamental problem in the quantum communication theory. We think that our analysis will help us to make more general recipe to optimize LOCC.

Keywords Quantum communication · Analytical optimization · Entanglement distillation · Local operations and classical communication · Post-selection

1 Introduction

Recently, development of quantum technology encourages the construction of practical systems which use quantum physical properties. One promising usage is the quantum communication, which is the communication by using quantum systems, e.g. Quantum key distribution (QKD). In the near future, the quantum communication will be used in many occasions.

The quantum communication can provide functions which can not be realized by using current communication (*classical communication*). In fact, such functions

---

G. Kato (✉)
NTT Communication Science Laboratories, 3-1, Morinosato Wakamiya,
Kanagawa Prefecture Atsugi-shi, Japan
e-mail: kato.go@lab.ntt.co.jp

M. Wakayama et al. (eds.), *The Impact of Applications on Mathematics*,
Mathematics for Industry 1, DOI: 10.1007/978-4-431-54907-9_15,

deeply relate to *entanglement* [5], which is a certain correlation in a quantum state. Strictly speaking, we can implement such a function by a physical channel if and only if the physical channel has an ability to generate a quantum state which has entanglement. The quantity of the entanglement (*entanglement monotone* [5]) for a given quantum state can be evaluated [11]. As a result, we can quantitatively evaluate the quality of a physical channel as an expedient of a quantum communication.

In realistic situations a quantum communication is implemented by a physical channel. In general, there exists a non-trivial upper bound of the entanglement monotone in a state that we can make by using such a physical channel. Such upper bound is an important index in order to design quantum communication protocols. Therefore, evaluation of the quality of the standard physical channel, i.e., a weak laser pulse with optical fiber, is a fundamental problem. Furthermore, generating a quantum state which has entanglement (entanglement generating protocol in short) is thought to be a fundamental task for the quantum communication. Various pragmatic protocols which implement this fundamental task are proposed [1, 3, 7, 10].

We must mention about the evaluation of the quality of a physical channel more precisely. It is well known that, even with local quantum operation and classical communication (LOCC), i.e. without quantum communication, we can increase the entanglement monotone probabilistically by post selection. This means that, there is a tradeoff between the probability P_s for the selected events and the average of the entanglement monotone $\bar{E}$ for those events [11]. Therefore, in order to evaluate the quality of a physical channel as an expedient of quantum communication, the function $\bar{E}^{(\max)}(P_s)$ is an appropriate index of the quality, where $\bar{E}^{(\max)}(P_s)$ is the maximum of $\bar{E}$ with respect to LOCC and an input state of the physical channel under the constraint that the success probability P_s is fixed. However, there is no analysis to derive $\bar{E}^{(\max)}(P_s)$ itself, though previous papers [3] show non-trivial lower bound of $\bar{E}^{(\max)}(P_s)$ for the standard physical channel. The difficulty in the derivation of $\bar{E}^{(\max)}(P_s)$ is the maximization with respect to LOCC. In order to identify LOCC, we have to fix infinite number of parameters. Therefore, for the maximization, we have to optimize such infinite number of parameters. In fact, a general recipe to treat optimization of LOCC exists only if the input state is a *pure state* [6], which is a special case of quantum states. However, most of the states transmitted by a physical channel are not pure states in general. Therefore, we have to extend analytical techniques of the optimization for non-pure input state in order to derive $\bar{E}^{(\max)}(P_s)$.

In this manuscript, we treat the most difficult analysis to derive $\bar{E}^{(\max)}(P_s)$. That is, for a given quantum state, we maximize the $\bar{E}$ with respect to LOCC for a fixed success probability P_s. Note that, in order to evaluate the quality of a physical channel, we have to execute optimization with respect to the input state too, which is not so difficult. As a result, the results in this manuscript enable us to derive the $\bar{E}^{(\max)}(P_s)$ for the cases of the standard physical channels [2].

It is fair to mention that the entanglement monotone is not uniquely defined [4, 9]. In this manuscript, we suppose extra conditions for the entanglement monotone which is not necessarily demanded. Though not all the entanglement monotone satisfy the extra conditions, many of them satisfy these conditions.

This manuscript is written as follows. In the next section, we mathematically define pure states, LOCC, etc. which are used in order to express our results, and we also write down the assumptions with respect to the entanglement monotone. In the third section, our main results are written. In 4th and 5th sections, we show abstracts of the proof of our results. And, the last section is devoted to the summary.

2 Notation

In this manuscript, we fix the parameters d and n as an positive integers and Ω as a finite set. b indicates 0 or 1. m, k, k', k'' and d' indicate integers. $d_{m\in\{0,1,\ldots\}}$, $k_{m\in\{0,1,\ldots\}}$ and $k'_{m\in\{0,1,\ldots\}}$ also indicate integers. j, j' and $j_{m\in\{1,2,\ldots\}}$ indicate elements in Ω. $\mathbf{j}_{m\in\{0,1,2,\ldots\}}$ indicates a vector $(j_1, j_2, \ldots, j_m)$, especially $\mathbf{j} := \mathbf{j_n}$. Furthermore, we use the following notations.

A bipartite quantum pure state (pure state in short) $|\phi\rangle$: a vector in a complex linear space $\mathbb{C}^{\otimes d} \otimes \mathbb{C}^{\otimes d}$ whose length is 1, i.e. $|||\phi\rangle|| = 1$.

$\langle\phi|$: The vector conjugate to $|\phi\rangle$.
Id:The identity operator on $\mathbb{C}^{\otimes d}$.
$|e_k\rangle_k$: an orthonormal basis of $\mathbb{C}^{\otimes d}$.

A bipartite quantum mixed state (mixed state in short): a Hermitian positive operator on $\mathbb{C}^{\otimes d_0} \otimes \mathbb{C}^{\otimes d_1}$ whose trace is 1.

A partial trace Tr_b: a map from $Hom(\mathbb{C}^{\otimes d} \otimes \mathbb{C}^{\otimes d})$ into $Hom(\mathbb{C}^{\otimes d})$ such that,

$$
\begin{aligned}
&\mathrm{Tr}_b\left(\sum_{k_0,k'_0,k_1,k'_1} \alpha_{k_0,k'_0,k_1,k'_1} |e_{k_0}\rangle\langle e_{k'_0}| \otimes |e_{k_1}\rangle\langle e_{k'_1}|\right) \\
&:= \sum_{k_0,k'_0,k_1,k'_1} \delta(k_b, k'_b)\alpha_{k_0,k'_0,k_1,k'_1} |e_{k_{1-b}}\rangle\langle e_{k'_{1-b}}|.
\end{aligned}
$$

A quantum operator from $\mathbb{C}^{\otimes d_0}$ to $\mathbb{C}^{\otimes d_1}$: a set of linear operator M_ω from$\mathbb{C}^{\otimes d_0}$ to $\mathbb{C}^{\otimes d_1}$ which is identified by an element ω in a certain finite set Ω' and satisfies $0 < \sum_{\omega\in\Omega'} M^\dagger_\omega M_\omega \leq Id_{d_0}$, where $M^\dagger_\omega$ is the Hermitian conjugate operator of M_ω.

LOCC: A list of quantum operators $\{M_{j'|\mathbf{j}_m}\}_{j'}$ from $\mathbb{C}^{d_m}$ into $\mathbb{C}^{d_{m+2}}$ which are identified by a vector $\mathbf{j}_m$ for $0 \leq m < n$, where d_0 and d_1 is equal to d.

A random unitary channel Λ_U: A map from a mixed state into a mixed state defined by a probability distribution $\{q_k\}_k$ and a set of unitary operators $\{U_k\}_k$ on $\mathbb{C}^{\otimes d}$ such that $\rho \mapsto \sum_k q_k(Id \otimes U_k)\rho(Id \otimes U^\dagger_k)$ where $\rho \in Hom(\mathbb{C}^{\otimes d} \otimes \mathbb{C}^{\otimes d})$.

Entanglement monotone E: a convex function from a mixed state into a real number which satisfies the following relations. For any mixed state ρ and any quantum operator $\{M_\omega\}_{\omega\in\Omega'}$,

$$E(\rho) \geq \sum_{\omega \in \Omega'} p_\omega E(\rho_\omega) \tag{1}$$

where $p_\omega := \mathrm{Tr}\tilde{\rho}_\omega$, $\tilde{\rho}_\omega := (M_\omega \otimes Id)\rho(M_\omega^\dagger \otimes Id)$, $\rho_\omega := p_\omega^{-1}\tilde{\rho}_\omega$, if $p_\omega \neq 0$, and ρ_ω is equal to an arbitrary mixed state if $p_\omega = 0$. This is standard constraint for E. In this manuscript, we assume the following extra condition: There is a convex and monotonically increasing function $G : \mathbb{R} \to \mathbb{R}$ and a map $C : Hom(\mathbb{C}^{\otimes d_0} \otimes \mathbb{C}^{\otimes d_1}) \to \mathbb{R}$ such that $E(\rho') = G(C(\rho'))$ and

$$C(M_0 \otimes M_1 \rho M_0^\dagger \otimes M_1^\dagger) = (\prod_b det(M_b^\dagger M_b)^{1/d})C(\rho) \tag{2}$$

for any mixed states ρ' on $\mathbb{C}^{\otimes d_0} \otimes \mathbb{C}^{\otimes d_1}$ and ρ on $\mathbb{C}^{\otimes d} \otimes \mathbb{C}^{\otimes d}$, any linear operator M_b from $\mathbb{C}^{\otimes d}$ into $\mathbb{C}^{\otimes d_b}$.

3 Main Results

Theorem 1 *If a mixed state ρ can be expressed by the form $\Lambda_U(|\phi\rangle\langle\phi|)$ for an appropriate pure state $|\phi\rangle$ and a random unitary channel Λ_U, there exists a positive operator $M \leq Id$ such that the "success probability" P_s is equal to $\mathrm{Tr}\tilde{\rho}$ and the "average of entanglement monotone" $\bar{E}$ is bounded by $E(P_s^{-1}\tilde{\rho})$ where $\tilde{\rho} := (M \otimes Id)\rho(M^\dagger \otimes Id)$, for any LOCC Λ and a set of success events $\mathscr{S} \subset \Omega^{\otimes n}$.*

The success probability P_s and the average of entanglement monotone $\bar{E}$ are defined by Λ, $\mathscr{S}$ and ρ as follows,

$$P_s := \sum_{\mathbf{j} \in \mathscr{S}} p_{\mathbf{j}}, \quad \bar{E} := P_s^{-1} \sum_{\mathbf{j} \in \mathscr{S}} p_{\mathbf{j}} E(\rho_{\mathbf{j}}) \tag{3}$$

$$p_{\mathbf{j}} := \mathrm{Tr}\tilde{\rho}_{\mathbf{j}}, \quad \rho_{\mathbf{j}} := p_{\mathbf{j}}^{-1}\tilde{\rho}_{\mathbf{j}}, \quad \tilde{\rho}_{\mathbf{j}} := M_{\mathbf{j}}\rho M_{\mathbf{j}}^\dagger \tag{4}$$

$$M_{\mathbf{j}_m} := \left(\prod_{m'=0}^{\lfloor (m-1)/2 \rfloor} M_{j_{2m'+1}|\mathbf{j}_{2m'}} \right) \otimes \left(\prod_{m'=1}^{\lfloor m/2 \rfloor} M_{j_{2m'}|\mathbf{j}_{2m'-1}} \right), \tag{5}$$

where $M_{j_{m+1}|\mathbf{j}_m}$ is an operator identified by Λ. Note that, if $p_{\mathbf{j}}$ is equal to 0, $\rho_{\mathbf{j}}$ is defined to be an arbitrary mixed state.

Lemma 1 *If $\bar{E}$, P_S are defined by Λ, $\mathscr{S}$ and $\rho = \Lambda_U(|\phi\rangle\langle\phi|)$ in the same way as (3),*

$$\max_{\Lambda, \mathscr{S} \subset \Omega^{\otimes n}} \bar{E} = G\left(\frac{(P_S - \sum_{j=k+1}^{d} \lambda_j^\downarrow)^{\frac{k}{d}}}{P_S k^{\frac{k}{d}} \prod_{j=1}^{k} \lambda_j^{\downarrow \frac{1}{d}}} C(\rho) \right) \tag{6}$$

holds under the constraint that P_s *and* $k \in \{1, \ldots, d\}$*are fixed so that*

$$k\lambda^{\downarrow}_{k+1} + \sum_{k'=k+1}^{d} \lambda^{\downarrow}_{k'} < P_s \leq (k-1)\lambda^{\downarrow}_{k} + \sum_{k'=k}^{d} \lambda^{\downarrow}_{k'}. \tag{7}$$

$\{\lambda^{\downarrow}_{k'}\}_{k'\in\{1,\ldots,d\}}$ *is a set of eigenvalues for* $\mathrm{Tr}_2\rho$ *such that* $\lambda^{\downarrow}_{j} \geq \lambda^{\downarrow}_{j+1}$.

Note that, for any $0 < P_s \leq 1$, there is a k which satisfies (7).

4 Proof of the Theorem

In this section, some characters written in the followings are used as fixed values. $|\phi\rangle$ is an arbitrary pure state on $\mathbb{C}^d \otimes \mathbb{C}^d$, Λ_U is a random unitary channel identified by $\{q_k\}_k$ and $\{U_k\}_k$, Λ is a LOCC identified by $\{\{M_{j'|\mathbf{j}_m}\}_{j'}|m < n\}$, and $\mathscr{S}$ is a subset of $\Omega^{\otimes n}$.

In order to prove the theorem, we explicitly give the matrix M.

For any linear operator M', there are an isometry V and an operator $\tilde{M}'$ such that $\tilde{M}' = \sqrt{M'^{\dagger}M'}$ and $V\tilde{M}' = M'$. By using this property recursively, we can define a POVM $\tilde{\Lambda} = \{\{\tilde{M}_{j'|\mathbf{j}_m}\}_{j'}|m < n\}$ so that $\tilde{M}_{j_m|\mathbf{j}_{m-1}} = \sqrt{V^{\dagger}_{\mathbf{j}_{m-2}} M^{\dagger}_{j_m|\mathbf{j}_{m-1}} M_{j_m|\mathbf{j}_{m-1}} V_{\mathbf{j}_{m-2}}}$ and $V_{\mathbf{j}_m}\tilde{M}_{j_m|\mathbf{j}_{m-1}} = M_{j_m|\mathbf{j}_{m-1}} V_{\mathbf{j}_{m-2}}$ where $V_{\mathbf{j}_{-1}}$ and $V_{\mathbf{j}_0}$ are defined to be Id_d. Note that, $\tilde{M}_{j'|\mathbf{j}_m}$ is a linear map on $\mathbb{C}^d$, i.e. a linear map from $\mathbb{C}^d$ into $\mathbb{C}^d$. From this definition and the property of the entanglement monotone (2), we can easily check that, P_S and $\bar{E}$ for the LOCC Λ are equivalent to those for the LOCC $\tilde{\Lambda}$ for the same input state ρ and the same set of success events $\mathscr{S}$. As a result, without loss of generality, we can suppose $M_{j_m|\mathbf{j}_{m-1}}$ is a linear map on $\mathbb{C}^d$ and we investigate such a restricted case in the following.

In order to make this manuscript self-contained, we explicitly explain the fundamental procedure *Schmidt decomposition* [8] used in the quantum information society. That is a decomposition $|\psi\rangle = \sum_k \sqrt{\lambda_k}|\mu^{(0)}_k\rangle \otimes |\mu^{(1)}_k\rangle$ of a given vector $|\psi\rangle$ in $\mathbb{C}^{\otimes d} \otimes \mathbb{C}^{\otimes d}$ where $\{\lambda_k\}_k$ is a set of positive numbers and $\{|\mu^{(b)}_k\rangle\}_k$ are orthonormal bases. Note that, from this expression, we know that $\{\lambda_k\}_k$ is equal to the set of eigenvalues for $\mathrm{Tr}_b|\psi\rangle\langle\psi|$.

A proof of the existence of the decomposition is as follows. We write the vector as $|\psi\rangle = \sum_{k,k'} \alpha_{k,k'}|e_k\rangle \otimes |e_{k'}\rangle$. We define $\{\lambda_k\}_k$ as a set of all the eigenvalues of $\mathrm{Tr}_1|\psi\rangle\langle\psi|$, and we can define the set of orthonormal basis $\{|\mu^{(0)}_k\rangle\}_k$ such that $|\mu^{(0)}_k\rangle$ is an eigenvector whose eigenvalue is λ_k since $\mathrm{Tr}_1|\psi\rangle\langle\psi|$ is a Hermitian operator. Because $\left\langle\tilde{\mu}^{(1)}_k|\tilde{\mu}^{(1)}_{k'}\right\rangle = \lambda_k\delta(k,k')$ where $|\tilde{\mu}^{(1)}_k\rangle := \sum_{k',k''} \alpha_{k',k''}\left\langle\mu^{(0)}_k|e_{k'}\right\rangle|e_{k''}\rangle$, we can define an orthonormal basis $\{|\mu^{(1)}_k\rangle\}_k$ such that $\sqrt{\lambda_k}|\mu^{(1)}_k\rangle = |\tilde{\mu}^{(1)}_k\rangle$. These bases give the Schmidt decomposition, i.e. $\alpha_{k',k''} = \sum_k \sqrt{\lambda_k}\left\langle e_{k'}|\mu^{(0)}_k\right\rangle\left\langle e_{k''}|\mu^{(1)}_k\right\rangle$, since

$\sum_{k'} \alpha_{k',k''} \left\langle \mu_k^{(0)} | e_{k'} \right\rangle = \sum_k \sqrt{\lambda_k} \left\langle e_{k''} | \mu_k^{(1)} \right\rangle$ and $\sum_k \left\langle e_{k'} | \mu_k^{(1)} \right\rangle \left\langle \mu_k^{(1)} | e_{k''} \right\rangle = \delta(k', k'')$ hold by definition.

The Schmidt decomposition gives a simple property written as the *proposition* 1 in [6]. That is, for any vector $|\psi\rangle$ in $\mathbb{C}^{\otimes d} \otimes \mathbb{C}^{\otimes d}$ and any quantum operator $\{M_j^{(1)}\}_j$ on $\mathbb{C}^d$, there are unitary operator $V_j^{(b)}$, and a quantum operator $\{M_j^{(0)}\}_j$ such that $V_j^{(0)} \otimes M_j^{(1)} |\psi\rangle = M_j^{(0)} \otimes V_j^{(1)} |\psi\rangle$ and $\det(M_j^{(b)\dagger} M_j^{(b)})$ does not depend on b.

A sketch of a proof is as follows. The Schmidt decomposition of $|\psi\rangle$ gives a set of real number $\{\lambda_k\}_k$ and orthonormal bases $\{|\mu_k^{(b)}\rangle\}_k$. We define $M_j^{(0)} := W^\dagger M_j^{(1)} W$ where $W := \sum_k |\mu_k^{(0)}\rangle\langle\mu_k^{(1)}|$ is a unitary operator . From the definition, $\{M_j^{(0)}\}_j$ is a quantum operator and $\det(M_j^{(b)\dagger} M_j^{(b)})$ does not depend on b. We can check that the set of eigenvalues $\{\lambda'_k\}_k$ for $\sum_k \lambda_k M_j^{(b)} |\mu_k^{(b)}\rangle\langle\mu_k^{(b)}| M_j^{(b)\dagger}$ does not depend on b. Therefore, the Schmidt decompositions of $Id \otimes M_j^{(1)} |\psi\rangle$ and $M_j^{(0)} \otimes Id |\psi\rangle$ give orthonormal bases $\{|\mu_k^{(b,b')}\rangle\}_k$ such that the two vectors are equal to $\sum_k \sqrt{\lambda'_k} |\mu_k^{(0,b')}\rangle \otimes |\mu_k^{(1,b')}\rangle$ for $b' = 0$ and 1 respectively. These bases give unitary operators $V_j^{(b)} := \sum_k |\mu_k^{(b,1-b)}\rangle\langle\mu_k^{(b,b)}|$. We can easily check that these variables satisfy the conditions in the proposition 1.

By using the proposition 1, we can give a quantum operator $\{M'_{j_{2m}|\mathbf{j}_{2m-1},k}\}_{j_{2m}}$ and unitary operators $V_{\mathbf{j}_{2m},k}$ which satisfy

$$Id \otimes M_{j_{2m}|\mathbf{j}_{2m-1}} |\phi_{\mathbf{j}_{2m-1}|k}\rangle = M'_{j_{2m}|\mathbf{j}_{2m-1},k} \otimes V_{\mathbf{j}_{2m},k} |\phi_{\mathbf{j}_{2m-1}|k}\rangle, \tag{8}$$

and $\det M_{j_{2m}|\mathbf{j}_{2m-1}} = \det M'_{j_{2m}|\mathbf{j}_{2m-1},k}$ where $|\phi_{\mathbf{j}_m|k}\rangle := M_{\mathbf{j}_m}(Id_d \otimes U_k)|\phi\rangle$. By using these notations, we can define a quantum operator $\{M_{\mathbf{j},k} := \prod_{m=1}^n M'_{j_m|\mathbf{j}_{m-1},k}\}_{\mathbf{j}\in\Omega^{\otimes n}}$ where $M'_{j_{2m+1}|\mathbf{j}_{2m},k} := M_{j_{2n+1}|\mathbf{j}_{2m}}$. Note that, the definition of $M_{\mathbf{j},k}$ gives the relation

$$\prod_{m=1}^{n} \det(M^\dagger_{j_m|\mathbf{j}_{m-1}} M_{j_m|\mathbf{j}_{m-1}}) = \det(M^\dagger_{\mathbf{j},k} M_{\mathbf{j},k}). \tag{9}$$

Now, we give the operator M explicitly as $M := \sqrt{\sum_k q_k \sum_{\mathbf{j}\in\mathscr{S}} M^\dagger_{\mathbf{j},k} M_{\mathbf{j},k}}$. The rest of works in this section is to prove the property expressed in the theory 1.

We can check $0 < M \leq Id$, because $\{q_k\}_k$ is a probability distribution, $\{M_{\mathbf{j},k}\}_\mathbf{j}$ is a quantum operator and $\mathscr{S}$ is a subset of $\Omega^{\otimes n}$.

Since Λ_U affects only on the second space, $|||\phi_{\mathbf{j}|k}\rangle||^2 = \mathrm{Tr}\tilde{\rho}_\mathbf{j}^{(k)} =: p_\mathbf{j}^{(k)}$ holds where $\tilde{\rho}_\mathbf{j}^{(k)} := (M_{\mathbf{j},k} \otimes Id)\rho(M^\dagger_{\mathbf{j},k} \otimes Id)$. This relation gives

$$(p_\mathbf{j} =)\mathrm{Tr}\rho_\mathbf{j} = \sum_k q_k \mathrm{Tr}\tilde{\rho}_\mathbf{j}^{(k)} (= \sum_k q_k p_\mathbf{j}^{(k)}) \tag{10}$$

where we use the relation $\rho_{\mathbf{j}} = \sum_k q_k |\phi_{\mathbf{j}|k}\rangle\langle\phi_{\mathbf{j}|k}|$ which is given by definition. As a result, by considering the definitions of $\tilde{\rho}_{\mathbf{j}}^{(k)}$, $\tilde{\rho}$ and M, we can show that the success probability P_s is equal to $\sum_{\mathbf{j}\in\mathscr{S}} \mathrm{Tr}\rho_{\mathbf{j}} = \sum_k q_k \sum_{\mathbf{j}\in\mathscr{S}} \mathrm{Tr}\tilde{\rho}_{\mathbf{j}}^{(k)} = \mathrm{Tr}\tilde{\rho}$.

Next, we define a Hermitian operator $\bar{M}$ such that M and $\bar{M}$ have the same span and $\bar{M}M$ is a projection P_M onto the span. If we define $M'_{(\mathbf{j},k)} := q_k^{\frac{1}{2}} M_{\mathbf{j},k}\bar{M}$, $\{M'_\omega\}_{\omega\in(\mathscr{S}\otimes\{k\})}$ is a quantum operator. By using the quantum operator, we can define p_ω and $\rho_\omega$by $p_\omega := \mathrm{Tr}\tilde{\rho}_\omega$, $\rho_\omega := p_\omega^{-1}\tilde{\rho}_\omega$ and $\tilde{\rho}_\omega := (M'_\omega \otimes Id)P_s^{-1}\tilde{\rho}(M'^{\dagger}_\omega \otimes Id)$. Note that, If $p_\omega = 0$, ρ_ω is defined to be an arbitrary quantum mixed state.

By using these notations and relations, we give

$$\begin{aligned} P_s E(P_s^{-1}\tilde{\rho}) &\geq P_s \sum_{\omega\in(\mathscr{S},\{k\})} p_\omega E(\rho_\omega) = \sum_k q_k \sum_{\mathbf{j}\in\mathscr{S}} p_{\mathbf{j}}^{(k)} G\left(C\left(\rho_{(\mathbf{j},k)}\right)\right) \\ &\geq \sum_{\mathbf{j}\in\mathscr{S}'} p_{\mathbf{j}} G\left(\sum_k \frac{q_k p_{\mathbf{j}}^{(k)}}{p_{\mathbf{j}}} C\left(\rho_{(\mathbf{j},k)}\right)\right) = \sum_{\mathbf{j}\in\mathscr{S}} p_{\mathbf{j}} G\left(C\left(\rho_{\mathbf{j}}\right)\right) = P_s\,\bar{E}. \end{aligned} \quad (11)$$

where $\mathscr{S}'$ is a set of all elements $\mathbf{j}$ in $\mathscr{S}$ which satisfy $p_{\mathbf{j}} \neq 0$. The first inequality comes from Eq. (1) and the fact that $\{M'_\omega\}_\omega$ is a quantum operator. The second equality comes from the fact that $P_s p_{(\mathbf{j},k)} = q_k p_{\mathbf{j}}^{(k)}$ for $\mathbf{j} \in \mathscr{S}$, which is justified by the relation $M_{\mathbf{j},k}P_M = M_{\mathbf{j},k}$ and the definitions of $p_{\mathbf{j},k}$ and $p_{\mathbf{j}}^{(k)}$. The third inequality comes from the convexity of G and Eq. (10). The fourth equality comes from the following three relations: The first relation is $\alpha C\left(\tilde{\rho}_{\mathbf{j}}^{(k)}\right) = C\left(\alpha\tilde{\rho}_{\mathbf{j}}^{(k)}\right)$, which is justified by (2). The second one is $p_{\mathbf{j}}^{(k)}\rho_{(\mathbf{j},k)} = \tilde{\rho}_{\mathbf{j}}^{(k)}$. The last one is $C\left(\tilde{\rho}_{\mathbf{j}}^{(k)}\right) = C\left(\tilde{\rho}_{\mathbf{j}}\right)$, which is justified by the facts that (5), (9), $C\left(\tilde{\rho}_{\mathbf{j}}\right) = \det(M_{\mathbf{j}}^{\dagger}M_{\mathbf{j}})^{\frac{1}{d}} C\left(\rho\right)$ and $C\left(\tilde{\rho}_{\mathbf{j}}^{(k)}\right) = \det(M_{\mathbf{j},k}^{\dagger}M_{\mathbf{j},k})^{\frac{1}{d}} C\left(\rho\right)$, where the last two relations come from Eq. (2). The last equality of Eq. (11) comes from the definition of $\bar{E}$. The inequality (11) guarantees that $\bar{E}$ is bounded by $E(P_s^{-1}\tilde{\rho})$.

5 Proof of the Lemma

From the theorem, for a fixed $0 < P_s$,

$$\max_{\Lambda,\mathscr{S}\subset\Omega^{\otimes n}} \bar{E} = \max_{0<M\leq Id_d} E(P_s^{-1}\tilde{\rho}) = \max_{0<M\leq Id_d} G(P_s^{-1}(\det M)^{\frac{1}{d}} C(\rho))$$

under the constraint $P_s = \mathrm{Tr}\tilde{\rho}$ where $\tilde{\rho} := (M \otimes Id)\rho(M^{\dagger} \otimes Id)$ in the second and third values. Note that, If $0 < M \leq Id_d$ and $0 < P_s$, $\max_{\Lambda,\mathscr{S}\subset\Omega^{\otimes n}} \bar{E} \geq E(P_s^{-1}\tilde{\rho})$ is trivial since we can easily construct LOCC Λ' and a set of success event $\mathscr{S}'$ by using M such that the success probability and the average of entanglement are respectively

P_s and $E(P_s^{-1}\tilde{\rho})$ for the same initial state ρ. Since the G is a monotonically increasing function, the only thing we have to do is to show the relation

$$\max_{0<M\le Id_d} \det M = (P_s - \sum_{k'=k+1}^{d} \lambda_{k'}^{\downarrow})^k k^{-k} \prod_{k'=1}^{k} \lambda_{k'}^{\downarrow -1} \tag{12}$$

in case k satisfies (7) under the constraint that $\mathrm{Tr}(M\bar{\rho}) = P_s$ where $\bar{\rho} := \mathrm{Tr}_1\rho$.

By definition of $\lambda_j^{\downarrow}$, we can give an orthonormal basis $\{|\mu_j\rangle\}$ such that $\rho = \sum_j \lambda_j^{\downarrow}|\mu_j\rangle\langle\mu_j|$. If M is a matrix which satisfies the constraint and gives the maximum, $\sum_{\{\theta_j\}} p(\{\theta_j\})U_M M U_M^{\dagger}$ is also such a matrix, where $p(\{\theta_j\})$ is a probability distribution with respect to $\{\theta_j\}$ and $U_M := \sum_j e^{\theta_j i}|\mu_j\rangle\langle\mu_j|$. Therefore, we can restrict a positive operator M into diagonal positive operator under the basis $\{|\mu_j\rangle\}$. That means, the left hand side of (12) is equal to $\max_{0\le x_{k'}\le 1}\prod_{k'=1}^{d} x_{k'}$ under the modified constraint $\sum_{k'=1}^{d} x_{k'}\lambda_{k'}^{\downarrow} = P_s$. We define $\nu_0 := (P_s - \sum_{k''=k+1}^{d}\lambda_{k''}^{\downarrow})^{-1}k$ and $\nu_{k'} := \nu_0\lambda_{k'}^{\downarrow} - 1$ for $k' \neq 0$. From the assumption (7), $\nu_{k'}$ are non-positive numbers if $k' \ge k+1$. Therefore, when $f(\{x_{k'}\}_{k'}) := \sum_{k'=1}^{d}\log(x_{k'}) - \sum_{k'=k+1}^{d}\nu_{k'}(1 - x_{k'}) - \nu_0(\sum_{k'=1}^{d} x_{k'}\lambda_{k'}^{\downarrow} - P_s)$, the relation $\max_{0\le x_{k'}\le 1}\sum_{k'=1}^{d}\log(x_{k'}) \le \max_{0\le x_{k'}} f(\{x_{k'}\}_{k'})$ holds, if the maximization in the left hand side is executed under the modified constraints. Since $\frac{d}{dx_{k''}} f(\{x_{k'}\}_{k'})$ is monotonically decreasing function of $x_{k''} > 0$, we can easily evaluate maximum point of the function $f(\{x_{k'}\}_{k'})$. That is $x_{k'}^{(max)} = \nu_0^{-1}\lambda_{k'}^{\downarrow -1}$ in the case of $k' \le k$, and $x_{k'}^{(max)} = 1$ in the case of $k+1 \le k'$. We can easily check that if $x_{k'} = x_{k'}^{(max)}$, the relation $0 \le x_{k'} \le 1$ and the modified constraint is satisfied and $\sum_{j=1}^{d}\log(x_j^{(max)}) = f(\{x_j^{(max)}\}_j)$. Therefore, the left hand side of (12) is equal to $\exp(f(\{x_j^{(max)}\}_j))$ and we can easily check that this value is equal to the right hand side of (12).

6 Summary and Physical Meanings

In this manuscript, we analyze the optimization of the average of the entanglement monotone $\bar{E}$ with respect to LOCC for non-pure input states. In this analysis, we show that some properties of $\bar{E}$ guarantee that optimal point of LOCC is in the finite dimensional space. Therefore, the difficulty of the optimization for the infinite number of variables disappears.

Though the optimization with LOCC is needed in many investigations in the quantum communication and computation theory, the optimization is very difficult and there has been no recipe to treat the problem except for the case that the input state is a pure state. Our results suggest that some extension of the recipe will exist, and construction of such a general recipe is a future work.

Physical meanings of this analysis is as follows: In case of $d = 2$, our results enable us to evaluate $\bar{E}^{(max)}(P_s)$ for the standard physical channel [2] which have been deeply investigated before. Furthermore, our results give us some hints to make efficient design of entanglement generating protocols [2].

References

1. Azuma, K.: Phys. Rev. A **85**, 062309 (2012)
2. Azuma, K., Kato, G.: Phys. Rev. A **85**, 060303(R) (2012)
3. Childress, L., et al.: Phys. Rev. Lett. **96**, 070504 (2006)
4. Gour, G.: Phys. Rev. A **71**, 012318 (2005)
5. Horodecki, R., Horodecki, P., Horodecki, M.: K. Horodecki. Rev. Mod. Phys. **81**, 865 (2009)
6. Lo, H.K., Popescu, S.: Phys. Rev. A **63**, 022301 (2001)
7. Munro, W.J., et al.: Phys. Rev. Lett. **101**, 040502 (2008)
8. Nielsen, M.A., Chuang, I.L.: Quantum Computation and Quantum Information. Cambridge University Press, Cambridge (2000)
9. Streltsov, A., Kapermann, H.: D. Bru. New J. Phys. **12**, 123004 (2010)
10. van Look, P., et al.: Phys. Rev. Lett. **96**, 240501 (2006)
11. Vidal, E.: J. Mod. Opt. **47**, 355 (2000)

Cellular Networks with α-Ginibre Configurated Base Stations

Naoto Miyoshi and Tomoyuki Shirai

Abstract We consider a cellular network model with base stations configurated according to the α-Ginibre point process with $\alpha \in (0, 1]$, which is one of the determinantal point processes. In this model, we focus on the asymptotic behavior of the so-called coverage probability (or link success probability) as the threshold value tends to 0 and ∞, and discuss the Padé approximation of the coverage probability at 0 and the dependence on $\alpha \in (0, 1]$ of the asymptotic constant at ∞ both numerically and theoretically.

Keywords Cellular network · Ginibre point process · α-Ginibre point process · Determinantal point process · SINR · Coverage probability · Padé approximation · Stochastic geometry

1 Introduction

Recent studies on cellular networks are based on the theory of stochastic geometry and random geometric graphs by assuming that the base stations of a network are randomly distributed in space (cf. [2, 3, 5, 6]). Cellular networks are usually modeled by the ingredients ($\Phi = \{X_i\}_{i=1}^{\infty}, \ell(r), \{F_i\}_{i=1}^{\infty}, W$), spatially distributed base stations, a path-loss function, the effect of fading, and noise. A configuration $\Phi = \sum_{i=1}^{\infty} \delta_{X_i}$ is a simple, stationary point process on $\mathbb{R}^d$ and it

N. Miyoshi
Department of Mathematical and Computing Sciences, Tokyo Institute of Technology, 2-12-1-W8-52 Ookayama, Tokyo 152-8552, Japan
e-mail: miyoshi@is.titech.ac.jp

T. Shirai (✉)
Institute of Mathematics for Industry, Kyushu University, 744 Motooka, Nishi-ku, Fukuoka 819-0395, Japan
e-mail: shirai@imi.kyushu-u.ac.jp

M. Wakayama et al. (eds.), *The Impact of Applications on Mathematics*,
Mathematics for Industry 1, DOI: 10.1007/978-4-431-54907-9_16,

stands for a realization of spatial distribution of base stations of a cellular network. A decreasing function $\ell\colon (0,\infty) \to [0,\infty)$ is a *path-loss function*, which represents the attenuation of signals at distance r. What we have in mind is, for example, $\ell(r) = ar^{-d\beta}$ or $\ell(r) = a(r^{-d\beta} \wedge 1)$ with $a > 0$ and $\beta > 1$. A random variable F_i independent of Φ represents a random effect of fading from the base station X_i to the typical user. Here we assume the so-called *Rayleigh fading*, i.e., $\{F_i\}_{i=1}^{\infty}$ are i.i.d. exponential random variables with mean 1. W is a random variable representing thermal noise independent of $\{F_i\}_{i=1}^{\infty}$ and Φ. Suppose that a typical user is located at the origin and she/he is associated with the nearest base station X_B from the origin, where B is the index corresponding to the nearest base station. SINR (signal-to-interference-plus-noise-ratio) at the origin is defined by

$$\mathsf{SINR}_o = \frac{F_B\, \ell(|X_B|)}{W + I(B)} \left(= \frac{\text{signal}}{\text{noise}}\right),$$

where $I(B) = \sum_{i \neq B} F_i\, \ell(|X_i|)$ is the cumulative interference signal from all the base stations other than B.

One of the main concerns of the study of wireless networks is to analyze the *coverage probability* $P(\mathsf{SINR}_o > \theta)$ with given threshold $\theta > 0$ as a functional of $(\Phi, \ell(r), \{F_i\}_{i=1}^{\infty}, W)$. The coverage probability plays a role of natural metric for measuring the performance of a wireless network. For the first attempt of such an analysis, one often considers a Poisson point process because of its tractability and computability by its spatial independence. The coverage probability for the stationary Poisson point process has been computed explicitly in [1] (see Example 1 below). However, sometimes, it does not seem to be plausible as a real world model since the Poisson points have some clusters due to spatial independence. In our previous paper [9], we treated the cellular network whose base stations are configurated according to the Ginibre point process, which might be more natural as base stations than the Poisson point process since it has repulsion or negative correlation. In the study, we derive an integral representation of the coverage probability by using random infinite products to obtain an asymptotic behavior and perform numerical computation. It is observed in our numerical computation that the coverage probability for the Ginibre point process is larger than that of the Poisson point process in wide range of threshold values θ. In [10], the α-Ginibre point process is considered as a one-parameter interpolation between Poisson ($\alpha = 0$) and Ginibre ($\alpha = 1$), and the numerical computation of the coverage probabilities for them is performed.

In this paper, we give some discussions on the asymptotics of the coverage probability of a cellular network model based on stationary point processes. In Sect. 2, we consider the asymptotics as $\theta \to 0$ and propose an idea of the Padé approximation of coverage probability (Fig. 1). In Sect. 3, we focus on the coverage probability for the α-Ginibre point processes (Fig. 2) and give an asymptotic behavior as $\theta \to \infty$ (Theorem 1). Also we discuss the asymptotic constant which appears in the limit (Figs. 3, 4 and Proposition 5).

2 Cellular Network with Stationary Base Stations

For simplicity, we only consider the interference limited case $W = 0$.

Proposition 1 *Suppose that base stations are distributed according to a simple point process $\Phi = \sum_{i=1}^{\infty} \delta_{X_i}$. Then, the coverage probability is given by the formula*

$$P(\mathsf{SINR}_o > \theta) = E\left[\prod_{j \neq B} \left(1 + \theta \frac{\ell(|X_j|)}{\ell(|X_B|)}\right)^{-1}\right], \tag{1}$$

where X_B is of the least modulus.

Proof Straightforward (cf. [9]). □

In what follows, we assume that the spatial dimension $d = 2$, Φ is simple and stationary, and the path-loss function is given by $\ell(r) = ar^{-2\beta}$ with $a > 0$ and $\beta > 1$. We denote the coverage probability $P(\mathsf{SINR}_o > \theta)$ by $p_c(\theta, \beta)$, which does not depend on a in this case and is given by

$$p_c(\theta, \beta) = E\left[\prod_{j \neq B} \left(1 + \theta \left|\frac{X_B}{X_j}\right|^{2\beta}\right)^{-1}\right]. \tag{2}$$

Example 1 If Φ is the stationary Poisson point process on $\mathbb{R}^2$ with intensity $\pi^{-1} dxdy$, then

$$p_c^{(Poi)}(\theta, \beta) = \frac{1}{1 + \rho(\theta, \beta)}, \quad \rho(\theta, \beta) = \theta^{1/\beta} \int_{\theta^{-1/\beta}}^{\infty} \frac{du}{1 + u^{\beta}}. \tag{3}$$

In particular, as $\theta \to \infty$,

$$p_c^{(Poi)}(\theta, \beta) \sim \frac{\sin(\pi/\beta)}{\pi/\beta} \theta^{-1/\beta}.$$

See [1], for example.

We have another expression for the coverage probability in terms of the number of base stations inside the disk.

Proposition 2 *The base stations are distributed according to a point process $\Phi = \sum_{i=1}^{\infty} \delta_{X_i}$. Then, the coverage probability is expressed as*

$$p_c(\theta, \beta) = E\left[\exp\left(-\theta \int_1^{\infty} \frac{\tilde{N}_{\Phi}(s^{\frac{1}{2\beta}}|X_B|)}{s(\theta + s)} \mathrm{d}s\right)\right], \tag{4}$$

where $\tilde{N}_{\Phi}(t)$ is the number of $|X_i|$'s less than or equal to t except $|X_B|$.

Proof Let $Z_j = |X_j/X_B|^{2\beta}$. Then, for $s > 0$, we have

$$\#\{j \geq 1;\, j \neq B, Z_j \leq s\} = \#\{j \geq 1;\, j \neq B, |X_j| \leq s^{\frac{1}{2\beta}}|X_B|\} = \tilde{N}_\Phi(s^{\frac{1}{2\beta}}|X_B|).$$

From (2) and Lemma 1 below, we obtain (4). □

Lemma 1 *Let z_j, $j = 1, 2, \dots$ be an increasing sequence of positive reals. Then, for $T > 0$,*

$$\prod_{j=1}^{\infty}(1 + \frac{T}{z_j}) = \exp\left(T\int_0^\infty \frac{N(s)}{s(T+s)}\mathrm{d}s\right), \tag{5}$$

where $N(s)$ is the number of z_j's less than or equal to s. Both sides are finite if and only if $\sum_{j=1}^\infty z_j^{-1} (= \int_0^\infty \frac{N(s)}{s^2}ds) < \infty$.

Proof By summation by parts, we see that

$$\begin{aligned}\log\prod_{j=1}^{\infty}(1 + \frac{T}{z_j}) &= \sum_{j=1}^{\infty}\log(1 + \frac{T}{z_j}) = \sum_{j=1}^{\infty} j\left(\log(1 + \frac{T}{z_j}) - \log(1 + \frac{T}{z_{j+1}})\right)\\ &= \sum_{j=1}^{\infty} j\int_{z_j}^{z_{j+1}} \frac{T}{s(T+s)}\mathrm{d}s = T\int_0^\infty \frac{N(s)}{s(T+s)}\mathrm{d}s.\end{aligned}$$

The last equality follows from $\sum_{j=1}^\infty z_j^{-1} = \int_0^\infty \frac{dN(s)}{s}$ and integration by parts. □

Remark 1 The formula (5) is also valid for a finite sequence of z_j, $j = 1, 2, \dots, M$ by replacing $\prod_{j=1}^\infty$ with $\prod_{j=1}^M$ on the left-hand side.

By using Proposition 2, we can compute the Taylor expansion at $\theta = 0$.

Proposition 3 *For $t > 0$, let*

$$\begin{aligned}\kappa_1(t,\beta) &= E[\tilde{N}_\Phi(t^{\frac{1}{2\beta}}|X_B|)],\\ \kappa_2(t,s,\beta) &= E[\tilde{N}_\Phi(t^{\frac{1}{2\beta}}|X_B|)\tilde{N}_\Phi(s^{\frac{1}{2\beta}}|X_B|)],\end{aligned}$$

where $\tilde{N}_\Phi(s)$ is the number of $|X_i|$'s except $|X_B|$ less than or equal to s. Suppose that $\kappa_1(t,\beta)$ and $\kappa_2(t,s,\beta)$ are finite. Then, we have

$$\begin{aligned}&p_c(\theta,\beta)\\ &= 1 - \theta\int_1^\infty \frac{\kappa_1(t,\beta)}{t^2}\mathrm{d}t + \frac{\theta^2}{2}\left\{\int_1^\infty\int_1^\infty \frac{\kappa_2(t,s,\beta)}{t^2 s^2}\mathrm{d}t\mathrm{d}s + 2\int_1^\infty \frac{\kappa_1(t,\beta)}{t^3}\mathrm{d}t\right\}\\ &\quad + O(\theta^3) \quad (\theta \to 0).\end{aligned}$$

Proof It follows from the Taylor expansion at $\theta = 0$ in (4). □

In the case of the stationary Poisson point process on $\mathbb{R}^2$ with intensity $\pi^{-1}dxdy$, it is easy to see that

$$\kappa_1(t,\beta) = (t^{\frac{1}{\beta}} - 1)I_{[1,\infty)}(t), \quad \kappa_2(t,s,\beta) = \kappa_1(t \wedge s, \beta) + 2\kappa_1(t,\beta)\kappa_1(s,\beta)$$

and that

$$p_c^{(Poi)}(\theta,\beta) = 1 - \frac{\theta}{\beta - 1} + \frac{\beta^2\theta^2}{(2\beta-1)(\beta-1)^2} + O(\theta^3) \quad (\theta \to 0)$$

by Proposition 3, although it can also be computed directly from the expression (3) for the Poisson case. In order to apply Proposition 3 to the α-Ginibre point processes (see Sect. 3.2 for the definition), we need to compute the conditional product moments of random variables $\{\tilde{N}_\Phi(u|X_B|), u > 0\}$ given $|X_B| = r$, i.e., the condition that there are no points except X_B within the disk of radius r. For example, the asymptotic behavior of the conditional first moment (slightly different version) for $\alpha = 1$ can be found in [13].

A Padé approximant is a rational function whose power series expansion agrees with a prescribed Taylor series to the highest possible order. Once one knows the second order Taylor expansion of a function, one can compute its $(1,1)$-Padé approximant by the formula

$$\frac{1 + p_1\theta}{1 + q_1\theta} = 1 - a_1\theta + a_2\theta^2 + O(\theta^3) \quad (\theta \to 0).$$

Then, if $p_c(\theta,\beta) = 1 - a_1\theta + a_2\theta^2 + O(\theta^3)$ as $\theta \to 0$, the $(1,1)$-Padé approximant is given by

$$p_c(\theta,\beta) = \frac{1 + \frac{a_2 - a_1^2}{a_1}\theta}{1 + \frac{a_2}{a_1}\theta} + O(\theta^3).$$

For example, in the Poisson case above, we have

$$p_c^{(Poi)}(\theta,\beta) = \frac{1 + \frac{\beta-1}{2\beta-1}\theta}{1 + \frac{\beta^2}{(\beta-1)(2\beta-1)}\theta} + O(\theta^3).$$

The $(1,1)$-Padé approximant provides a better approximation of the coverage probability (3) in wide range near $\theta = 0$ than the second order Taylor expansion. See Fig. 1, in which $\theta = 0$ corresponds to $-\infty$ dB.

More details on the Padé-approximation for coverage probability will be discussed elsewhere.

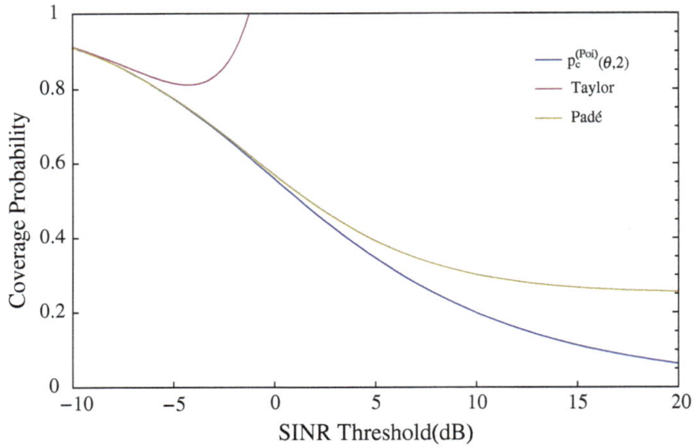

Fig. 1 The second order Taylor expansion and the $(1, 1)$-Padé approximant of the coverage probability $p_c^{(Poi)}(\theta, 2)$. Here "dB" means that $\theta(\text{dB}) = 10^{\theta/10}$

3 Cellular Network with α-Ginibre Configurated Base Stations

We recall the definition of determinantal point processes. The α-Ginibre point process is defined as a determinantal point process (cf. [4, 10]). In Sects. 3.3 and 3.4, we discuss the asymptotic behavior of the coverage probability for the α-Ginibre point process as $\theta \to \infty$.

3.1 Determinantal Point Processes

Let ν be a Radon measure on $\mathbb{R}^d$ and $K(\cdot, \cdot)\colon \mathbb{R}^d \times \mathbb{R}^d \to \mathbb{C}$ a continuous kernel which defines a self-adjoint integral operator K on $L^2(\mathbb{R}^d, \nu)$. Under the assumption that (i) K is of locally trace class, i.e., the restriction of K on a compact set is of trace class, and (ii) the spectrum of K is contained in [0, 1], there exists a unique simple point process Φ on $\mathbb{R}^d$ such that the n-th correlation function with respect to the reference measure ν is given by

$$\rho_n(x_1, x_2, \dots, x_n) = \det(K(x_i, x_j))_{i,j=1}^n.$$

Here the n-th correlation function (if exists) is defined by the formula

$$E\Big[\sum_{\substack{x_1,\dots,x_n \in \Phi \\ \text{distinct}}} f(x_1, x_2, \dots, x_n)\Big] = \int_{(\mathbb{R}^d)^n} f(x_1, x_2, \dots, x_n)\rho_n(x_1, x_2, \dots, x_n) \prod_{i=1}^n \nu(\mathrm{d}x_i)$$

for every symmetric continuous function f on $(\mathbb{R}^d)^n$ with compact support. Such a point process $\Phi = \sum_i \delta_{X_i}$ is called the *determinantal point process on* $\mathbb{R}^d$ *associated with* K *and* ν. Determinantal point processes, a.k.a. fermion point processes, form a nice class of point processes with repulsion and they have been investigated from various points of view over the last 20 years. We refer the reader to [7, 14, 15] for more details on determinantal point processes and their properties.

Here we just mention one property for later use. For a compact set $\Lambda \subset \mathbb{R}^d$, let $K_\Lambda = I_\Lambda K I_\Lambda$ be the restriction of K onto the compact set Λ. By the assumption for K, the operator K_Λ has eigenvalues $\{\lambda_i(\Lambda), i \in \mathbb{N}\}$ satisfying $\lambda_i(\Lambda) \in [0, 1]$ for every i. It is known that the number of points inside Λ, say $\Phi(\Lambda)$, is expressed as the sum of independent Bernoulli random variables, that is,

$$\Phi(\Lambda) \stackrel{d}{=} \sum_{i \in \mathbb{N}} \xi_i(\Lambda), \tag{6}$$

where $\xi_i(\Lambda)$ is a Bernoulli random variable with $P(\xi_i(\Lambda) = 1) = \lambda_i(\Lambda)$ and $P(\xi_i(\Lambda) = 0) = 1 - \lambda_i(\Lambda)$. In particular,

$$E[\Phi(\Lambda)] = \sum_{i \in \mathbb{N}} \lambda_i(\Lambda) = \mathrm{Tr} K_\Lambda, \tag{7}$$

$$\mathrm{Var}(\Phi(\Lambda)) = \sum_{i \in \mathbb{N}} \lambda_i(\Lambda)(1 - \lambda_i(\Lambda)) = \mathrm{Tr} K_\Lambda(1 - K_\Lambda). \tag{8}$$

We note that $\mathrm{Var}(\Phi(\Lambda)) \le E[\Phi(\Lambda)]$ and the assumption (i) for K guarantees that $E[\Phi(\Lambda)]$ is finite when Λ is compact.

3.2 α-Ginibre Point Processes and Their Properties

We assume that $\alpha \in [0, 1]$ and $\alpha = 0$ is understood to be the limit $\alpha \to 0$ as we will see in Remark 3. For $\alpha \in (0, 1]$, let μ_α be the α-Ginibre point process, i.e., the determinantal point process on $\mathbb{C}(\cong \mathbb{R}^2)$ associated with

$$K_\alpha(z, w) = e^{z\bar{w}/\alpha}, \quad \mathsf{g}_\alpha(dz) = \pi^{-1} e^{-|z|^2/\alpha} m(dz),$$

where $m(dz)$ is the Lebesgue measure on $\mathbb{C}$. The integral operator K_α acting on $L^2(\mathbb{C}, \mathsf{g}_\alpha)$ has eigenvalues α and 0. Indeed, the functions

$$\phi_j(z) = \frac{z^{j-1}}{\sqrt{(j-1)!\alpha^j}}, \quad j = 1, 2 \ldots$$

are the normalized eigenfunctions corresponding to the eigenvalue α, and the functions $z^n \overline{z^m}, n = 0, 1, 2, \ldots, m = 1, 2, \ldots$ are the eigenfunctions corresponding to 0. Hence, the spectral decomposition of the kernel K_α is given by

$$K_\alpha(z, w) = \sum_{j=1}^{\infty} \alpha \cdot \phi_j(z)\overline{\phi_j(w)}.$$

Lemma 2 *Let $(K_\alpha)_{D_r}$ be the restriction of K_α on the disk D_r of radius r. Then, for each $j \in \mathbb{N}$, $\phi_j(z)$ is also an eigenfunction of $(K_\alpha)_{D_r}$ corresponding to the eigenvalue*

$$\lambda_j(r) = \alpha \int_0^{r^2/\alpha} \frac{s^{j-1}e^{-s}}{(j-1)!} ds = \alpha P(Y_j \le \frac{r^2}{\alpha}),$$

where $Y_j \sim \mathrm{Gamma}(j, 1)$, i.e., $P(Y_j \le t) = \int_0^t \frac{s^{j-1}e^{-s}}{(j-1)!} ds$.

Proof We can show it by direct computation (cf. [12] for $\alpha = 1$). □

This lemma is closely related to the following remark.

Remark 2 When $\Phi = \sum_{i \in \mathbb{N}} \delta_{X_i}$ is the original Ginibre ($\alpha = 1$), it is known that $\{|X_i|, i \in \mathbb{N}\} \stackrel{d}{=} \{\sqrt{Y_j}, j \in \mathbb{N}\}$, where $\{Y_j, j \in \mathbb{N}\}$ are independent and $Y_j \sim \mathrm{Gamma}(j, 1)$ ([7, 8]). This fact is useful for computation of the coverage probability since the path-loss function only depends on the distance in our setting.

The n-th correlation function with respect to $\mathfrak{g}_\alpha(dz)$ is given by

$$\rho_n(z_1, \dots, z_n) = \det(K_\alpha(z_i, z_j))_{i,j=1}^n$$

for each $n \in \mathbb{N}$. For example, the first and second correlation measures are the following.

$$\begin{aligned}
\rho_1(z)\mathfrak{g}_\alpha(dz) &= K_\alpha(z, z)\mathfrak{g}_\alpha(dz) = \pi^{-1}m(dz), \\
\rho_2(z, w)\mathfrak{g}_\alpha(dz)\mathfrak{g}_\alpha(dw) &= (K_\alpha(z, z)K_\alpha(w, w) - K_\alpha(z, w)K_\alpha(w, z))\mathfrak{g}_\alpha(dz)\mathfrak{g}_\alpha(dw) \\
&= \pi^{-2}e^{-|z-w|^2/\alpha}m(dz)m(dw).
\end{aligned}$$

Both measures are motion invariant, i.e., invariant under translation and rotation. Moreover, the n-th correlation measure is also motion invariant for every $n \in \mathbb{N}$. Hence, the α-Ginibre point process is motion invariant.

Remark 3 As remarked above, the α-Ginibre point process is motion invariant and the intensity is here normalized to be π^{-1} for all $\alpha \in (0, 1]$. One can show that μ_α converges weakly to the Poisson point process with the same intensity as $\alpha \to 0$ so that μ_0 can be regarded as the Poisson point process, which itself is not determinantal.

Proposition 4 *Let N_r be the number of points inside D_r. Then, under μ_α, we have*

$$E[N_r] = r^2, \quad Var(N_r) \sim (1-\alpha)r^2$$

as $r \to \infty$.

Proof We give a sketch of proof. It follows from (7) that

$$E[N_r] = \sum_{i=1}^{\infty} \lambda_i(r) = \alpha \cdot \frac{r^2}{\alpha} = r^2.$$

Note that by Lemma 2 and the law of large numbers, we have

$$\lambda_n(r) \sim \begin{cases} \alpha + o(1), & n \leq (1-\epsilon)\frac{r^2}{\alpha}, \\ o(1), & n \geq (1+\epsilon)\frac{r^2}{\alpha} \end{cases}$$

for any $\epsilon > 0$ and $o(1)$ is exponentially small by the large deviations result. Hence it follows from (8) that

$$\mathrm{Var}(N_r) = \sum_{i=1}^{\infty} \lambda_i(r)(1-\lambda_i(r)) \sim \sum_{i=1}^{r^2/\alpha} \alpha(1-\alpha) = (1-\alpha)r^2.$$

□

Remark 4 When $\alpha = 1$, the variance is of $o(r^2)$ by Proposition 4, which is called small fluctuation property of the (1-)Ginibre point process. Indeed, it is known that $\mathrm{Var}(N_r) \sim \pi^{-1/2} r$ under μ_1 (see [12]).

Remark 5 The α-Ginibre point process can be constructed from the 1-Ginibre by scaling and independent *thinning*. We scale the 1-Ginibre point process by factor $\sqrt{\alpha}$, and for each point in the scaled point process, independently, retain it with probability α and delete it otherwise. Then the resultant point process is equal in law to the α-Ginibre point process. The thinning operation makes the variance large for $\alpha \in (0, 1)$.

3.3 Asymptotics of $p_c^{(\alpha)}(\theta, \beta)$

The coverage probability for α-Ginibre point process, denoted by $p_c^{(\alpha)}(\theta, \beta)$, was discussed in [9] when $\alpha = 1$ and in [10] for general $\alpha \in (0, 1]$. It is given by

$$p_c^{(\alpha)}(\theta, \beta) = \alpha \sum_{k=1}^{\infty} E\Big[\prod_{j \in \mathbb{N} \setminus \{k\}} q\Big(\alpha, \theta \Big(\frac{Y_k}{Y_j}\Big)^{\beta}\Big) 1_{\{Y_j \geq Y_k\}} \Big], \tag{9}$$

where $Y_j \sim \mathrm{Gamma}(j, 1)$, $j = 1, 2, \ldots$ and $q(\alpha, x) = 1 - \alpha x(1+x)^{-1}$. Note that $q(\alpha, x)$ is decreasing in x and α. The formula (9) can be shown by Proposition 1, Remark 2 and Remark 5.

For $k \in \mathbb{N}$ and $\beta > 1$, let

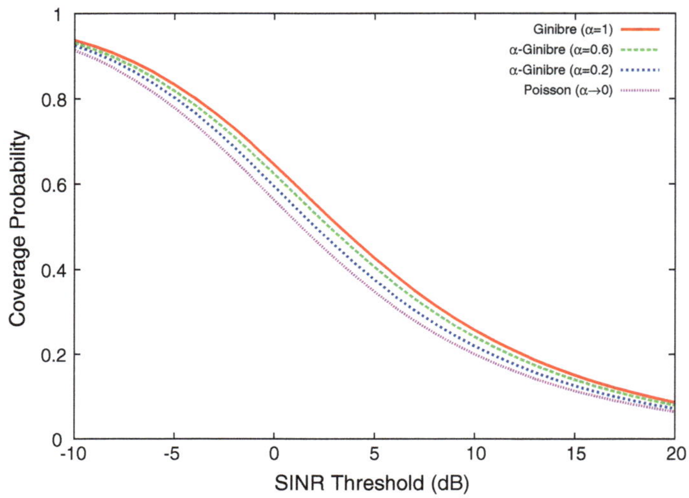

Fig. 2 The coverage probability for the α-Ginibre point process for $\beta = 2$ [10]

$$A_k(\alpha) = A_{k,\beta}(\alpha) := \alpha \int_0^\infty \frac{v^{k-1}}{(k-1)!} \prod_{j \in \mathbb{N} \setminus \{k\}} E\Big[q\Big(\alpha, \Big(\frac{v}{Y_j}\Big)^\beta\Big)\Big] \mathrm{d}v, \tag{10}$$

where Y_j's and $q(\alpha, x)$ are defined as above. Then, we can show the following.

Theorem 1 *Fix $\beta > 1$. Then, for $\alpha \in (0, 1]$,*

$$p_c^{(\alpha)}(\theta, \beta) \sim A_1(\alpha)\theta^{-1/\beta} \tag{11}$$

as $\theta \to \infty$.

We note that the decay rate $\theta^{-1/\beta}$ is the same as that of the Poisson point process as in Example 1. As we will see in Lemma 4 and Proposition 5, the asymptotic constant $A_1(\alpha)$ is finite (see Fig. 3), and $A_1(\alpha)$ converges to that of the Poisson point process as $\alpha \to 0+$, as was naturally expected from Remark 3.

We also observe that $A_1(\alpha)$ is not increasing in α for $\beta = 1.25$ (near $\beta = 1$) in Fig. 3, and it seems that $A_{1,\beta}(\alpha)$ is increasing in β and the asymptotic constant is bounded below by $A_1(0+)$ in Fig. 4.

Lemma 3 *Let $k \in \mathbb{N}$ be fixed. Then,*

$$\lim_{\theta \to \infty} \theta^{k/\beta} \cdot \alpha E\Big[\prod_{j \in \mathbb{N} \setminus \{k\}} \Big\{q\Big(\alpha, \theta\Big(\frac{Y_k}{Y_j}\Big)^\beta\Big) \mathbf{1}_{\{Y_j \geq Y_k\}}\Big\}\Big] = A_k(\alpha). \tag{12}$$

Proof By a change of variables and the monotone convergence theorem, we see that

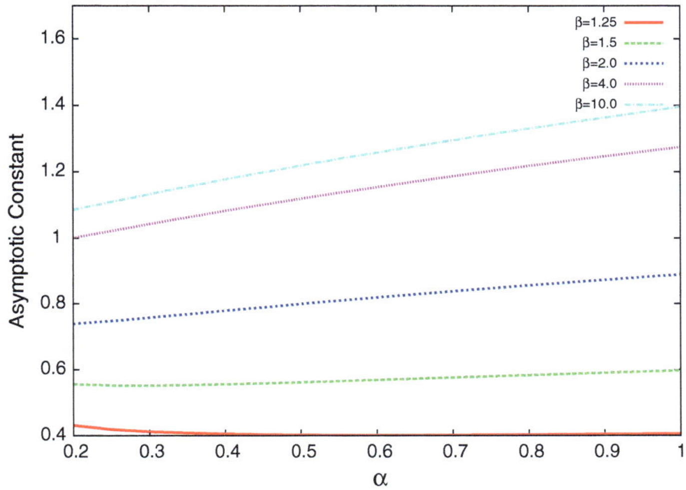

Fig. 3 The asymptotic constant $A_1(\alpha)$

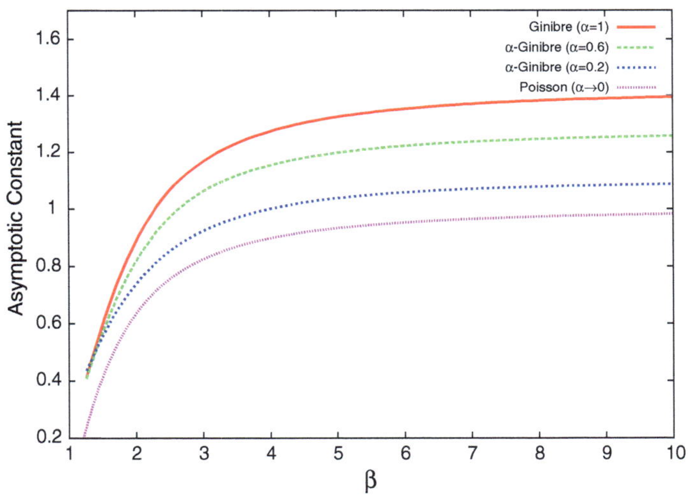

Fig. 4 The asymptotic constant $A_1(\alpha)$ as a function of $\beta > 1$

$$\theta^{k/\beta} \cdot \alpha E\left[\prod_{j\in\mathbb{N}\setminus\{k\}} \left\{q\left(\alpha, \theta\left(\frac{Y_k}{Y_j}\right)^{\beta}\right)\mathbf{1}_{\{Y_j\geq Y_k\}}\right\}\right]$$

$$= \theta^{k/\beta} \cdot \alpha \int_0^{\infty} \frac{u^{k-1}e^{-u}}{(k-1)!} \prod_{j\in\mathbb{N}\setminus\{k\}} E\left[q\left(\alpha, \theta\left(\frac{u}{Y_j}\right)^{\beta}\right)\mathbf{1}_{\{Y_j\geq u\}}\right]\mathrm{d}u$$

$$= \alpha \int_0^\infty \frac{v^{k-1} e^{-\theta^{-1/\beta} v}}{(k-1)!} \prod_{j \in \mathbb{N} \setminus \{k\}} E\Big[q\Big(\alpha, \Big(\frac{v}{Y_j}\Big)^\beta\Big) \mathbf{1}_{\{Y_j \ge \theta^{-1/\beta} v\}}\Big] \mathrm{d}v$$

$$\nearrow \alpha \int_0^\infty \frac{v^{k-1}}{(k-1)!} \prod_{j \in \mathbb{N} \setminus \{k\}} E\Big[q\Big(\alpha, \Big(\frac{v}{Y_j}\Big)^\beta\Big)\Big] \mathrm{d}v = A_k(\alpha)$$

as $\theta \to \infty$. □

Lemma 4 *Let $k \ge 1$. Then $A_k(\alpha) < \infty$. Moreover, for every $N \ge 1$,*

$$\lim_{\theta \to \infty} \theta^{1/\beta} \cdot \alpha \sum_{k=1}^{N} E\Big[\prod_{j \in \mathbb{N} \setminus \{k\}} \Big\{ q\Big(\alpha, \theta \Big(\frac{Y_k}{Y_j}\Big)^\beta\Big) \mathbf{1}_{\{Y_j \ge Y_k\}} \Big\} \Big] = A_1(\alpha). \tag{13}$$

Proof Since $q(\alpha, x) \le \exp(-\alpha \frac{x}{1+x})$, we have

$$A_k(\alpha) \le \alpha \int_0^\infty \frac{v^{k-1}}{(k-1)!} E\Big[\prod_{\substack{j=c_1 v \\ j \ne k}}^{c_2 v} \exp\Big(-\alpha \frac{(\frac{v}{Y_j})^\beta}{1 + (\frac{v}{Y_j})^\beta}\Big)\Big] \mathrm{d}v$$

$$= \alpha \int_0^\infty \frac{v^{k-1}}{(k-1)!} E\Big[\exp\Big(-\alpha \sum_{\substack{j=c_1 v \\ j \ne k}}^{c_2 v} \frac{v^\beta}{Y_j^\beta + v^\beta}\Big)\Big] \mathrm{d}v$$

for $0 < c_1 < c_2$. For $\epsilon > 0$, let

$$B_v := \bigcap_{c_1 v \le j \le c_2 v} \{Y_j \le (1+\epsilon) j\}.$$

Now we estimate

$$E\Big[\exp\Big(-\alpha \sum_{\substack{j=c_1 v \\ j \ne k}}^{c_2 v} \frac{v^\beta}{Y_j^\beta + v^\beta}\Big)\Big]$$

$$= E\Big[\exp\Big(-\alpha \sum_{\substack{j=c_1 v \\ j \ne k}}^{c_2 v} \frac{v^\beta}{Y_j^\beta + v^\beta}\Big); B_v\Big] + E\Big[\exp\Big(-\alpha \sum_{\substack{j=c_1 v \\ j \ne k}}^{c_2 v} \frac{v^\beta}{Y_j^\beta + v^\beta}\Big); B_v^c\Big]$$

$$= (I) + (II).$$

For (I), we have

$$(I) \le E\Big[\exp\Big(-\alpha \sum_{\substack{j=c_1 v \\ j \ne k}}^{c_2 v} \frac{v^\beta}{\{(1+\epsilon) j\}^\beta + v^\beta}\Big); B_v\Big] \le \exp\Big(-\alpha \frac{(c_2 - c_1) v}{\{(1+\epsilon) c_2\}^\beta + 1}\Big).$$

For (II), by large deviations result, we see that

$$
\begin{aligned}
(II) &\le P(B_v^c) = P\Big(\bigcup_{c_1 v \le j \le c_2 v} \{Y_j > (1+\epsilon)j\}\Big) \le \sum_{c_1 v \le j \le c_2 v} P(Y_j > (1+\epsilon)j) \\
&\le \sum_{c_1 v \le j \le c_2 v} \exp\big(- I(1+\epsilon)j\big) \\
&\le (c_2 - c_1)v \exp\big(- c_1 I(1+\epsilon)v\big),
\end{aligned}
$$

where $I(x) = x - 1 - \log x$ $(x > 0)$ is the rate function with which the large deviation principle holds for sum of i.i.d. exponential random variables with mean 1. Hence, for some $c, C > 0$ independent of v, we have

$$(I) + (II) \le c(1+v)e^{-Cv}.$$

Therefore,

$$A_k(\alpha) \le \alpha \int_0^\infty \frac{v^{k-1}}{(k-1)!} \cdot c(1+v)e^{-Cv} \mathrm{d}v < \infty.$$

The second part of the assertion immediately follows from the above and Lemma 3. □

Proof of Theorem 1 We observe that

$$
\begin{aligned}
&\alpha \sum_{k=N+1}^{\infty} E\Big[\prod_{j \in \mathbb{N}\setminus\{k\}} \Big\{ q\Big(\alpha, \theta \Big(\frac{Y_k}{Y_j}\Big)^{\beta}\Big) \mathbf{1}_{\{Y_j \ge Y_k\}} \Big\}\Big] \\
&\le \alpha q(\alpha,\theta)^{-1} \sum_{k=N+1}^{\infty} \int_0^\infty \frac{u^{k-1}e^{-u}}{(k-1)!} \prod_{j\in\mathbb{N}} E\Big[q\Big(\alpha, \theta\Big(\frac{u}{Y_j}\Big)^{\beta}\Big)\Big] \mathrm{d}u \\
&\le \alpha q(\alpha,\theta)^{-1} \int_0^\infty \frac{u^N}{N!} \prod_{j\in\mathbb{N}} E\Big[q\Big(\alpha, \theta\Big(\frac{u}{Y_j}\Big)^{\beta}\Big)\Big] \mathrm{d}u \\
&\le q(\alpha,\theta)^{-1} A_{N+1}(\alpha) \theta^{-\frac{N+1}{\beta}} \\
&= \begin{cases} O(\theta^{-\frac{N+1}{\beta}}) & 0 < \alpha < 1, \\ O(\theta^{1-\frac{N+1}{\beta}}) & \alpha = 1. \end{cases}
\end{aligned}
$$

The last equality follows from Lemma 3 since $q(\alpha,\theta) \ge 1 - \alpha$ when $0 < \alpha < 1$ and $q(\alpha,\theta) = (1+\theta)^{-1}$ when $\alpha = 1$. Therefore, letting $N = \lfloor \beta \rfloor + 1$ when $\alpha = 1$, we have the asymptotic formula

$$\lim_{\theta\to\infty} \theta^{1/\beta} p_c^{(\alpha)}(\theta,\beta) = A_1(\alpha). \tag{14}$$

This together with (13) in Lemma 4 completes the proof. □

3.4 A Remark on the Asymptotic Constant $A_1(\alpha)$

In this subsection, we give a probabilistic representation of $A_1(\alpha)$ and its asymptotic behavior as $\alpha \to 0+$.

Fix $\beta > 1$ and let $f_j(v) = E\left[\frac{v^\beta}{Y_j^\beta + v^\beta}\right]$. Then, it is easy to see that

$$A_1(\alpha) = \alpha \int_0^\infty \prod_{j\ge 2}(1-\alpha f_j(v))\mathrm{d}v, \tag{15}$$

$$A_1'(\alpha) = \int_0^\infty \Big\{\prod_{j\ge 2}(1-\alpha f_j(v)) - \sum_{k\ge 2}\alpha f_k(v) \prod_{j\ge 2, j\ne k}(1-\alpha f_j(v))\Big\}\mathrm{d}v.$$

Note that $f_j(v) \in [0,1]$ is increasing in v with $f_j(0) = 0$ and $f_j(\infty) = 1$ for every j. We consider independent Bernoulli random variables $\{\xi_{j,\alpha}(v), j \ge 2\}$ such that $P(\xi_{j,\alpha}(v) = 1) = \alpha f_j(v)$ and $P(\xi_{j,\alpha}(v) = 0) = 1 - \alpha f_j(v)$, and set $X_\alpha(v) = \sum_{j\ge 2}\xi_{j,\alpha}(v)$. Then,

$$E[X_\alpha(v)] = \alpha \int_0^\infty \frac{v^\beta}{t^\beta + v^\beta}(1-e^{-t})\mathrm{d}t = \alpha v \int_0^\infty \frac{1}{s^\beta + 1}(1-e^{-vs})\mathrm{d}s. \tag{16}$$

It follows from (16) that as $v \to \infty$

$$E[X_\alpha(v)] \sim \alpha v \int_0^\infty \frac{\mathrm{d}s}{s^\beta + 1} = \alpha \frac{\pi/\beta}{\sin(\pi/\beta)} v \quad (\beta > 1)$$

and as $v \to 0$

$$E[X_\alpha(v)] \sim \alpha \begin{cases} -\Gamma(1-\beta)v^\beta & 1 < \beta < 2, \\ -v^2 \log v & \beta = 2, \\ \frac{\pi/\beta}{\sin(2\pi/\beta)} v^2 & \beta > 2. \end{cases}$$

In terms of $X_\alpha(v)$, the quantities $A_1(\alpha)$ and $A_1'(\alpha)$ can be rewritten as

$$A_1(\alpha) = \alpha \int_0^\infty P(X_\alpha(v) = 0)\mathrm{d}v,$$

$$A_1'(\alpha) = \int_0^\infty \{P(X_\alpha(v) = 0) - P(X_\alpha(v) = 1)\}\mathrm{d}v.$$

Although the convergence of the numerical computation is too slow to observe the values near $\alpha = 0$ in Fig. 3, we can show the following limiting behavior as $\alpha \to 0+$.

Proposition 5 *For every $\beta > 1$, it holds that*

$$\lim_{\alpha \to 0+} A_1(\alpha) = \frac{\sin(\pi/\beta)}{\pi/\beta}.$$

The right-hand side is the asymptotic constant for the Poisson case as in Example 1.

Proof For every $\delta > 0$ there exists $0 < x_\delta < 1$ such that

$$e^{-(1+\delta)x} \le 1 - x \le e^{-x} \quad (0 \le \forall x \le x_\delta). \tag{17}$$

Fix $\delta > 0$. From (15) and (17), since $f_j(v) \in [0, 1]$ for all j and v, we see that for any sufficiently small $\alpha > 0$

$$\alpha \int_0^\infty e^{-(1+\delta)E[X_\alpha(v)]} \mathrm{d}v \le A_1(\alpha) \le \alpha \int_0^\infty e^{-E[X_\alpha(v)]} \mathrm{d}v,$$

and hence we have

$$\int_\epsilon^\infty e^{-(1+\delta)E[X_\alpha(u/\alpha)]} \mathrm{d}u \le A_1(\alpha) \le \epsilon + \int_\epsilon^\infty e^{-E[X_\alpha(u/\alpha)]} \mathrm{d}u$$

for any $\epsilon > 0$. From (16) we see that $E[X_\alpha(u/\alpha)] \ge (1 - e^{-\epsilon})u \int_1^\infty \frac{ds}{s^\beta + 1}$ uniformly in $u \in [\epsilon, \infty)$ and that $E[X_\alpha(u/\alpha)] \nearrow \frac{\pi/\beta}{\sin(\pi/\beta)} u$ as $\alpha \searrow 0$. Therefore, we obtain the assertion by the monotone convergence theorem since ϵ and δ are arbitrary. □

Acknowledgments The authors would like to thank the referee for his/her comments. The first author (NM)'s work was supported in part by JSPS (Japan Society for the Promotion of Science) Grant-in-Aid for Scientific Research (C) 25330023. The second author (NM)'s work was supported in part by JSPS Grant-in-Aid for Scientific Research (B) 22340020.

References

1. Andrews, J.G., Baccelli, F., Ganti, R.K.: A tractable approach to coverage and rate in cellular networks. IEEE Trans. Commun. **59**, 3122–3134 (2011)
2. Andrews, J.G., Ganti, R.K., Haenggi, M., Jindal, N., Weber, S.: A primer on spatial modeling and analysis in wireless networks. IEEE Commun. Mag. **48**, 156–163 (2010)
3. Baccelli, F., Błaszczyszyn, B.: Stochastic Geometry and Wireless Networks, Vol. I: Theory/Volume II: Applications. Foundations and Trends(R) in Networking 3, 249–449/ 4, 1–312 (2009)
4. Goldman, A.: The palm measure and the Voronoi tessellation for the Ginibre process. Ann. Appl. Probab. **20**, 90–128 (2010)
5. Haenggi, M.: Stochastic Geometry for Wireless Networks. Cambridge University Press, Cambridge (2013)
6. Haenggi, M., Andrews, J.G., Baccelli, F., Dousse, O., Franceschetti, M.: Stochastic geometry and random graphs for the analysis and design of wireless networks. IEEE J. Select. Areas Commun. **27**, 1029–1046 (2009)
7. Hough, J.B., Krishnapur, M., Peres, Y., Virág, B.: Zeros of Gaussian Analytic Functions and Determinantal Point Processes. American Mathematical Society, Providence, RI (2009)

8. Kostlan, E.: On the spectra of Gaussian matrices. Directions in matrix theory (Auburn, AL, 1990). Linear Algebra Appl. **162**(164), 385–388 (1992)
9. Miyoshi, N., Shirai, T.: A cellular network model with Ginibre configurated base stations. To appear in Advances in Applied Probability (2014)
10. Nakata, I., Miyoshi, N.: Spatial stochastic models for analysis of heterogeneous cellular networks with repulsively deployed base stations. To appear in Performance Evaluation (2014)
11. Nagamatsu, H., Miyoshi, N., Shirai, T.: Padé approximation for coverage probability in cellular networks. Proc. 12th Int'l Symposium on Modeling and Optimization in Mobile, Ad Hoc, and Wireless Networks (WiOpt), pp. 699–706, Hammamet, Tunisia, May 2014
12. Shirai, T.: Large deviations for the fermion point process associated with the exponential kernel. J. Stat. Phys. **123**, 615–629 (2006)
13. Shirai, T.: Ginibre-type point processes and their asymptotic behavior. To appear in J. Math. Soc. Japan. http://mathsoc.jp/publication/JMSJ/inpress.html
14. Shirai, T., Takahashi, Y.: Random point fields associated with certain Fredholm determinants I: fermion, poisson and boson processes. J. Funct. Anal. **205**, 414–463 (2003)
15. Soshnikov, A.: Determinantal random point fields. Russ. Math. Surv. **55**, 923–975 (2000)

Nucleation Rate Identification in Binary Phase Transition

Dietmar Hömberg, Shuai Lu, Kenichi Sakamoto and Masahiro Yamamoto

Abstract In this chapter, we study a PDE-ODE system arising from binary phase transition coupled with an energy balance to account for recalescence effects. The phase transition is described by classic arguments on nucleation and growth process. The main novelty of our work is the identification of temperature dependent nucleation rates from measurements in a subdomain. We prove the uniqueness of the parameter identification problem and numerical results support the theoretical results.

Keywords Inverse problem · Optimal control · Coupled system · Parameter identification

D. Hömberg
Weierstrass Institute for Applied Analysis and Stochastics, Mohrenstr. 39, 10117 Berlin, Germany
e-mail: hoemberg@wias-berlin.de

S. Lu (✉)
School of Mathematical Sciences, Fudan University, Shanghai 200433, China
e-mail: slu@fudan.edu.cn

K. Sakamoto
Mathematical Science and Technology Research Laboratory,
Advanced Technology Research Laboratories, Technical Development Bureau,
Nippon Steel & Sumitomo Metal Corporation, 20-1 Shintomi,
Futtsu, Chiba 293-8511, Japan
e-mail: sakamoto.a2c.kenichi@jp.nssmc.com

M. Yamamoto
Department of Mathematical Sciences, The University of Tokyo, Komaba Meguro,
Tokyo 153-8914, Japan
e-mail: myama@ms.u-tokyo.ac.jp

M. Wakayama et al. (eds.), *The Impact of Applications on Mathematics*,
Mathematics for Industry 1, DOI: 10.1007/978-4-431-54907-9_17,

1 Introduction

According to [6], phase transition may occur in all metastable systems and the initial or final phase may be solid, liquid, or gaseous. The new phase grows at the expense of the old one by the migration of the interphase boundary via nucleation and growth process. At a fixed temperature the reaction proceeds isothermally and will continue until it completes. Hence the final amount of transformation is independent of temperature as long as the equilibrium phase fraction is so.

To become more specific let us consider a test volume $V \subset I\!R^3$ in which a transformation from a phase A to a phase B happens. We call $V^A(t)$ and $V^B(t)$ the sub-volumes occupied by binary phases A and B at time t, respectively, i.e.

$$V = V^A(t) + V^B(t) \quad \text{for all } t \in [0, T].$$

Moreover, we define the phase volume fraction of the product phase,

$$P(t) = \frac{V^B(t)}{V}.$$

We introduce the growth rate ρ, which we assume to be a constant throughout the context. In many cases such a linear isotropic growth is well justified. However, especially in solid-solid phase transitions with an underlying grain structure one would observe rather an anisotropic growth perpendicular to the grain boundary. When the composition of the matrix also changes during the transformation, a parabolic growth corresponding to $\rho \sim t^{-1/2}$ can be expected.

Assuming spherical growth, the volume of a new phase B region originating from a nucleus born at time τ is given by

$$v(t, \tau) = \frac{4\pi}{3}\rho^3(t - \tau)^3. \tag{1}$$

In the sequel we use the abbreviation $\gamma := 4\pi\rho^3$. The way to derive the nucleation and growth model is to start with an extended volume V^B_{ext} of the new phase B disregarding impingement of different B sub-regions. To this end, multiplying the single grain volume (1) with the number of nuclei born at time τ, i.e., $\alpha(\theta(\tau))V$, we obtain the extended volume fraction

$$V^B_{ext}(t) = \frac{\gamma V}{3}\int_0^t \alpha(\theta(\tau))(t - \tau)^3 d\tau. \tag{2}$$

Here, α is the temperature θ dependent nucleation rate which denotes the number of stable nuclei formed per unit time and space. After some time the B sub-regions will first impinge and then grow into each other. Moreover, new nuclei will be born in already transformed regions. In reality, the new phase grows either until the growing process ceases locally due to impingement of sub-regions or until an equilibrium volume $V^B_{eq}(\theta)$ is reached with corresponding equilibrium volume fraction

$$P_{eq}(\theta) = \frac{V^B_{eq}(\theta)}{V}.$$

Usually, the equilibrium value is temperature dependent and can be extracted from the respective equilibrium phase diagram. Then, we may assume only that fraction of an incremental extended volume fraction dV^B_{ext} contributes to the growth of the really transformed fraction dV^B, which previously has not been transformed. In other words we conjecture that

$$dV^B = \left(1 - \frac{V^B}{V^B_{eq}(\theta)}\right) dV^B_{ext}. \tag{3}$$

This so-called Avrami correction has been investigated independently by Avrami [1–3] and Kolmogorov [17], see also [15]. Tacitly assuming that θ is a constant, we integrate (3) using (2) to obtain

$$-\ln\left(1 - \frac{P}{P_{eq}(\theta)}\right) = \frac{\gamma}{3}\frac{1}{P_{eq}(\theta)} \int_0^t \alpha(\theta(\tau))(t-\tau)^3 d\tau \tag{4}$$

from which we conclude

$$P(t) = P_{eq}(\theta)\left(1 - e^{-\frac{\gamma}{3}\frac{1}{P_{eq}(\theta)}\int_0^t \alpha(\theta(\tau))(t-\tau)^3 d\tau}\right). \tag{5}$$

In the case of a constant nucleation rate and $P_{eq} \equiv 1$, (5) boils down to the classical Johnson-Mehl-Avrami-Kolmogorov equation

$$P(t) = 1 - e^{-\frac{\pi}{3}\rho^3 \alpha t^4}. \tag{6}$$

Note that the latter is still often used to quantify phase transitions in steel, especially in the engineering sciences, see, e.g., [9]. Our interest is to identify the temperature dependent nucleation rate $\alpha(\theta)$ in the generalized Avrami model (5). To simplify the exposition in the sequel we assume $P_{eq} \equiv 1$.

Phase transitions are known to be accompanied by the release or consumption of latent heat, which is usually assumed to be proportional to the phase growth rate P_t. To incorporate this effect it is convenient to take the derivative of (4) with respect to time (recall that we assume ($P_{eq} \equiv 1$) and replace (5) with the integro-differential equation

$$P_t(t) = \gamma(1 - P(t)) \int_0^t \alpha(\theta(\tau))(t-\tau)^2 d\tau \tag{7a}$$

$$P(0) = 0. \tag{7b}$$

To account for the release of latent heat during the phase change we couple the phase kinetics with the balance of internal energy, which reads

$$\bar{\rho}\frac{\partial e}{\partial t} - \nabla \cdot (\kappa \nabla \theta) = 0,$$

where we have employed Fourier's law of heat conduction. Here, $\bar{\rho}$ is the mass density, e the specific internal energy and κ the heat conductivity. Now we proceed as in [23] and assume that there exists a differentiable material function $\hat{e}$ such that the internal energy takes the form

$$e(x,t) = \hat{e}(\theta, P),$$

with the partial derivatives

$$\frac{\partial \hat{e}}{\partial \theta} = c, \quad \frac{\partial \hat{e}}{\partial P} = -L, \tag{8}$$

where L denotes the latent heat and c the specific one, respectively. Then the energy balance reads as

$$\bar{\rho}c\theta_t - \nabla \cdot (\kappa \nabla \theta) = \bar{\rho}LP_t. \tag{9}$$

The goal of this paper is to study the system (7a), (7b) together with (9). We investigate the solvability of the state system and study the inverse problem of identifying the temperature dependent nucleation rate $\alpha(\theta)$. To this end we also establish a uniqueness result. We refer to Choulli et al. [5], DuChateau and Rundell [7], Egger et al. [8], Isakov [14], Klibanov [16], Lorenzi [20], Pilant and Rundell [21]. Those papers discuss parabolic equations without integral term, and proved the uniqueness with boundary measurements and the key is the maximum principle. In our recent work [12], we have proved a uniqueness result on the parameter identification problem arising in the PDE-ODE coupled system (7a), (7b), (9) and developed an optimal control approach for its numerical computation.

The paper is organized as follows. Section 2 contains the well-posedness of our coupled model with appropriate boundary and initial conditions. In Sect. 3 we show that indeed the nucleation rate α can be uniquely determined from measurements in a subdomain. Numerical examples are presented in Sect. 4 to identify the nucleation rate α by minimizing a cost functional defined on a subdomain.

2 Well-posedness of the Forward Model

For sake of simplicity, we skip most of the physical-based constants and obtain the simplified forward problem in the following PDE-ODE coupled system. By assuming that $\Omega \subset I\!R^3$ is a domain with $C^{1,1}$ boundary, we consider a transition from phase A stable at high temperature to a low temperature phase B. Accordingly, the cooling processes is observed if the initial temperature θ_0 is greater than the coolant temperature θ_w, assumed to be constant. Then the governing parabolic system for the temperature distribution θ is

$$\theta_t - \kappa \Delta\theta = L(\theta)P_t \quad \text{in} \quad \Omega \times (0, T); \tag{10a}$$

$$\kappa \partial_\nu \theta + \sigma(\theta - \theta_w) = 0 \quad \text{on} \quad \partial\Omega \times (0, T); \tag{10b}$$

$$\theta(x, 0) = \theta_0 \quad \text{in} \quad \Omega \tag{10c}$$

where ν is the normal vector, $\sigma > 0$ is the constant heat exchange coefficient and $\kappa > 0$ the constant heat conductivity. The governing ODE system for the phase volume fraction P is

$$P_t(x, t) = \gamma(1 - P) \int_0^t \alpha(\theta(x, t))(t - \tau)^2 d\tau \quad \text{in} \quad \Omega \times (0, T); \tag{11a}$$

$$P(0) = 0 \quad \text{in} \quad \Omega. \tag{11b}$$

Changing of variables $\eta = \ln\left(\frac{1}{1-P}\right)$ and taking additional initial conditions, we can reformulate an equivalent PDE-ODE coupled system

$$\theta_t - \kappa \Delta\theta = L(\theta)e^{-\eta}\eta_t \quad \text{in} \quad \Omega \times (0, T); \tag{12a}$$

$$\kappa \partial_\nu \theta + \sigma(\theta - \theta_w) = 0 \quad \text{on} \quad \partial\Omega \times (0, T); \tag{12b}$$

$$\theta(x, 0) = \theta_0 \quad \text{in} \quad \Omega \tag{12c}$$

and

$$\frac{d^4\eta}{dt^4} = 2\gamma\alpha(\theta) \quad \text{in} \quad \Omega \times (0, T); \tag{13a}$$

$$\eta^{(i)}(0) = 0, \quad i = 0, \ldots, 3, \quad \text{in} \quad \Omega. \tag{13b}$$

The following assumptions are important in the sequel:

(A1) θ_0 and θ_w are positive constants satisfying $\theta_0 > \theta_w$.
(A2) $L(\theta)$ is in $C^{1,1}(I\!R)$ and $L(\theta) = 0$ if $\theta \leq \theta_-$ or $\theta \geq \theta_+$, and $L(\theta) > 0$ if $\theta_- < \theta < \theta_+$, where $\theta_{+,-}$ are chosen such that $\theta_w \leq \theta_- < \theta_+ \leq \theta_0$.

(A3) The admissible set for $\alpha(\theta)$ is

$$\mathscr{A}_{ad} := \left\{ \alpha \in C^{1,\zeta}(I\!R) \colon \|\alpha\|_{C^{1,\zeta}} \leq M_0,\ \operatorname{supp}\alpha \subset (\theta_-, \theta_+),\ \alpha(s)|_{s\in I\!R} \geq 0 \right\}$$

where $\zeta \in (0, 1)$.

(A4) The measurement data satisfy $\theta_m \in L^p(0, T; L^p(\omega))$, where ω is an interior open domain such that $\omega \subset \Omega$.

In (A2), we note that $\operatorname{supp}\alpha \subset (\theta_-, \theta_+)$ means that we can choose numbers a, b such that $\theta_- < a < b < \theta_+$ and $\operatorname{supp}\alpha \subset (a, b)$. According to (A1)–(A3) we consider a cooling process from high initial temperature to quenchant temperature. The phase transition happens in the subdomain $(\theta_-, \theta_+) \subset [\theta_w, \theta_0]$. We note well-posedness of the forward model (12a), (12b), (12c)–(13a), (13b) holds true with lower regularity on L and $\alpha(\theta)$. Assumptions (A2)–(A3) are presented here to unify the arguments in the forthcoming discussion, namely, on inverse problems.

In order to proceed further, we recall a standard parabolic regularity result for linear parabolic equations in the space $W_p^{2,1}(Q) := W^{1,p}(0, T; L^p(\Omega)) \cap L^p(0, T; W^{2,p}(\Omega))$ where $Q := \Omega \times (0, T)$ is the space-time cylinder.

Lemma 1 ([18, Theorem 9.1]) *Assume that assumption (A1) holds. Then for any $f \in L^p(\Omega)\,(p \in (1, \infty))$, there exists a unique solution in $W_p^{2,1}(Q)$ for the parabolic system*

$$\begin{aligned} \theta_t - \kappa\Delta\theta &= f \text{ in } \Omega \times (0, T); \\ \kappa\partial_\nu\theta + \sigma(\theta - \theta_w) &= 0 \text{ on } \partial\Omega \times (0, T); \\ \theta(x, 0) &= \theta_0 \text{ in } \Omega \end{aligned}$$

and we have the following a priori estimate

$$\|\theta\|_{W_p^{2,1}(Q)} \leq C_1 + C_2\|f\|_{L^p(Q)}.$$

with constants $C_{1,2}$ and $C_1 = 0$ if $\theta_0 = \theta_w = 0$. If in addition $p > 5/2$, then for $\varepsilon \in (0, 2 - 5/p)$ the solution θ is in $C^{0,\varepsilon}(\bar{Q})$ and the same estimate holds for the $C^{0,\varepsilon}(\bar{Q})$-norm.

Meanwhile, the *a priori* estimates for the ODE system are carried out by changing of variables $\eta := \ln\left(\frac{1}{1-P}\right)$.

Lemma 2 *Assume (A2)–(A3), $\theta \in L^1(Q)$, and fix a finite final time T. Then there holds $\eta(t) \in [0, \eta_{\max}]$ with $t \in [0, T]$ and a constant $\eta_{\max} < \infty$. Moreover, there exists a constant M independent of θ such that*

$$\|\eta\|_{W^{1,\infty}(0,T;L^\infty(\Omega))} \leq M.$$

At the same time, assume that there exist $\theta_1, \theta_2 \in L^p(Q)$ with $p \in [2, \infty)$ with solutions η_1, η_2, then the following estimate holds with a constant $C > 0$

$$\|\eta_1 - \eta_2\|_{W^{1,p}(0,T;L^p(\Omega))} \leq C\|\theta_1 - \theta_2\|_{L^p(Q)}.$$

Proof The proof follows by changing of variables $\eta := \ln\left(\frac{1}{1-P}\right)$ from the original ODE system on P in (11a), (11b). Notice

$$\begin{cases} \eta_t = \gamma \int\limits_0^t \alpha(\theta)(t-\tau)^2 d\tau; \\ \eta(0) = 0. \end{cases}$$

Assuming $\theta \in L^1(Q)$ and the initial condition, we conclude that $\eta(t)$ is increasing and finite in the time interval $[0, T]$ such that $0 \leq \eta(t) \leq \eta_{\max} = \frac{\gamma}{12}M_0T^4$. Moreover, η_t satisfies $0 \leq \eta_t \leq \frac{\gamma}{3}M_0T^3$. The rest of the proof follows by testing the difference of η_1, η_2 by $|\eta_1 - \eta_2|^{p-2}(\eta_1 - \eta_2)$ and applying the Gronwall's and Young's inequalities.

Remark 1 We emphasize that by adding appropriate initial conditions, the original ODE system (11a), (11b) is equivalent to the 4-th order ODE system (13a), (13b). The *a priori* estimates in Lemma 2 are adjusted, respectively, in the following estimates

$$\|\eta\|_{W^{4,\infty}(0,T;L^\infty(\Omega))} \leq M;$$
$$\|\eta_1 - \eta_2\|_{W^{4,p}(0,T;L^p(\Omega))} \leq C\|\theta_1 - \theta_2\|_{L^p(Q)}.$$

In the sequel, we denote by η the solution of the 4-th order ODE system (13a), (13b) where the standard estimates in Lemma 2 are sufficient for the well-posedness of the forward model.

Corollary 1 *Let $\theta \in L^1(Q)$ and fix a finite final time T. Then the term $e^{-\eta}\eta_t$ is nonnegative and bounded with an a priori estimate*

$$\|e^{-\eta}\eta_t\|_{L^\infty(0,T;L^\infty(\Omega))} \leq M$$

where the constant M is independent of η and θ.

Now, we are ready to present the main existence theorem for the PDE-ODE coupled system (12a), (12b), (12c)–(13a), (13b).

Theorem 1 *Assume that assumptions (A1)–(A3) hold true. Then the PDE-ODE coupled system (12a),* (12b), *(12c)–(13a),* (13b) *admits a unique solution (θ, η) such that $\theta \in W_p^{2,1}(Q)$ for $p \in (2, \infty)$ and $\eta \in W^{1,\infty}(0, T; L^\infty(\Omega))$.*

Proof Fix a finite final time $T > 0$, we consider the following closed set

$$K_T := \{\theta \in W_p^{2,1}(Q) : \theta(x,0) = \theta_0\}.$$

Choose $\hat{\theta} \in K_T$, and define the solution η of (13a), (13b) to the governing ODE by

$$\frac{d^4\eta}{dt^4} = 2\gamma\alpha(\hat{\theta}). \tag{14}$$

The solution η uniquely exists and satisfies the *a priori* estimates in Lemma 2.

Now define θ as the solution to (12a), (12b), (12c), where the right-hand side of the governing parabolic equation is replaced by the solution η to (14). Since the *a priori* estimates in Lemma 1 and Corollary 1 are independent of $\hat{\theta}$, we can infer that the operator $S: \hat{\theta} \to \theta$ maps K_T into itself.

At the same time, defining $S(\hat{\theta}_i) = \theta_i$, $i = 1, 2$ with $\hat{\theta}_{1,2} \in K_T$, we can obtain, for $\theta = \theta_1 - \theta_2$ and $\hat{f} := L(\hat{\theta}_1)e^{-\eta_1}\eta_{1,t} - L(\hat{\theta}_2)e^{-\eta_2}\eta_{2,t}$, where η_i is the solution to (14) with respect to $\hat{\theta}_i$ and $\eta_{i,t}$ is the time derivative of each η_i,

$$\begin{cases} \theta_t - \kappa\Delta\theta = \hat{f} \text{ in } (0,T) \times \Omega; \\ \kappa\partial_\nu\theta + \sigma\theta = 0 \text{ on } (0,T) \times \partial\Omega; \\ \theta(x,0) = 0 \text{ in } \Omega. \end{cases}$$

Lemmas 1, 2, (A1), and Hölder's inequality then yield

$$\|\theta_1 - \theta_2\|_{W_p^{2,1}(Q)} \le C\|\hat{f}\|_{L^p(Q)} \le C\|\hat{\theta}_1 - \hat{\theta}_2\|_{L^p(Q)} \le CT^{\frac{p-1}{p}}\|\hat{\theta}_1 - \hat{\theta}_2\|_{W_p^{2,1}(Q)}.$$

Thus, S is a contraction map if we choose $T := T^+$ sufficiently small. The existence of a unique local solution then follows from the Banach fixed point theorem. The global *a priori* estimates in Lemma 1 and Corollary 1 guarantee that such an estimate holds true on the whole interval $[0, T]$.

Remark 2 If we additionally let $p > 5/2$ in Theorem 1, θ and P are provided with additional regularity $\theta \in C^{0,\varepsilon}(\bar{Q})$ and $P \in C^1([0,T], C(\bar{\Omega}))$ consequently.

3 Uniqueness for the Inverse Problems

On in this section, we consider a generalized form of (11a), (11b) for an inverse problem, because the arguments do not depend on the concrete form of the integrand factor $(t - \tau)^2$ in (11a), (11b). More precisely, we consider the system (10a), (10b), (10c) together with

$$P_t(x,t) = (1-P)\int_0^t \alpha(\theta(x,t))\psi(x,t,\tau)d\tau, \quad \text{in} \quad \Omega\times(0,T) \tag{11'a}$$

and

$$P(x,0) = 0 \quad \text{in} \quad \Omega. \tag{11'b}$$

Here for a natural number m we assume

(A5) $\psi(x,t,\tau) \in C(\overline{\Omega}\times[0,T]^2)$ and $\psi(x,\cdot,\tau) \in C^{m+1}[0,T]$ for each $x \in \overline{\Omega}$, $0 \le \tau \le T$. Moreover, we assume

$$\frac{\partial^j \psi}{\partial t^j}(x,t,t) = 0, \quad x \in \overline{\Omega},\ 0 \le t \le T \quad j = 0,1,\dots,m-1 \tag{15}$$

and

$$\frac{\partial^m \psi}{\partial t^m}(x,t,t) \neq 0, \quad x \in \overline{\Omega},\ 0 \le t \le T.$$

Alternatively, if $\psi(x,\cdot,\tau) \in C^m[0,T]$ for each $x \in \overline{\Omega}$ and (15) is satisfied, then we can also consider a particular set of functions such that $\frac{\partial^m \psi}{\partial t^m}(x,t,t)$ is a non-zero constant. Obviously the function $(t-\tau)^2$ satisfies the latter conditions with $m = 2$.

For concentrating on the inverse problem, we assume the unique existence of solution $(\theta(\alpha), P(\theta))$ to the forward problem (10a), (10b), (10c) and (11') such that $\theta(\alpha) \in W_p^{2,1}(Q)\cap C(\overline{Q})$ and $P(\alpha) \in C^1([0,T]; C(\overline{\Omega}))$ with $p > \frac{5}{2}$. For the case of $\psi(x,t,\tau) = \gamma(t-\tau)^2$, as for such solutions, see Remark 2.

Now we formulate our inverse problem and are concerned with whether we can uniquely determine α from temperature measurements in an arbitrarily chosen subdomain $\omega \subset \Omega$ with non-zero measure, that is, we consider the problem

(IP) determine α by $\theta|_{\omega\times(0,T)}$.

Our inverse problem is over-determined. That is, an unknown function α depends on a single variable, while observation data are taken in $x \in \omega$ and $0 < t < T$, and are a function with multi-variables (x,t). However the arguments below are not trivial because we have to discuss the range of $\theta(\alpha)$ in $\omega\times(0,T)$ taking into consideration the integral term of $\theta(\alpha)$ in (11a), (11b).

We are ready to state the main result on the inverse problem.

Theorem 2 (Uniqueness) *Assume (A1)–(A3) and (A5). If $\theta(\alpha^1)(x,t) = \theta(\alpha^2)(x,t)$ for $x \in \omega$ and $0 < t < T$, then $I := \{\theta(\alpha^1)(x,t) : x \in \omega,\ 0 < t < T\}$ contains a non-empty open interval and $\alpha^1(\eta) = \alpha^2(\eta)$ for $\eta \in \overline{I}$.*

In general, $I \cap (\operatorname{supp}\alpha^1 \cup \operatorname{supp}\alpha^2)$ may not have an interior point. Then both sides of the conclusion of the theorem are zero and the theorem is trivial. For example, if $T > 0$ is too small, then by (A2) and $\theta(\cdot, 0) = \theta_0 \geq \max\{\max \operatorname{supp} \alpha^1, \max \operatorname{supp} \alpha^2\}$, we see that $\alpha(\theta) \equiv 0$ in $\omega \times (0, T)$. That is, the conclusion is still trivial.

Remark 3 (1) For our inverse problem, we cannot expect any maximum principle or monotonicity property of θ with respect to α, and we use interior observation data in $\omega \times (0, T)$.

(2) If we can prove that I contains the whole range of $\theta(\alpha)$, that is,

$$\{\theta(\alpha^1)(x, t): x \in \omega, 0 < t < T\} \supset \{\theta(\alpha^1)(x, t): x \in \Omega, 0 < t < T\} := I_{\max},$$

then the theorem gives the uniqueness in determining α^1, α^2 over $I_{\max}$. We notice that (10a), (10b), (10c)–(11a), (11b) does not give any information of α outside of $I_{\max}$ and we cannot expect determination of α outside $I_{\max}$. If we can apply a suitable maximum principle, similarly to DuChateau and Rundell [7] we can expect that our data in $\omega \times (0, T)$ give information of α in $I_{\max}$, and the identification of α over $I_{\max}$ may be possible. However for our system (10a), (10b), (10c)–(11a), (11b) the maximum principle does not work and the uniqueness of α over I cannot be considered as reasonable.

(3) As is seen by the proof, the local uniqueness also follows, if we replace assumption (A1) by one of the following conditions:

- θ_0 is not a constant function in ω.
- θ_0 is a constant in ω and $\theta_w(x, t) \not\equiv \theta_0$ on $\partial\Omega \times (0, T)$.

For the proof we need the following

Lemma 3 *Let $z \in W_2^{2,1}(\Omega \times (0, T))$ satisfy*

$$\partial_t z - \kappa \Delta z = \int_0^t A(x, t, \tau) z(x, \tau) d\tau, \quad x \in \Omega, 0 < t < T$$
$$z(x, 0) = 0, \quad x \in \Omega$$

with $A \in L^\infty(\Omega \times (0, T)^2)$. If $z = 0$ in $\omega \times (0, T)$, then $z = 0$ in $\Omega \times (0, T)$.

The proof of Lemma 3 can be found in Lemma 4.6 in [12] and is done by a Carleman estimate [24].

Proof (Proof of Theorem 2) We divide the proof into two steps.

First Step.

We set $u = \theta(\alpha^1)$, $v = \theta(\alpha^2)$, $p = P(\alpha^1)$, $q = P(\alpha^2)$ and

$$y = u - v, \quad R = p - q.$$

In this step, we will prove

$$\alpha^1(u(x,t)) = \alpha^2(u(x,t)) \tag{16}$$

for $(x,t) \in \omega \times (0,T)$.

By (A3), we can choose a, b such that

$$\theta_- < a < b < \theta_+, \quad \operatorname{supp} \alpha \subset (a,b)$$

for any α under consideration. Thus we have $\alpha^1(u(x,t)) = \alpha^2(u(x,t)) = 0$ if $u(x,t) \le a$. Therefore we can assume

$$u(x,t) > a, \quad (x,t) \in \omega \times (0,T). \tag{17}$$

Since $u = v$ in $\omega \times (0,T)$, we have $y = 0$ in $\omega \times (0,T)$. Since

$$\partial_t y = \kappa \Delta y + L(u)\partial_t R + (L(u) - L(v))\partial_t q \quad \text{in } \Omega \times (0,T),$$

we have

$$L(u(x,t))\partial_t R(x,t) = 0 \quad \text{in } \omega \times (0,T) \tag{18}$$

and

$$\begin{aligned}\partial_t R &= (1-p)\int_0^t \alpha^1(u)\psi(x,t,\tau)d\tau - (1-q)\int_0^t \alpha^2(v)\psi(x,t,\tau)d\tau \\ &= -R\int_0^t \alpha^1(u(x,\tau))\psi(x,t,\tau)d\tau + (1-q)\int_0^t (\alpha^1 - \alpha^2)(u(x,\tau))\psi(x,t,\tau)d\tau, \end{aligned} \tag{19}$$

$$\text{for all } (x,t) \in \omega \times (0,T).$$

Let $x \in \omega$ be arbitrarily fixed. Since $u(x,0) = \theta_0 = \theta_+$ and $\alpha^k(\theta_0) = 0, k = 1,2$ by (A3), we have (16) for $t = 0$. Starting at $t = 0$, we will prove (16) for all $(x,t) \in \omega \times (0,T)$ by increasing t. The proof can be done by using (18) and (19) in the cases $u(x,t) > b$ for $0 \le t \le T$ and $u(x,t') \le b$ for some $t' > 0$ respectively. Since (19) is involved with the integral term from 0 to t, the extension argument in t is necessary. More precisely, we will do as follows. Let $u(x,t) > b$ for $0 \le t \le T$. Then $\alpha^1(u(x,t)) = \alpha^2(u(x,t)) = 0$ for $0 \le t \le T$, and we already have proved (16) for $0 \le t \le T$. Next we assume that there exists $\widetilde{t} \in (0,T]$ such that $u(x,\widetilde{t}) = b$. We set $t_0 = t_0(x) = \min\{t \in [0,T] \colon u(x,t) = b\}$. By the continuity of u, such t_0 exists. Moreover $t_0 > 0$ by $u(x,0) = \theta_0 > b$. Hence we have $u(x,t) > b$ for $0 \le t < t_0$. Therefore $\alpha^1(u(x,t)) = \alpha^2(u(x,t)) = 0$ for $0 \le t < t_0$ and

$$\partial_t R(x,t) = 0, \quad 0 \le t \le t_0 \tag{20}$$

by the first equation in (19). By $u \in C(\overline{Q})$, there exists $\delta_0 > 0$ such that $a < u(x,t) < \theta_+$ for $t_0 - \delta_0 \le t \le t_0 + \delta_0$. Therefore (A2) implies $L(u(x,t)) \ne 0$ for $t_0 - \delta_0 \le t \le t_0 + \delta_0$. Hence (18) yields $\partial_t R(x,t) = 0$ for $t_0 \le t \le t_0 + \delta_0$ and it follows from (20) that $\partial_t R(x,t) = 0$ for $0 \le t \le t_0 + \delta_0$. Consequently, by $R(x,0) = P(\alpha^1(x,0)) - P(\alpha^2(x,0)) = 0$, we have

$$R(x,t) = 0, \quad 0 \le t \le t_0 + \delta_0. \tag{21}$$

On the other hand, we can solve (11') to verify $1 - q(x,t) > 0$ for $x \in \overline{\Omega}$ and $0 \le t \le T$. By (19), we see

$$\int_0^t (\alpha^1 - \alpha^2)(u(x,\tau))\psi(x,t,\tau)d\tau = 0, \quad 0 \le t \le t_0 + \delta_0.$$

Differentiating $(m+1)$-times, we have

$$(\alpha^1 - \alpha^2)(u(x,t))\frac{\partial^m \psi}{\partial t^m}(x,t,t) + \int_0^t (\alpha^1 - \alpha^2)(u(x,\tau))\frac{\partial^{m+1}\psi}{\partial t^{m+1}}(x,t,\tau)d\tau = 0$$

for $0 \le t \le t_0 + \delta_0$. Since $\frac{\partial^m \psi}{\partial t^m}(x,t,t) \ne 0$ for $x \in \overline{\Omega}$ and $0 \le t \le T$ by assumption (A5), the Gronwall inequality yields (16) for $0 \le t \le t_0 + \delta_0$. Moreover (19) yields

$$\partial_t R(x,t) = R(x,t)H(x,t) + (1 - q(x,t))\int_{t_0+\delta_0}^t (\alpha^1 - \alpha^2)\psi(x,t,\tau)d\tau, \qquad t \ge t_0 + \delta_0, \tag{22}$$

where we set $H(x,t) = -\int_0^t \alpha^1(u(x,\tau))\psi(x,t,\tau)d\tau$.

Next we extend (16) for $t > t_0 + \delta_0$. We consider two cases $u(x, t_0 + \delta_0) < b$ and $u(x, t_0 + \delta_0) \ge b$. First we assume that $u(x, t_0 + \delta_0) < b$. By the continuity of u, we can choose sufficiently small $\delta_1 > 0$ such that $u(x,t) < b$ for $t_0 + \delta_0 \le t \le t_0 + \delta_0 + \delta_1$. By (17) we can assume that $u(x,t) > a$. Hence by (A2) we have $L(u(x,t)) \ne 0$ for $t_0 + \delta_0 \le t \le t_0 + \delta_0 + \delta_1$, and by (18) we have $\partial_t R(x,t) = 0$ for $t_0 + \delta_0 \le t \le t_0 + \delta_0 + \delta_1$. By (21), we have $R(x, t_0 + \delta_0) = 0$, so that

$$R(x,t) = 0, \quad t_0 + \delta_0 \le t \le t_0 + \delta_0 + \delta_1. \tag{23}$$

Then using (23) and differentiating (22) $(m+1)$-times, in view of $1 - q > 0$ on $\overline{Q}$, we obtain (16) for $t_0 + \delta_0 \le t \le t_0 + \delta_0 + \delta_1$ in the case of $u(x, t_0 + \delta_0) < b$.

Next we assume that $u(x, t_0 + \delta_0) \ge b$. Again by the continuity of u, we see that there exists small $\delta_1 > 0$ such that $u(x,t) > \max_{k=1,2} \operatorname{supp} \alpha^k$ for $t_0 + \delta_0 \le t \le t_0 + \delta_0 + \delta_1$. Hence we have $\alpha^1(u(x,t)) = \alpha^2(u(x,t)) = 0$ for $t_0 + \delta_0 \le t \le t_0 + \delta_0 + \delta_1$ with small $\delta_1 > 0$. Therefore (22) yields

$$\partial_t R(x,t) = R(x,t)H(x,t), \quad t_0+\delta_0 \le t \le t_0+\delta_0+\delta_1. \tag{24}$$

Hence, since $R(x, t_0+\delta_0) = 0$ by (21), the uniqueness for the initial value problem (24) with $R(x, t_0+\delta_0) = 0$ yields (23) and (16) for $t_0+\delta_0 \le t \le t_0+\delta_0+\delta_1$. Thus in both cases of $u(x, t_0+\delta_0) < b$ and $u(x, t_0+\delta_0) \ge b$, we have (16) for $0 \le t \le t_0+\delta_0+\delta_1$.

We will extend (16) for $t > t_0+\delta_0+\delta_1$. Let $u(x,t)$ first gain the value $u_{\min} := \min_{0\le t\le T} u(x,t)$ at t_1:

$$u(x,t_1) = u_{\min}, \quad u(x,t) > u_{\min} \quad \text{if } 0 \le t < t_1.$$

Such t_1 exists by the continuity of u. Moreover $t_1 > 0$ by $u(x,0) = \theta_0 > b$. We can repeat the above argument and obtain (16) for $0 \le t \le t_1$. In fact, we set $\widehat{t} = \max\{t \in [0,t_1]: \alpha^1(u(x,\xi)) = \alpha^2(u(x,\xi)),\ 0 \le \xi \le t\}$ and we assume that $\widehat{t} < t_1$. In view of (17), repeating the previous argument in extending (16) on $[0, t_0+\delta_0]$ to $[0, t_0+\delta_0+\delta_1]$, we can see that there exists small $\widetilde{\delta} > 0$ such that (16) holds for $0 \le t \le \widehat{t}+\widetilde{\delta}$. This contradicts the definition of $\widehat{t}$. Thus (16) holds for $0 \le t \le t_1$.

Thus, varying $x \in \omega$, we obtain (16) for all $x \in \omega$ and $0 < t < T$.

Second Step.

In this step, we will prove that I contains a non-empty open interval. To this end, in terms of the intermediate value theorem, it suffices that $I = \{u(x,t): x \in \omega, 0 \le t \le T\}$ contains at least two points. Assume that $u(x,t) = \theta(\alpha^1)(x,t)$ is constant for $x \in \omega$ and $0 < t < T$. Then $u \equiv \theta_0$ in $\omega \times (0,T)$ by (10a), (10b), (10c). We set $z = u - \theta_0$. Therefore $z = 0$ in $\omega \times (0,T)$. On the other hand, by (A3) and $\theta_0 > b$, we obtain $\alpha^1(\theta_0) = 0$. The mean value theorem yields $\alpha^1(u(x,\tau)) = \alpha^1(z+\theta_0) = \alpha^1(\theta_0) + (\alpha^1)'(\mu)z = (\alpha^1)'(\mu)z(x,\tau)$ for $(x,\tau) \in \Omega \times (0,T)$, where μ is between θ_0 and $u(x,\tau)$. Hence, choosing $A \in L^\infty(\Omega \times (0,T)^2)$ suitably, we can rewrite (10a), (10b), (10c) and (11') in terms of $z := u - \theta_0$:

$$\partial_t z - \kappa\Delta z = \int_0^t A(x,t,\tau)z(x,\tau)d\tau, \quad x \in \Omega,\ 0 < t < T$$

with

$$z(x,0) = 0, \quad x \in \Omega.$$

In view of Lemma 3, we have $u = \theta_0$ in $\Omega \times (0,T)$. Therefore the boundary condition of θ^1 yields $\theta_0 = \theta_w$. This contradicts $\theta_0 > \theta_w$. Thus the proof is completed.

Remark 4 We can prove the uniqueness if we replace $\operatorname{supp}\alpha \subset (\theta_-, \theta_+)$ by a weaker condition $\operatorname{supp}\alpha \subset [\theta_-, \theta_+]$. However we here assume the former by the physical background.

4 Numerical Illustration

In this section we present a numerical example with a 2D problem as an illustration. The minimization approach is realized by the following cost functional

$$J(\alpha) := \frac{1}{2}\int_0^T \int_\omega (F(\alpha) - \theta_m)^2 \mathrm{d}x\mathrm{d}t$$

with the measurement data θ_m in a small interior domain $\omega \subset \Omega$ satisfying (A4). Thus we aim at seeking a minimizer α satisfying

$$\min_{\alpha \in \mathscr{A}_{ad}} J(\alpha). \tag{25}$$

For simplicity's sake the following PDE-ODE coupled system will be considered

$$\begin{cases} \theta_t - \kappa \Delta\theta = L P_t \text{ in } \quad \Omega \times (0, T); \\ \kappa \partial_\nu \theta + \sigma(\theta - \theta_w) = 0 \text{ on } \quad \partial\Omega \times (0, T); \\ \theta(x, 0) = \theta_0 \text{ in } \quad \partial\Omega \end{cases} \tag{26}$$

and

$$\begin{cases} P_t = \gamma(1 - P) \int_0^t \alpha(\theta)(t - \tau)^2 d\tau \text{ in } \quad \Omega \times (0, T); \\ P(0) = 0 \text{ in } \quad \Omega \end{cases} \tag{27}$$

where $\Omega = (-1, 1) \times (-1, 1)$, $L = 151.099$, $\kappa = 0.125$, $\sigma = 1$, $\theta_w = 20$, $\theta_0 = 800$ and $\gamma = 4\pi$. Our choice of these data reflects the cooling of an eutectoid carbon steel, which is known to exhibit one diffusive phase transition below the temperature θ_0, see, e.g., [13]. Moreover, we assume a uniform growth rate $\rho = 1$. To realize the forward problem we let the nucleation rate $\alpha(\theta) = 6\exp(-0.02(\theta - 700)^2)$ and discretize the coupled system with the finite element method by the Matlab PDE toolbox. The measurement domain ω is a circle centered at $(-0.4, 0.6)$ with a radius 0.2. In Figs. 1 and 2, we collect the complete domain, the measurement θ_m at $T = 3$ and the temperature distribution $\theta(x, t)$, phase volume fraction $P(x, t)$ at $x = (-0.4, 0.6)$. As one can observe in the left panel of Fig. 2 the cooling process is disturbed by the latent heat induced by the phase volume fraction P especially at $t \in (0.5, 1.5)$.

In order to identify the nucleation rate $\alpha(\theta)$ with respect to the measured temperature distribution on ω we define the support of α $\mathrm{supp}(\alpha) = (\theta_-, \theta_+)$ with $\theta_- = 650$ and $\theta_+ = 750$. We then discretize the domain $[\theta_-, \theta_+]$ with equal-distance distributed points $\theta_- := \tau_0 \le \tau_1 < \cdots < \tau_N := \theta_+$ and approximate α with cubic B-splines basis functions $\varphi_i(\tau)$ such that

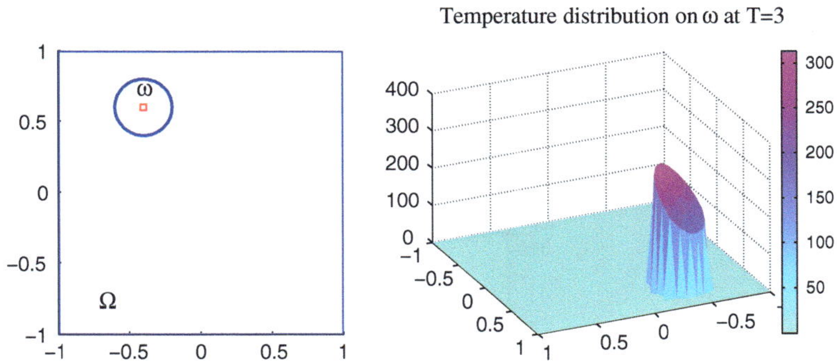

Fig. 1 *Left* The whole domain Ω and the observation domain ω. Temperature distribution and phase volume fraction at $\square$ ($x = (-0.4, 0.6)$) are presented in Fig. 2. *Right* Measurement $\theta_m(x, T)$ for $x \in \omega$ and $T = 3$

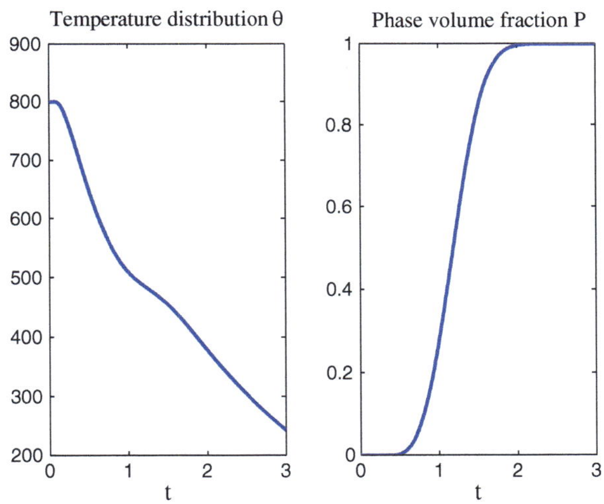

Fig. 2 *Left* Temperature distribution $\theta(x, t)$. *Right* Phase volume fraction $P(x, t)$ for $x = (-0.4, 0.6)$ and $t \in [0, 3]$

$$\alpha^N(\tau) = \sum_{i=1}^{N} \alpha_i \varphi_i(\tau), \quad \tau \in [\theta_-, \theta_+]$$

with $N = 9$.

We thus define a finite-dimensional admissible set

$$\alpha_{ad}^N = \{\alpha^N = (\alpha_1, \ldots, \alpha_N)^T \in I\!R^N\colon \ 0 \leq C_m \leq \alpha_i \leq C_M \quad \text{for} \quad i = 1, \ldots, N\}$$

with the upper and lower constraints C_M and C_m. The original (infinite-dimensional) minimization problem (25) is reduced into a finite form such that $J_{dis}(\alpha^N) =$

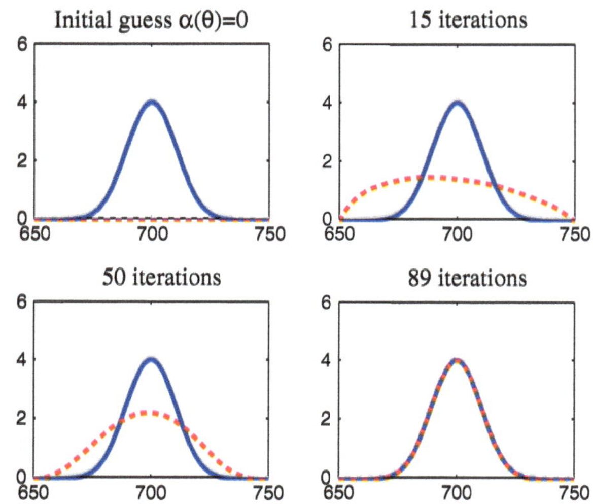

Fig. 3 4 snapshots of the approximated solution towards the exact measured data. The *solid line* is the exact solution, the *dotted line* is the approximated one

$J(\theta(\alpha^N), \alpha^N)$ and define $\bar{\alpha}^N$ to be the variable. The parameterized discrete system allows us to solve the minimization problem heuristically with a quasi-Newton method by calling Matlab command *fmincon*. In Fig. 3, we displayed four snapshots of the iterative solutions towards the exact measured data. In this example, the *fmincon* iterates 89 times and provides a small functional value $J_{dis}(\alpha^N)$ approximately at 1.8×10^{-12}.

5 Conclusions

In the present chapter we have investigated the identification of the temperature dependent nucleation rate in binary phase transition. We have shown its unique identifiability from measurements in a subdomain. Numerical results with model data support the feasibility of our approximation schemes.

We note that justified by the uniqueness result we also have employed an optimal control approach to the realization of (25) which can be found in our recent work [12]. The optimal control approach there is done in the spirit of [22], where the identification of a nonlinear heat transfer law is studied. In [11] a similar approach has been taken to identify a temperature dependent rate law for the coagulation of cancerous tissue. In addition we note that optimal control problems for nucleation and growth models related to the crystallization of polymers have been studied in [4, 10]. In [19] a simplified version of the generalized Avrami model has been developed.

References

1. Avrami M.: Kinetics of phase change. I general theory. J. Chem. Phys. **7**, 110311.12. (1939)
2. Avrami, M.: Kinetics of phase change. II transformation time relations for random distribution of nuclei. J. Chem. Phys. **8**, 212–224 (1940)
3. Avrami, M.: Kinetics of phase change. III granulation, phase change, and microstructure kinetics of phase change. J. Chem. Phys. **9**, 177–184 (1941)
4. Burger, M., Capasso, V., Micheletti, A.: Optimal control of polymer morphologies. J. Eng. Math. **49**, 339–358 (2004)
5. Choulli, M., Ouhabaz, E.M., Yamamoto, M.: Stable determination of a semilinear term in a parabolic equation. Commun. Pure Appl. Anal. **5**, 447–462 (2006)
6. Christian, J.W.: The Theory of Transformations in Metals and Alloys, Part I. Pergamon, New York (2002)
7. DuChateau, P., Rundell, W.: Unicity in an inverse problem for an unknown reaction term in a reaction-diffusion equation. J. Differ. Equ. **59**, 155–164 (1985)
8. Egger, H., Engl, H., Klibanov, M.V.: Global uniqueness and Hölder stability for recovering a nonlinear source term in a parabolic equation. Inverse Prob. **21**, 271–290 (2005)
9. Eisenhüttenleute, V.D. (ed): Steel—A Handbook for Materials Research and Engineering. Vol. 1: Fundamentals. Springer, Berlin (1992)
10. Götz, Th, Rinnau, R., Struckmeier, J.: Optimal control of crystallization processes. Math. Models Methods Appl. Sci. **16**, 2029–2045 (2006)
11. Hömberg, D., Liu, J., Togobytska, N.: Identification of the thermal growth characteristics of coagulated tumor tissue in laser-induced thermotherapy. Math. Meth. Appl. Sc. **35**, 497–509 (2012)
12. Hömberg, D., Lu, S., Sakamoto, K., Yamamoto, M.: Parameter identification in non-isothermal nucleation and growth processes. Inverse Prob. **30**, 035003 (24pp) (2014)
13. Hömberg, D., Togobytska, N., Yamamoto, M.: On the evaluation of dilatometer experiments. Appl. Anal. **88**, 669–681 (2009)
14. Isakov, V.: On uniqueness in inverse problems for semilinear parabolic equations. Arch. Rat. Mech. Anal. **124**, 1–12 (1993)
15. Johnson, W.A., Mehl, R.F.: Reaction kinetics in processes of nucleation and growth. Trans. Amer. Inst. Min. Metallurg. Eng. Iron Steel Div. **135**, 416–458 (1939)
16. Klibanov, M.V.: Global uniqueness of a multidimensional inverse problem for a nonlinear parabolic equation by a Carleman estimate. Inverse Prob. **20**, 1003–1032 (2004)
17. Kolmogorov, A.N.: A statistical theory for the recrystallization of metals. Izvestia Akademia Nauk Serie Mathematica SSSR **1**, 355–359 (1937)
18. Ladyženskaja, O.A., Solonnikov, V.A., Ural'ceva, N.N.: Linear and Quasilinear Equations of Parabolic Type. Amer. Math. Soc. Transl. **23**, Providence (1968)
19. Leblond, J.B., Devaux, J.: A new kinetic model for anisothermal metallurgical transformations in steels including effect of austenite grain size. Acta Met. **32**, 137–146 (1984)
20. Lorenzi, A.: An inverse problem for a semilinear parabolic equation. Annali di Mat. Pura ed Appl. **131**, 145–166 (1982)
21. Pilant, M.S., Rundell, W.: An inverse problem for a nonlinear parabolic equation. Comm. Partial Differ. Equ. **11**, 445–457 (1986)
22. Rösch, A., Tröltzsch, F.: An optimal control problem arising from the identification of nonlinear heat transfer law. Arch. Control Sci. **1**(3–4), 183–195 (1992)
23. Visintin, A.: Mathematical models of solid-solid phase transitions in steel. IMA J. Appl. Math. **39**, 143–157 (1987)
24. Yamamoto, M.: Carleman estimates for parabolic equations and applications. Inverse Prob. **25**, 123013 (75pp) (2009)

Multi-scale Problems, High Performance Computing and Hybrid Numerical Methods

G. Balarac, G.-H. Cottet, J.-M. Etancelin, J.-B. Lagaert, F. Perignon and C. Picard

Abstract The turbulent transport of a passive scalar is an important and challenging problem in many applications in fluid mechanics. It involves different range of scales in the fluid and in the scalar and requires important computational resources. In this work we show how hybrid numerical methods, combining Eulerian and Lagrangian schemes, are natural tools to address this mutli-scale problem. One in particular shows that in homogeneous turbulence experiments at various Schmidt numbers these methods allow to recover the theoretical predictions of universal scaling at a minimal cost. We also outline how hybrid methods can take advantage of heterogeneous platforms combining CPU and GPU processors.

Keywords High performance computing · Particle method · Hybrid computing · Turbulence · Transport equations

1 Introduction

Numerical simulations have become a routine tool to develop, prototype and/or validate products and processes in industry. Applications encompass virtually all sectors of activity from Aeronautics, Automotive industry and Oil exploration to Circuit

G. Balarac
LEGI, CNRS and Université de Grenoble, BP 53, 38041 Grenoble Cedex 9, France
e-mail: guillaume.balarac@grenoble-inp.fr

G.-H. Cottet (✉), J.-M. Etancelin, F. Perignon and C. Picard
Laboratoire Jean Kuntzmann, CNRS and Université de Grenoble, BP 53,
38041 Grenoble Cedex 9, France
e-mail: georges-henri.cottet@imag.fr

J.-B. Lagaert
Laboratoire de Mathématiques, Université Paris 11, 91405 Orsay Cedex, France
e-mail: jean-baptiste.lagaert@math.u-psud.fr

M. Wakayama et al. (eds.), *The Impact of Applications on Mathematics*,
Mathematics for Industry 1, DOI: 10.1007/978-4-431-54907-9_18,

design, Biomechanics and Animations studios, to name a few. With the need to perform more and more realistic simulations and the advent of supercomputers, available in national or regional centers, the field of High Performance Computing (HPC) is not anymore restricted to academia and scientific grand challenges but starts to reach SMEs.

HPC requires easy and flexible access to HPC facilities, obviously, and to master the appropriate programming language, but also to question the numerical methods and algorithms that are used in the simulations. These methods and algorithms should be adapted both to the physics of the problems to solve and to the architecture of the simulation platforms. Moreover since these platforms are often of an hybrid nature, that is combine different type of processors, typically CPU and GPU processors, one may also wish to develop or use methods which couple different types of algorithms that can be optimally distributed to different types of processors.

This is particularly desirable if the problem to solve is multi-level by nature. In that case the different scales that are to be represented can also be resolved on different types of processors. This is hybrid computing. In some sense the nature of the problem and of the hardware inspires the type of mathematical and numerical models that should be used for optimal efficiency.

The purpose of this paper is to describe ongoing work in our group towards hybrid computing for applications in turbulent transport of passive scalar. In the next section we briefly describe the physical context of this work. In Sect. 3 we describe a hybrid method coupling a semi-Lagrangian method for the scalar transport and a spectral method for incompressible flows and we show some results obtained with this method. Section 4 is devoted to the implementation of scalar transport on GPU processors.

2 Universal Scaling in Turbulent Transport

The prediction of the dynamics of a scalar advected by a turbulent flow is an important challenge in many applications. Some of these applications are illustrated in Figs. 1, 2 and 3. Figure 1 shows the dynamics of a pollutant ejected by a sewer in the Los Angeles bay at two different times. Figure 2 shows the atomization of a jet. In that case the transported quantity is the water-air interface [1]. Figure 3 shows a similar experiment but in the context of combustion. In this case the transported quantities are concentrations of chemical species [2]. All these illustrations share a common feature, namely that very small scales spontaneously appear and need to be captured if accurate predictions are needed for the location of the pollutant, the size of the droplets or the combustion efficiency, respectively, are sought.

The production of small scales in an advected scalar indeed reflects some fundamental turbulence properties and is driven by the value of the Schmidt number, Sc, the ratio between the viscosity of the fluid and the diffusivity of the scalar. If $Sc > 1$, the so-called Batchelor scale η_B which measures the size of the smallest scalar fluctuations is smaller than the smallest length scales of the turbulent flow

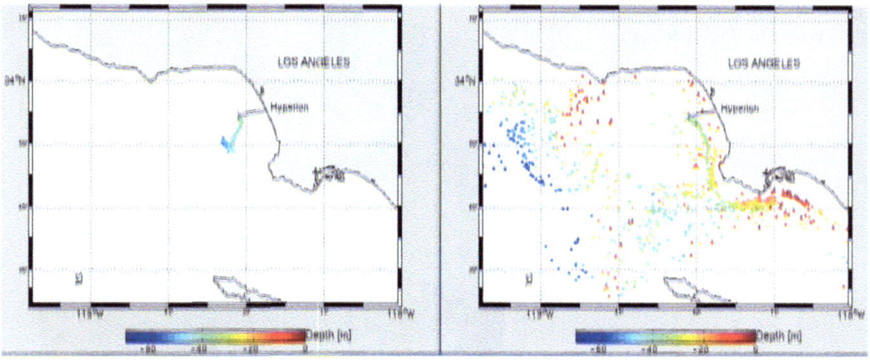

Fig. 1 Transport of a pollutant in the bay of Los Angeles at two different times. Courtesy of E. Blayo (Université Joseph Fourier, Grenoble)

Fig. 2 Atomization of a jet. Courtesy of S. Zaleski, Université Pierre et Marie Curie, Paris

(the Kolmogorov scale η_K). These scales are related by $\eta_B = \eta_K/\sqrt{Sc}$. More precisely, for $Sc > 1$, Batchelor [3] reports that the classical Corrsin-Obukhov cascade associated with a $k^{-5/3}$ law (where k is the wave number) for the scalar variance spectrum [4, 5] is followed by a viscous-convective range with a k^{-1} power law. This viscous-convective range is followed by the dissipation range, where various theoretical scalings have been proposed for the spectrum [3, 6]. A direct consequence of this fact is that, for $Sc > 1$, in numerical simulations the scalar is more demanding, in terms of grid resolution, than the flow itself. It is therefore natural to envision numerical approaches which use different grid resolutions for the scalar and the momentum.

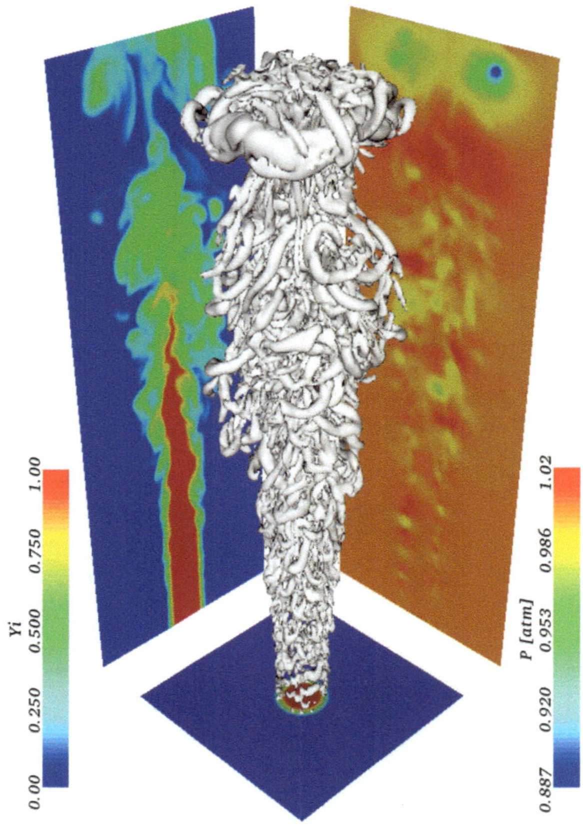

Fig. 3 Reacting jet. Courtesy of L. Vervisch, INSA Rouen

3 Hybrid Particle-Spectral Method

We consider in the following the scalar equation

$$\frac{\partial \theta}{\partial t} + \mathbf{u} \cdot \nabla \theta = \nabla \cdot (\kappa \nabla \theta) \tag{1}$$

coupled with the incompressible Navier-Stokes equation

$$\frac{\partial \mathbf{u}}{\partial t} + \mathbf{u} \cdot \nabla \mathbf{u} = \nabla \cdot (\nu \nabla \mathbf{u}) - \nabla p \ , \ \nabla \cdot \mathbf{u} = 0, \tag{2}$$

in a periodic box. κ is the molecular scalar diffusivity, ν the flow viscosity and $\mathbf{u}$ the flow velocity field. Using different grid resolutions for the scalar and the flow has already been considered for instance in [7, 8]. In the latter reference a compact finite-difference method was used for the scalar and a pseudo-spectral method for the flow. A significant speed-up over a pure spectral solver with high resolution for both the momentum and the scalar was obtained. Our choice here is to combine a particle method for the scalar and a spectral method for the flow.

Our motivation to choose a particle method for the scalar advection comes form the fact that, for large Schmidt numbers, the scalar dynamics is essentially driven by advection, a regime for which Lagrangian or semi-Lagrangian methods are well suited. Moreover in such method, the stability limits for the time-step are governed by the amount of strain in the flow, and not by the grid size. In the present context where high resolutions of the scalar are desired, this is definitely a feature that is expected lead to an important speed up.

More precisely, the method we use for the scalar is a semi-Lagrangian (or remeshed) particle method, where at every time step particles carrying the scalar values are moved along the streamlines of the velocity then remeshed on a regular cartesian grid. Remeshing particles on a regular grid is a way to guarantee the accuracy of particle methods. This approach has been systematically used and validated in a number of simulation of vortex flows [9–13] or in combination with level set methods for interface capturing [14–16]. Remeshing particles at every time-step also allows to easily couple the method to grid based methods, in particular when velocity values are computed on a grid. These methods can be summarized by the following formula

$$\theta_i^{n+1} = \sum_j \theta_j^n \Gamma\left(\frac{x_j^{n+1} - x_i}{\Delta x^\theta}\right), \tag{3}$$

where θ_j^n denotes the value of the scalar at the grid point x_j and at time $t_n = n\Delta t^\theta$, x_j^{n+1} is the location after one advection step of the particle initialized at time t_n on the grid point x_j, and Δx^θ and Δt^θ denote the grid size and the time-step. In the above formula Γ is an interpolating kernel, the smoothness and the moment properties of which govern the spatial overall accuracy of the method [17]. In this work we chose the following second order kernel

$$\Gamma(x) = \begin{cases} \frac{1}{12}(1-|x|)\left(25|x|^4 - 38|x|^3 - 3|x|^2 + 12|x| + 12\right) & \text{if } 0 \le |x| < 1 \\ \frac{1}{24}(|x|-1)(|x|-2)\left(25|x|^3 - 114|x|^2 + 153|x| - 48\right) & \text{if } 1 \le |x| < 2 \\ \frac{1}{24}(3-|x|)^3(5|x|-8)(|x|-2) & \text{if } 2 \le |x| < 3 \\ 0 & \text{if } 3 \le |x|. \end{cases}$$

The scalar time-step is given by $\Delta t^\theta = (|\nabla \mathbf{u}|_{max})^{-1}$. As already mentioned this value does not depend on the scalar grid size.

For the momentum equation we use a classical pseudo-spectral method, with the 3/2 rule to de-alias inertial terms and a second-order Runge-Kutta scheme is used both for the time-stepping of the spectral method and to advect particles. Precise descriptions of the methods and of the experimental set up are given in [17, 18].

Figure 4 shows a comparison of the scalar spectra obtained by the present coupling method and pure spectral method in an experiment of decaying homogeneous turbulence. In the hybrid method two different resolutions where used for the scalar.

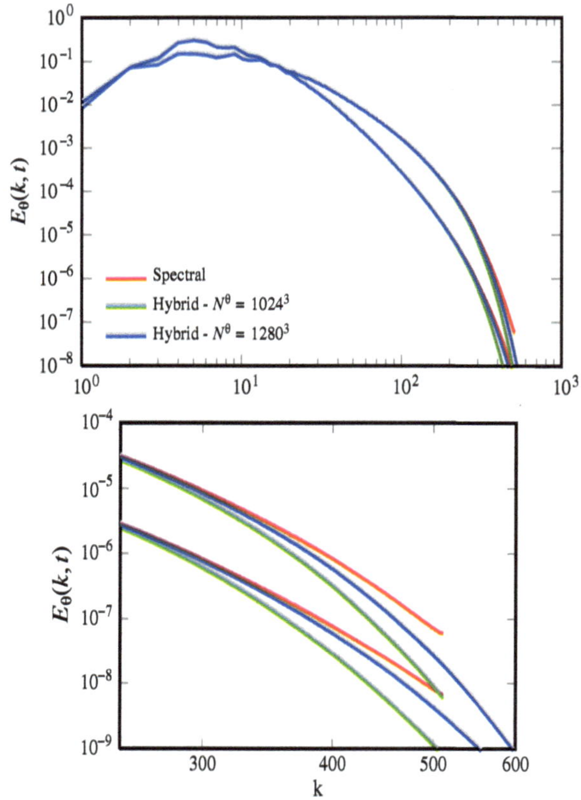

Fig. 4 Spectra of the scalar variance $E_\theta(k, t)$ at two two different times for $Sc = 50$. *The bottom* picture is a zoom of *the top* picture on the smallest scales

This experiment shows that, provided the particle method is used with slightly more grid points than needed by the spectral method, the scalar values are well recovered all the way to the dissipation scale. In this Direct Numerical Simulation, the Schmidt number was equal to 50 and the momentum equation was solved with 256 modes in each direction.

To evaluate the efficiency of the hybrid method, we show in Table 1 CPU times for the full spectral method and the hybrid method for $Sc = 50$. All runs correspond to fully resolved simulations for the Navier-Stokes equations. One can see that, because it can use much larger time-steps, the hybrid method, even when it uses slightly more points to accurately resolve the finest scales, leads to significant savings over the pure spectral method. Additional validation and diagnostics are give in [18].

The computational efficiency of the hybrid method allows to address more challenging cases and to investigate in a systematic fashion the universal scaling laws in the case of forced homogeneous turbulence. Table 2 summarizes the simulation set up corresponding to two values of the Reynolds number and several Schmidt numbers.

Table 1 Numerical efficiency of the different methods on a decaying turbulence experiment-Runs are performed on 2,048 cores of a Blue Gene Q

Method	N^u	N^θ	Δt^u ($\times 10^{-4}$)	Δt^θ ($\times 10^{-4}$)	Total CPU time (s)
Spectral	1024^3	$1{,}024^3$	2.5	2.5	43,590
Spectral	256^3	$1{,}024^3$	2.5	2.5	16,671
Hybrid	256^3	$1{,}024^3$	10	100	1,139
Hybrid	256^3	$1{,}280^3$	10	100	1,328

N^u, N^θ denotes the spatial resolution for velocity and scalar and Δt^u, Δt^θ are the numerical time steps for momentum and scalar equations. CPU times correspond to the simulation time $t = 6$

Table 2 Setup of simulations performed in forced homogeneous turbulence

R_λ	N^u	Δt^u	Sc	N^θ	Δt^θ (e^{-2})	$\Delta t^\theta_{\text{spec}}$ (e^{-3})
130	256^3	1.2 e^{-2}	0.7	512^3	8.6	6
			4	1024^3		3
			8	1024^3		3
			16	1536^3		2
			32	1536^3		2
			64	2048^3		1.5
			128	3064^3		1
210	512^3	3 e^{-3}	0.7	770^3	2	2
			4	1024^3		1.5

Δt^u is the time step used to solve the Navier-Stokes equation with a pseudo-spectral solver. Δt^θ is the time step used to solve the scalar transport equation with the particle method. $\Delta t^\theta_{\text{spec}}$ is the time step which would be needed if a pseudo-spectral method was used for the same number of scalar grid points [18]

For the highest Schmidt number, $Sc = 128$, the simulation used $3{,}064^3$ computational elements for the scalar equation, on a IBM Blue Gene supercomputer. The ratio between the time-step used in the present simulation and that which would have been required in a comparable spectral simulation is almost equal to 100. Figure 5 shows the compensated spectra of the scalar for a Reynolds number based on the Taylor mico-scale R_λ [21] equal to 130. These spectra do exhibit a k^{-1} decay on a range which increases with the Schmidt number. Beyond this viscous-convective range, the spectra follow an exponential decay coinciding with the scaling law proposed by Kraichnan [6].

Figure 6 shows vorticity and scalar contours in a cross section of the computational box for the simulation corresponding $R_\lambda \approx 130$ and $Sc = 128$. It illustrates the scale separation between the flow and the scalar for these parameters. The extension of the hybrid method to the coupling of particle methods with finite-volume methods to address engineering configurations is under way.

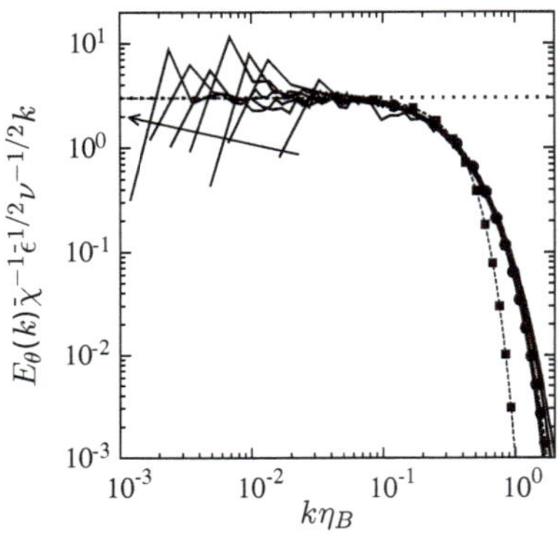

Fig. 5 Compensated spectra for the scalar variance at $R_\lambda \approx 130$. The *arrow* shows the direction of increasing Schmidt numbers. In the dissipative region, the *circles* show the law proposed by Kraichnan and the *squares* show the law proposed by Batchelor in the dissipative scales. The *vertical axis* shows the spectra compensated by the Batchelor law predicting a k^{-1} decay in the intermediate scale

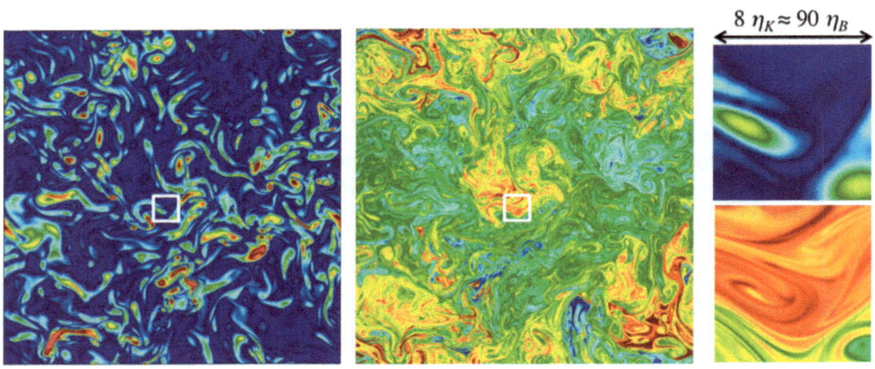

Fig. 6 Cross-section colored by the vorticity magnitude (*left*, *blue* regions are for the lowest vorticity values and *red* regions are for the highest vorticity values) and by the passive scalar (*middle*, *blue* regions are for the lowest scalar values and *red* regions are for the highest scalar values) for $R_\lambda \approx 130$ and $Sc = 128$. The zooms (*right*) for the vorticity magnitude (*top*) and the scalar (*bottom*) correspond to the *white box* with a length approximately equal to the Kolmogorov scale

4 Towards Hybrid Computing

As already mentioned the multi-scale nature of turbulent transport makes natural the idea of hybrid computing methodologies where different part of the problems are distributed to different types of hardware. To be able to implement hybrid algorithms on hybrid architectures, one needs to develop frameworks and libraries with a high level description which allows to distribute different solvers and grids to different parts of the clusters in a seamless fashion. Both particle advection and particle remeshing are local operations. This limits the communications between computational elements

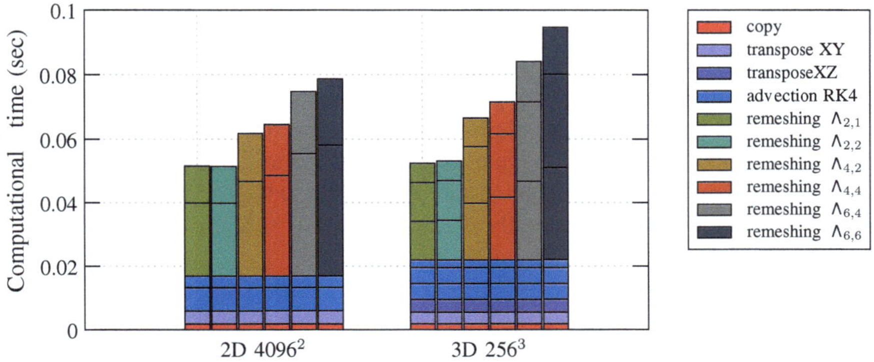

Fig. 7 Code profiling of one time step for different problem sizes in double precision

and makes semi-Lagrangian particle methods well suited to parallel implementations [20]. Such an implementation is described in [19] for the 2D Navier-Stokes equations and in [17] for 3D linear transport equations. In [17], to achieve good portability, the computational frameworks are written using OpenCL.

The efficiency of GPU algorithms is very much conditioned by memory access strategies. To minimize the resulting computational overhead, we use a directional splitting where particles are pushed and remeshed successively along each direction. This allows to send a given number of independent particle lines on a single work-group. This strategy requires to transpose data after each direction has been processes. However on modern GPU cards, transpositions can be achieved at a cost close to that of a simple copy operation. Figure 7 shows the computational cost of our GPU implementations [17] in double precision arithmetics for different remeshing kernels, for 2D and 3D experiments using about 16 million points. In these experiments a second order splitting was used to alternate one dimensional particle advection and remeshing. The number of points in the kernel stencils in each direction varied from 4 to 8 [17]. These calculations were done on a NVIDIA Tesla K20m. These performances reached between 20 and 50 % of the peak performance of the GPU, depending on the size of the stencil and represented a speed up of about 25 over a multi-threaded MPI implementation running on 8 `Xeon E5-2640` cores.

Hybrid computing would consist of combining the above implementation of scalar transport at high resolution with flow calculations on CPU processors. Based on timing obtained in our CPU and GPU implementations, in the case when the full scalar grid fits on a single GPU, up to resolutions of 512^3, a target toy configuration where computational times on CPU and GPU would be similar, consist of a 128^3 flow resolution running on 8 CPUs together with a 512^3 scalar resolution running on the GPU, for an overall computational time of about 1s per iteration. To obtain such performance it is essential that communications between velocity data processed on the CPUs and the GPU are processed in an optimal way. This is the object of ongoing research.

5 Conclusion

Combining high order semi-Lagrangian and Eulerian methods is an efficient strategy to address turbulent transport problems. It allows to describe accurately the viscous-convective range and dissipation scales of the scalar at a minimal cost. This is due to the fact that semi-Lagrangian methods are not subject to CFL conditions. The local nature of particle methods naturally opens the way to hybrid computations using heterogeneous hardware.

Acknowledgments This work was partially supported by the Agence Nationale pour la Recherche (ANR) under Contracts No. ANR-2010-JCJC-091601 and ANR-2010-COSI-0009. G.-H. C. is also grateful for the support from Institut Universitaire de France. Computations reported in Sect. 3 were performed using HPC resources from GENCI-IDRIS (Grant 2012-020611).

References

1. Tryggvason, G., Scardovelli, R.S.: Direct Numerical Simulations of Gas-Liquid Multiphase Flows. Cambridge University Press, Zaleski (2011)
2. Lodato, G., Domingo, P., Vervisch, L.: Three-dimensional boundary conditions for direct and large-eddy simulation of compressible viscous flows. J. Comput. Phys. **227**(1), 5105–5143 (2008)
3. Batchelor, G.K.: Small-scale variation of convected quantities like temperature in turbulent fluid part 1. General discussion and the case of small conductivity. J. Fluid Mech. **5**(01), 113–133 (1959)
4. Corrsin, S.: On the spectrum of isotropic temperature fluctuations in an isotropic turbulence. J. Appl. Phys. **22**(4), 469–473 (1951)
5. Obukhov, A.M.: The structure of the temperature field in a turbulent flow. Dokl. Akad. Navk. SSSR **39**, 391 (1949)
6. Kraichnan, R.: Small-scale structure of a scalar field convected by turbulence. Phys. Fluids **11**, 945–953 (1968)
7. Cottet, G.-H., Balarac, G., Coquerelle M.: Subgrid particle resolution for the turbulent transport of a passive scalar. In: Proceedings of the 12th EUROMECH European Turbulence Conference, Advances in Turbulence XII, vol. 132, pp. 779–782 (September 2009)
8. Gotoh, T., Hatanaka, S., Miura, H.: Spectral compact difference hybrid computation of passive scalar in isotropic turbulence. J. Comput. Phys. **231**(21), 7398–7414 (2012)
9. Koumoutsakos, P., Leonard, A.: High-resolution simulations of the flow around an impulsively started cylinder using vortex methods. J. Fluid Mech. **296**, 1–38 (1995)
10. Cottet, G.-H., Michaux, B., Ossia, S., Vanderlinden, G.: A comparison of spectral and vortex methods in three-dimensional incompressible flows. J. Comput. Phys. **175**(2), 702–712 (2002)
11. Cottet, G.-H., Poncet, P.: Advances in direct numerical simulations of 3d wall-bounded flows by vortex-in-cell methods. J. Comput. Phys. **193**(1), 136–158 (2004)
12. Ploumhans, P., Winckelmans, G., Salmon, J., Leonard, A., Warren, M.: Vortex methods for direct numerical simulation of three-dimensional bluff body flows: application to the sphere at Re=300, 500 and 1000. J. Comput. Phys. **178**(2), 427–463 (2002)
13. Poncet, P.: Topological aspects of the three-dimensional wake behind rotary oscillating circular cylinder. J. Fluid Mech. **517**, 27–53 (2004)
14. Hieber, S.E., Koumoutsakos, P.: A lagrangian particle level set method. J. Comput. Phys. **210**(1), 342–367 (2005)

15. Bergdorf, M., Koumoutsakos, P.K.: A Lagrangian particle-wavelet method. Multiscale Model Simul: SIAM Interdisc J **5**, 980–995 (2006)
16. Magni, A., Cottet, G.: Accurate, non-oscillatory, remeshing schemes for particle methods. J Comput Phys. **231**(1), 152–172 (2012)
17. Cottet, G.-H., Etancelin, J.-M., Perignon, F., Picard, C.: High-order Semi-Lagrangian particle methods for transport equation: numerical analysis and implementation issues. ESAIM: Math. Model. Numer. Anal.
18. Lagaert, J.-B., Balarac, G., Cottet, G.-H.: Hybrid spectral-particle method for the turbulent transport of a passive scalar. J. Comput. Phys. **260**(1), 127–142 (2014)
19. Rossinelli, D., Bergdorf, M., Cottet, G.-H., Koumoutsakos, P.: GPU accelerated simulations of bluff body flows using vortex methods. J. Comput. Phys. **229**(9), 33163333 (2010)
20. Sbalzarini, I.F., Walther, J.H., Bergdorf, M., Hieber, S.E., Kotsalis, E.M., Koumoutsakos, P.: PPM—a highly efficient parallel particle-mesh library for the simulation of continuum systems. J. Comput. Phys. **215**, 566–588 (2006)
21. Lesieur, M.: Turbul. Fluids. Fluid mechanics and its applications, Springer, Dordrecht (2008)

Multi-frequency Induction Hardening: A Challenge for Industrial Mathematics

Dietmar Hömberg, Thomas Petzold and Elisabetta Rocca

Abstract Multi-frequency induction hardening is a rather new technology to produce contour-hardened gears by applying ac current of two different frequencies to the inductor coil. The approach results in a number of additional control parameters as compared to the standard induction heating approach. Accordingly, there is a strong demand in industry for mathematical modelling and simulation of this process. This paper reports on the results of a collaborative project between partners from academia and industry. We describe a mathematical model of multi-frequency induction hardening and remark on its qualitative mathematical analysis, we derive a numerical approximation strategy, compare the results with experiments and conclude with a further validation in collaboration with one of our industrial partners.

Keywords Induction hardening · Joule heating · Maxwell's equations · Finite element simulation

1 Introduction

In most structural components in mechanical engineering, the surface is particularly stressed. Therefore, the aim of surface hardening is to increase the hardness of the boundary layers of a workpiece by rapid heating and subsequent quenching. This heat treatment leads to a change in the microstructure, which produces the desired hardening effect. Depending on the respective heat source, one can distinguish between

D. Hömberg (✉) · T. Petzold · E. Rocca
Weierstrass Institute, Mohrenstr. 39, 10117 Berlin, Germany
e-mail: dietmar.hoemberg@wias-berlin.de; thomas.petzold@wias-berlin.de

E. Rocca
Dipartimento di Matematica, Università di Milano,Via Saldini 50, 20133 Milan, Italy
e-mail: elisabetta.rocca@wias-berlin.de; elisabetta.rocca@unimi.it

M. Wakayama et al. (eds.), *The Impact of Applications on Mathematics*,
Mathematics for Industry 1, DOI: 10.1007/978-4-431-54907-9_19,

Fig. 1 The effect of medium-, high- and multi-frequency induction heating. MF (*left*): only the root of the gear is heated, HF (*middle*): only the tip of the gear is heated, MF + HF (*right*): tip and root of the gear are heated

different surface hardening procedures. Induction heat treatments can easily be integrated into a process chain. Moreover, they are energy efficient since the heat is generated directly in the workpiece. That is why induction hardening is still the most important surface treatment technology.

Its mode of operation relies on the transformer principle. A given current density in the induction coil induces eddy currents inside the workpiece. Because of the Joule effect, these eddy currents lead to an increase in temperature in the boundary layers of the workpiece. Then the current is switched off, and the workpiece is quenched by spray-water cooling producing the desired hard, martensitic microstructure in the boundary layer. Due to the skin effect, the eddy currents tend to distribute in a small surface layer. The penetration depth of these eddy currents depends on the material and essentially on the frequency. Therefore, it is difficult to obtain a uniform contour hardened zone for complex workpiece geometries such as gears using a current with only one frequency.

If, for example, a high frequency (HF) is applied, then the penetration depth is small, and it is possible to harden only the tip of the gear. With a medium frequency (MF), it is possible to heat the root of the gear, but not the tip. With a single frequency, a hardening of the complete tooth can only be achieved by increasing the heating time. But then, the complete tooth is heated beyond the austenitization temperature, which results in a complete martensitic structure of the tooth after quenching, which is not desirable since this will foster fatigue effects.

Recently, a new approach was developed, which amounts to supplying medium and high frequency powers simultaneously to the induction coil. This concept is called *multifrequency induction hardening*, see also Fig. 1.

The inductor current consists of a medium frequency fundamental oscillation superimposed by a high frequency oscillation. The amplitudes of both frequencies are independently controllable, which allows separate regulation of the respective shares of the output power of both frequencies according to the requirements of the workpiece. This fact provides the ability to control the depth of hardening at the root and the tip of the tooth individually [7]. Owing to the complex interplay of nonlinear material data and system parameters, there is a high demand for simulation and

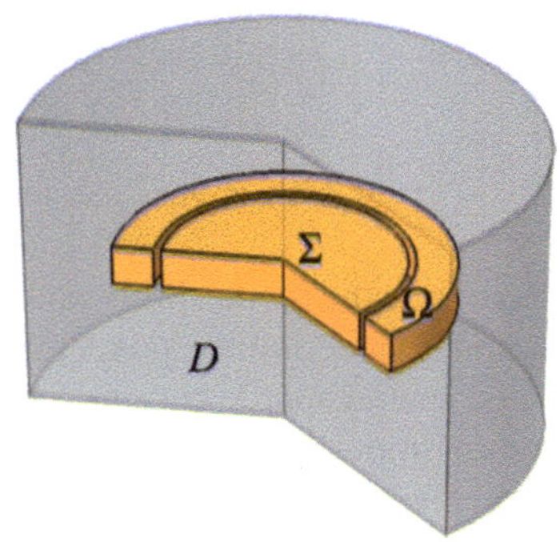

Fig. 2 Domain Dconsisting of the inductor Ω, the workpiece Σ and the surrounding air

control of this process. This was the starting point for the project "Modeling, simulation and optimization of multifrequency induction hardening" within the Federal Ministry of Education and Research's priority program "Mathematics for Innovations in Industry and Services", coordinated by WIAS. In the sequel, the main achievements of the WIAS sub-project will be presented.

2 The Model

The main parts of the model are a vector potential formulation of Maxwell's equations to describe the evolution of eddy currents, strongly coupled to the energy balance through the Joule heat term, and a rate law for the high-temperature phase, austenite, in the workpiece. It is assumed that during the quenching process following inductive heating, austenite transforms completely into martensite and is therefore an indicator of the hardening profile. The austenitization behaviour is directly linked to the temperature distribution by the transformation kinetics.

The following idealized geometric setting is considered (cf. Fig. 2), a hold-all domain D, containing the inductor Ω, the workpiece Σ, and the surrounding air. We call $G = \Omega \cup \Sigma$ the set of conductors and define the space-time domain as $Q = \Sigma \times (0, T)$, see Fig. 2.

Following [5], the mathematical model of multifrequency induction hardening amounts to finding the magnetic vector potential A, temperature ϑ, and austenite phase fraction z, satisfying the following nonlinear coupled boundary value problem:

$$\sigma A_t + \operatorname{curl}\left(\frac{1}{\mu}\operatorname{curl} A\right) = J_0(x)u(t) \quad \text{a.e. in } D \times (0, T), \tag{1}$$

$$\vartheta_t - \Delta\vartheta = -L(\vartheta, z)z_t + \sigma(x, z)|A_t|^2 \quad \text{a.e. in } Q, \tag{2}$$

$$z_t = \frac{1}{\tau(\vartheta)}\left(z_{eq}(\vartheta) - z\right)^+ \quad \text{a.e. in } Q, \tag{3}$$

$$\frac{\partial \vartheta}{\partial \nu} + \vartheta = g \quad \text{a.e. on } \partial\Sigma \times (0, T), \tag{4}$$
$$A \times n = 0 \quad \text{a.e. on } \partial D \times (0, T), \tag{5}$$
$$A(0) = A_0 \quad \text{a.e. in } D, \quad \vartheta(0) = \vartheta_0, \quad z(0) = 0 \quad \text{a.e. in } \Sigma. \tag{6}$$

Here, μ is the permeability and σ the electric conductivity. Since the latter vanishes in non-conducting regions, (1) is a degenerate parabolic equation. J_0 is a precomputed spatial source current density in the inductor, and $u(t)$ is the time dependent control imposing the different frequencies. The other physical parameters have been normalized to one.

Note that to ensure well-posedness of the above system in addition we have to impose the so-called Coulomb gauge, which amounts to demanding

$$\text{div} A = 0 \quad \text{a.e. in } D,$$

see [5] for details.

3 Analysis

Equations (1)–(6) are a strongly coupled system of evolution equations. The main analytical challenges are the quadratic Joule heating term in (2) and the nonlinearities in σ and μ. While the former depends on temperature and phase, the latter in addition also depends on the vector potential.

In two recent papers, the simpler frequency domain situation of Joule heating was studied. In [4], the Boccardo–Galluet approach was applied to prove existence of a weak solution, while in [3], new regularity results have been used to prove existence and stability in the frequency domain setting. In [5], the existence of a weak solution to a fully coupled electro-thermo-mechanical model was proven. Recently, existence and stability of solutions to (1)–(6) could be shown in the case that μ, σ depend on the phase fraction z, see [6]. Since the phase fraction grows with increasing temperature, still the effect of changing temperature is maintained.

4 Simulation

The system (1)–(6) is also numerically challenging. One has to deal with two different time scales, one for the heat equation and one for the rapidly oscillating magnetic vector potential. Owing to the skin effect, the eddy currents have to be resolved in a boundary layer, so one is also faced with two spatial scales. A further difficulty is imposed by the nonlinearities, especially the (ϑ, A)-dependent permeability.

While the temperature ϑ was approximated with standard P1 elements, the natural space for the vector potential A is the Hilbert space $H(\text{curl}, \text{D})$. To discretize A,

Fig. 3 Gear geometry

curl-conforming finite elements of Nédélec type were implemented in the finite element and finite volume toolbox `pdelib`. To account for the skin effect, the computational grid has to resolve the small surface layer of the workpiece in which eddy currents are distributed. Therefore, an adaptive grid was chosen for A governed by a residual based a posteriori error estimator developed in [1].

The temperature changes on a time scale much larger than that of its right hand side $\sigma|A_t|^2$, which is governed by the frequency of the source current. Hence, we can approximate the Joule heat term by its average over one period. Then we can solve the heat equation together with the ODE describing the phase transition using time steps $\Delta t \gg \delta t$, where δt denotes the time step for the time discretization of (1). We replace the rapidly varying Joule heat by an averaged Joule heat term, which is obtained from the solution of the vector potential equation.

To deal with the (ϑ, A)-dependency of the permeability, we proceed as in [2]. Assuming only a time averaged value of the permeability affects the magnetic field we first solve for the magnetic field with constant relative permeability $\hat{\mu}_r$. Since the magnetic field is periodic, this induces a periodic permeability $\mu(\vartheta(x,t), H(x,t)) = \mu_0 \mu_{\mathrm{r}}(\vartheta(x,t), H(x,t))$. Averaging over one period yields an effective permeability that is space dependent but independent of the magnetic field, i.e., we choose

$$\frac{1}{\mu_{\mathrm{r,av}}(x)} = \frac{1}{T} \int_0^T \frac{1}{\mu_{\mathrm{r}}(\vartheta(x,t), H(x,t))}\, dt.$$

5 Experimental Verification

The experimental verification has been performed by our project partner *Foundation Institute of Materials Science* (*IWT*), Bremen. Discs with different cross sections and a spur-gear with 21 teeth and a diameter of 47.7 mm were used, see Fig. 3. All the samples were heated by single frequency power, MF and HF separately, and by the multi-frequency approach in order to achieve a contour hardening. The temperature at the surface was measured by a pyrometer and compared to the simulation. In addition, metallographic analyses were performed and compared to the simulated

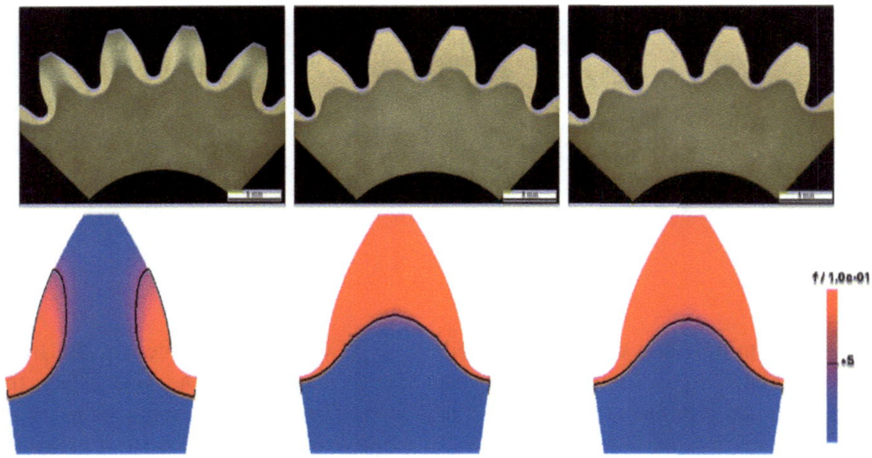

Fig. 4 Simulated and experimental hardening profile using MF (*left*), HF (*middle*), and multi-frequency approach, MF + HF (*right*)

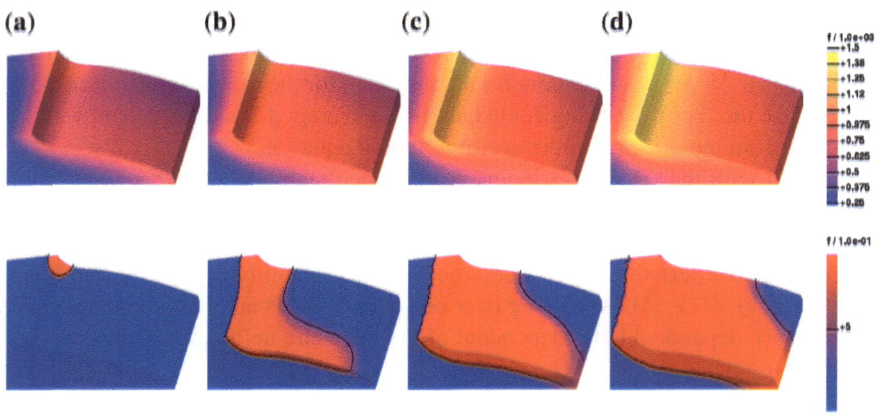

Fig. 5 Heating of a gear with the multi-frequency approach. Temperature profile (*top row*) and austenite profile (*bottom row*) for different time snapshots. For symmetry reasons only a quarter of the tooth is considered. **a** $t = 0.1$ s, **b** $t = 0.15$ s, **c** $t = 0.2$ s, **d** $t = 0.25$ s

austenite fraction, which by assumption transforms completely to martensite during the quenching process.

Figure 4 depicts a comparison for MF, HF and the multi-frequency approach. It shows a good coincidence between simulation and experiment, however, no contour hardening could be achieved due to technical limitations of the hardening equipment at IWT.

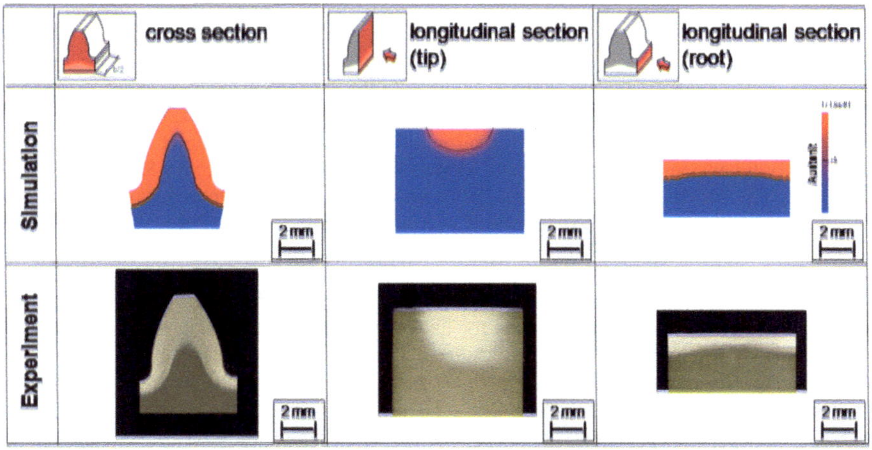

Fig. 6 Simulated and experimental hardening profile using the multi-frequency approach, MF + HF

6 Industrial Validation

To validate the developed code in an industrial setting, further experiments at our project partner *EFD Induction*, Freiburg were conducted. For the numerical validation of the experiments, at first the maximal power for MF and HF in the simulation code was adjusted to reproduce the corresponding mono-frequency experiments at EFD. The company's experiments showed that with machine specific power parameters of 53 % MF and 22 % HF, corresponding to approximately 300 kW MF and 100 kW HF, a contour hardening could be achieved with a heating time set to 0.25 s. Without any further fitting except for the mentioned fitting of the maximal mono frequency powers, the corresponding simulations were run. The heating phase simulation is shown in Fig. 5.

A comparison of experimental and simulated hardening profiles is depicted in Fig. 6, showing the desired contour hardening effect and an excellent accordance between experiment and simulation. Stimulated by these promising results, follow-up projects are in preparation aiming at the solution of related optimal control and shape design problems.

Acknowledgments The work of E. Rocca was supported by the FP7-IDEAS-ERC-StG Grant #256872 (EntroPhase). D. Hömberg and T. Petzold were partially supported by the Federal Ministry of Education and Research through the priority program "Mathematics for innovations in industry and services".

References

1. Beck, R., Hiptmair, R., Hoppe, R.H., Wohlmuth, B.: Residual based a posteriori error estimators for Eddy current computation. ESAIM Math. Model. Numer. Anal. **34**, 159–182 (2000)
2. Clain, S., Rappaz, J., Swierkosz, M., Touzani, R.: Numerical modeling of induction heating for two-dimensional geometries. Math. Models Methods Appl. Sci. **3**, 805–822 (1993)
3. Druet, P.E., Klein, O., Sprekels, J., Tröltzsch, F., Yousept, I.: Optimal control of three-dimensional state-constrained induction heating problems with nonlocal radiation effects. SIAM J. Control Optim. **49**, 1707–1736 (2011)
4. Montesinos González, M.T., Ortegón Gallego, F.: On an induction-conduction PDEs system in the harmonic regime. Nonlinear Anal. Real World Appl. **15**, 58–66 (2014)
5. Hömberg, D.: A mathematical model for induction hardening including mechanical effects. Nonlinear Anal. Real World Appl. **5**, 55–90 (2004)
6. Hömberg, D., Petzold, T., Rocca, E.: Analysis and simulations of multifrequency induction hardening. WIAS Preprint no. 1910, Berlin (2013)
7. Schwenk, W.R.: Simultaneous dual-frequency induction hardening. Heat Treat. Prog. **3**, 35–38 (2003)

Interactions in Mixed Lipid Bilayers

Sohei Tasaki

Abstract Fundamental interactions in mixed lipid bilayers are reviewed and discussed to clarify their influences on lipid microdomain formation. First, we describe a phase-separating elastic system of mixed lipid bilayers containing elastic and trans-bilayer interactions. The model can reflect characteristic properties of the bilayer, such as macroscopic elastic moduli and microscopic properties of the constituent molecules, so that we are able to analyze how the composition of the bilayer affects on the lateral morphology. Furthermore, it enables us to examine the interacting effects one by one. It is shown that the elastic interaction can stabilize intramembrane subdomain structures by secondary bifurcations of the steady states, even in simple situations with homogeneous and isotropic rigidity. On the other hand, the trans-bilayer coupling interaction may regulate the symmetry of the two leaflets of the bilayer. Indeed, simulations show us different mechanisms of synchronized lipid sorting and deformation of the bilayer. The fundamental interactions, together with further protein–protein and protein–lipid interactions, may be utilized depending on the situation to organize appropriate morphological structures.

Keywords Lipid bilayer · Phase separation · Elasticity · Pattern formation · Stability · Synchronization · Bifurcation

1 Introduction

Lateral heterogeneity in biomembranes is supposed to provide an infrastructure for a great variety of significant processes [46], such as signaling pathways [5, 47], membrane trafficking [14], entry sites for pathogens [7, 52], and cell adhesion

S. Tasaki (✉)
Frontier Research Institute for Interdisciplinary Sciences (FRIS), Tohoku University,
6-3 Aramaki-aza-Aoba, Aoba-ku, Sendai, Miyagi 980-8578, Japan
e-mail: tasaki@m.tohoku.ac.jp

M. Wakayama et al. (eds.), *The Impact of Applications on Mathematics*,
Mathematics for Industry 1, DOI: 10.1007/978-4-431-54907-9_20,

[31]. Artificial lipid membranes, studied as a simple model for the biomembrane raft hypothesis, also produce lateral heterogeneity. They often form intramembrane microdomains with specific lipid composition and local curvature. This phenomenon is observed in vesicles of bilayers [3, 40, 51, 53–55, 58], planar monolayers [11, 12, 51], and planar bilayers [8, 11].

The intramembrane lipid microdomains are formed by phase separation between liquid-ordered (L_o) and liquid-disordered (L_d) phases. The L_o phase is rich in saturated lipids and cholesterol, while the L_d phase is rich in unsaturated lipids. On the basis of this property, several models of mixed lipid bilayers are proposed by means of an order parameter which indicates the local composition. The interactions included in the models may be divided into two parts: elastic (order parameter—deformation) [20, 21, 23, 26, 30, 36, 42, 43, 48, 49] or trans-bilayer coupling (order parameter - order parameter) [1, 17, 27, 29, 49, 56]. The main purpose of this article is to review and discuss the influences of the two kinds of fundamental interations on lipid microdomain formation. In particular, we observe that stabilized and synchronized intramembrane molecule sorting and deformation in the bilayer may be driven by the fundamental interactions.

2 Model

Throughout this article, vectors (tensors of the first order) and tensors are indicated by bold letters. The inner product is denoted by a dot and the summation convention over repeated indices is used: For tensors $\mathbf{a} = (a_i)$, $\mathbf{b} = (b_i)$, $\mathbf{S} = (S_{ij})$, $\mathbf{R} = (R_{ij})$, $\mathbf{A} = (A_{ijkl})$, $\mathbf{B} = (B_{ijkl})$, we write $\mathbf{a} \cdot \mathbf{b} = a_i b_i$, $\mathbf{S} \cdot \mathbf{R} = S_{ij} R_{ij}$, $\mathbf{A} \cdot \mathbf{B} = A_{ijkl} B_{ijkl}$, $(\mathbf{Sa})_i = S_{ij} a_j$, $(\mathbf{AS})_{ij} = A_{ijkl} S_{kl}$, and so forth.

To model lateral morphological dynamics in mixed lipid bilayers, continuous models have been extensively studied. In general, the models are described by two order parameters and deformation vectors denoting the local lipid composition and displacement of the two monolayers. To examine the effects of the fundamental interactions, we consider a planar, nearly flat bilayer and ignore the thickness (molecular length), so that we use a single, real, scalar-valued function $u = u(\mathbf{x})$, denoting the vertical displacement of the membrane, and coupled order parameters $\phi = \phi(\mathbf{x}) = (\phi_1(\mathbf{x}), \phi_2(\mathbf{x}))$ for simplicity (Fig. 1). Here $\mathbf{x} = (x_1, x_2) \in \Omega$ is the space variable and $\Omega \subset \mathbb{R}^2$ a bounded domain occupied by a planar bilayer membrane in a fixed reference configuration. The component $\phi_1 = \phi_1(\mathbf{x})$ (resp. $\phi_2 = \phi_2(\mathbf{x})$) acts as an order parameter which expresses the relative composition of monolayer 1 (resp. monolayer 2), that is, the order parameter ϕ_n, $n = 1, 2$, stands for the difference between the density of the two phases, $\phi_n = c_n^{\mathrm{d}} - c_n^{\mathrm{o}}$. Here c_n^{d} and c_n^{o} respectively denote the concentration of the L_d phase and that of the L_o phase in monolayer n such that $c_n^{\mathrm{d}} + c_n^{\mathrm{o}} = 1$. Then $\phi_n \approx +1$ and $\phi_n \approx -1$ represents the L_d and L_o domains in monolayer n, respectively.

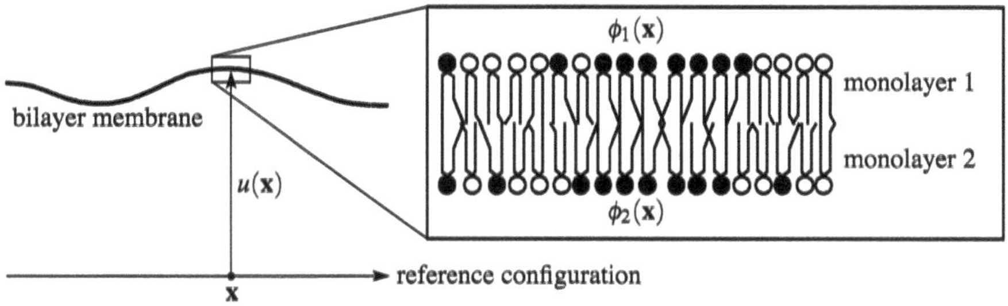

Fig. 1 Displacement $u = u(\mathbf{x})$ and relative composition $\phi = \phi(\mathbf{x}) = (\phi_1(\mathbf{x}), \phi_2(\mathbf{x}))$

The model is based on the minimization of the free energy consisting of three parts, $f = f_1 + f_2 + f_3$. Henceforth, $g_{,i} = \partial g/\partial x_i$ denotes the partial derivative of a function g with respect to the space variable x_i, $i = 1, 2$.

The first part is the Landau-Ginzburg-Cahn-Hilliard free energy [6] concerning lateral phase separations in the bilayer quenched below a critical temperature

$$f_1(\phi, \mathbf{D}\phi) = \frac{1}{2}\sum_{n=1}^{2}\left\{\frac{\gamma_n}{2}|\nabla\phi_n|^2 + W_n(\phi_n)\right\} \tag{1}$$

where $\mathbf{D}\phi = \nabla\phi = (\phi_{n,i})$ is the gradient of $\phi = \phi(\mathbf{x})$. Here, the coefficient $\gamma_n > 0$ is a constant, related to the line tension acting at the phase boundaries in monolayers n. The function W_n is double-well potential.

The second part is the following elastic energy of nearly flat membranes [16]

$$f_2\left(\mathbf{D}u, \mathbf{D}^2u, \phi\right) = \frac{\sigma}{2}|\nabla u|^2 + \frac{\kappa}{2}\left(\nabla^2 u - \overline{c}(\phi)\right)^2 \tag{2}$$

where $\mathbf{D}u = \nabla u = (u_{,i})$ and $\mathbf{D}^2u = \nabla\nabla u = (u_{,ij})$ are the gradient and the Hessian of $u = u(\mathbf{x})$, respectively. Here, we assume that the elastic interaction comes from only the spontaneous curvature. The first term in (2) is the contribution of the first gradient, which is related to the surface tension energy of the membrane with the modulus $\sigma > 0$. The second term in (2) represents the curvature energy of the membrane with the bending rigidity modulus $\kappa > 0$ and spontaneous curvature $\overline{c}(\phi)$.

The third part is the trans-bilayer coupling energy

$$f_3(\phi, \mathbf{D}\phi) = \frac{a_0}{2}(\phi_1 - s_0\phi_2)^2 + \frac{a_1}{2}|\nabla(\phi_1 - s_1\phi_2)|^2 \tag{3}$$

where the coefficient a_0, $a_1 \geq 0$ are constants, denoting the intensity of the trans-bilayer coupling interaction, and s_0, $s_1 = \pm 1$. The first term in (3) represents a local energy penalty for the compositional difference between the two monolayers [1, 27, 56]. The sign s_0 depends on the coupling mechanism. Similarly, the second term in (3) stands for an energy penalty for the difference between the compositional gradients.

Summing up, we obtain the free energy density

$$f\left(\mathbf{D}u, \mathbf{D}^2u, \phi, \mathbf{D}\phi\right) = f_1(\phi, \mathbf{D}\phi) + f_2\left(\mathbf{D}u, \mathbf{D}^2u, \phi\right) + f_3(\phi, \mathbf{D}\phi)$$

and the free energy $\mathscr{F} = \mathscr{F}(u, \phi)$ defined by

$$\mathscr{F}(u, \phi) = \int_\Omega f\left(\mathbf{D}u, \mathbf{D}^2u, \phi, \mathbf{D}\phi\right) \mathrm{d}\mathbf{x}.$$

One of the simplest systems is the following equations of the gradient flow type:

$$\tau_0 \frac{\partial u}{\partial t} = -\frac{\delta \mathscr{F}}{\delta u}(u, \phi), \quad \tau_n \frac{\partial \phi_n}{\partial t} = \nabla^2 \frac{\delta \mathscr{F}}{\delta \phi_n}(u, \phi) \quad \text{in } \Omega \times (0, T)$$

where $n = 1, 2$, and the coefficients $\tau_0, \tau_1, \tau_2 > 0$ are constants standing for the relaxation time. Here we assume that each order parameter ϕ_n is conserved by ignoring lipid flip-flop, chemical reactions, and any other external source. We impose the periodic boundary condition or the Neumann boundary condition

$$\frac{\partial}{\partial \mathbf{n}} u = \frac{\partial}{\partial \mathbf{n}} \nabla^2 u = 0, \quad \frac{\partial}{\partial \mathbf{n}} \phi_n = \frac{\partial}{\partial \mathbf{n}} \frac{\delta \mathscr{F}}{\delta \phi_n}(u, \phi) = 0 \quad \text{on} \quad \partial\Omega \times (0, T), \tag{4}$$

and the initial condition

$$u|_{t=0} = u^0, \quad \phi_n|_{t=0} = \phi_n^0 \quad \text{in } \Omega,$$

where $\mathbf{n}$ denotes the outer unit normal vector on $\partial\Omega$. Then the free energy $\mathscr{F} = \mathscr{F}(u, \phi)$ serves as a Lyapunov function and the mean value of each unknown variable is conserved:

$$\begin{aligned} \tfrac{1}{|\Omega|} \textstyle\int_\Omega u \mathrm{d}\mathbf{x} &= \overline{u^0}, \quad \text{(no vertical translation)} \\ \tfrac{1}{|\Omega|} \textstyle\int_\Omega \phi_n \mathrm{d}\mathbf{x} &= \overline{\phi_n^0} \quad \text{(composition conservation)} \end{aligned}$$

where $|\Omega|$ denotes the area of Ω and

$$\overline{u^0} = \frac{1}{|\Omega|} \int_\Omega u^0 \mathrm{d}\mathbf{x}, \quad \overline{\phi_n^0} = \frac{1}{|\Omega|} \int_\Omega \phi_n^0 \mathrm{d}\mathbf{x}.$$

By the change of the variable $u - \overline{u^0} \to u$, we can assume that $\overline{u^0} = 0$ without loss of generality. Henceforth, we set $m_1 = \overline{\phi_1^0}$ and $m_2 = \overline{\phi_2^0}$.

In the next section some simulation results [49] are reconstructed in the three dimensional space. In the simulations, for simplicity, the double-well potential and spontaneous curvature respectively take the form $W_n(\phi) = \phi^4/4 - \phi^2/2$, $\overline{c}(\phi) = c_1\phi_1 + c_2\phi_2$, where c_1, c_2 are constants. The other coefficients are homogeneous, i.e., γ_n, σ, and κ are positive constants. We also assume the symmetric property with respect to the membrane mid-surface: $\gamma_1 = \gamma_2$, $c_1 = -c_2$. The system is the square $\Omega = (-1, 1) \times (-1, 1)$, a 100×100 square lattice. The initial data is a homogeneous state with a small random perturbation, i.e., $u^0 \approx 0$, $\phi_1^0 \approx m_1$, and $\phi_2^0 \approx m_2$. The relaxation time coefficients are fixed: $\tau_0 = 0.5$, $\tau_1 = \tau_2 = 1$.

3 Interacting Effects

The model in the previous section enables us to examine the influences of the fundamental interactions one by one. The interactions may be divided into two parts: elastic and trans-bilayer interactions.

3.1 *Elastic Interaction*

The elastic energy is supposed to be a factor stabilizing intramembrane microdomain structures. By experiment and simulation, for instance, it has been predicted that curvature modulated lipid sorting can occur [15, 18, 32, 33, 50] particularly when lipid or protein components in the membrane differ in at least one of their macroscopic elastic moduli [28, 29]: the rigidity [3, 10, 32, 39, 41] or spontaneous curvature [4, 9, 10, 19, 22, 25, 34, 38, 44]. The resulting lipid microdomain structures may again deform the membrane, so that there is an elastic interaction acting between the deformation and phase separation. By simulation we can find that, due to the elastic interaction, the coarsening kinetics dramatically slows down (Fig. 2) and the lipid microdomain structures may be stabilized (Fig. 3). Here, we briefly review a mathematical structure corresponding to the stabilization effect of the elastic interaction derived only from the heterogeneous spontaneous curvature [49].

To ignore the trans-bilayer coupling interaction and to focus on the elastic interaction, the in-phase states, $\phi_1 = \phi_2$, (or the anti-phase states, $\phi_1 = -\phi_2$) has been formulated. The steady state problem of the resulting simplified system is described by

$$\begin{cases} -\kappa\nabla^2 u + \sigma u - b(\phi - m) = 0, \\ -\gamma\nabla^2\phi + W'(\phi) - \dfrac{1}{|\Omega|}\displaystyle\int_\Omega W'(\phi)\mathrm{d}\mathbf{x} + b\nabla^2 u = 0 \quad \text{in } \Omega, \\ \dfrac{\partial}{\partial \mathbf{n}}u = \dfrac{\partial}{\partial \mathbf{n}}\phi = 0 \quad \text{on } \partial\Omega, \quad \dfrac{1}{|\Omega|}\displaystyle\int_\Omega \phi\mathrm{d}\mathbf{x} = m \end{cases}$$

where the Neumann boundary condition is imposed. Let $\lambda = 1/\gamma$ and m be the bifurcation parameter and auxiliary one, respectively, and consider the one-dimensional case $\Omega = (0, l)$. In the absence of the interaction term, $b = 0$, there are in general

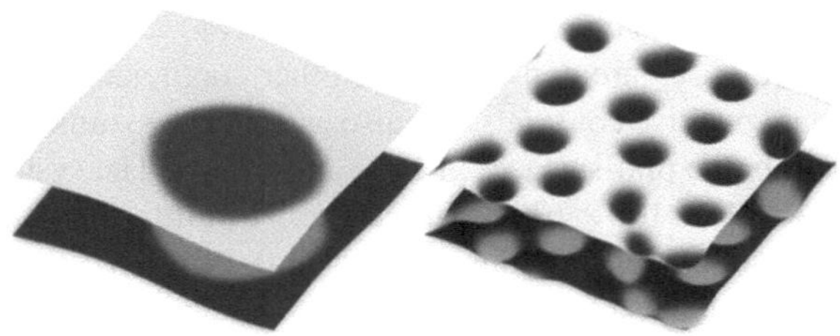

Fig. 2 Difference between the long-time coarsening kinetics in the absence (*left*) and presence (*right*) of the elastic interaction for $\gamma_i = 10^{-3}$ under the periodic boundary condition. We set $a_0 = a_1 = 0$ and give the initial data such that $\phi_1^0 = -\phi_2^0$ to ignore the trans-bilayer interaction effects and to focus on the elastic interaction. The other parameters: $\sigma = 10^{-5}$, $\kappa = 10^{-6}$, $c_1 = -c_2 = 8$, $\phi_1^0 = -\phi_2^0 \approx 0.46$

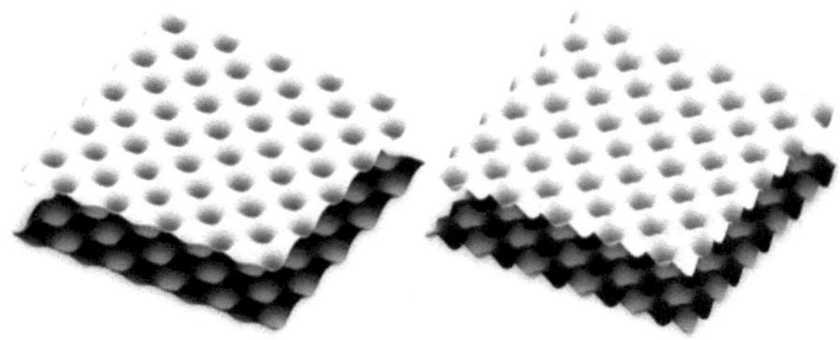

Fig. 3 Typical spatial configuration for small gradient energy coefficients, $\gamma_i = 4.7 \times 10^{-4}$. Similarly to Fig. 2, we impose the periodic boundary condition and $a_0 = a_1 = 0, \phi_1^0 = -\phi_2^0 \approx 0.46$. The other parameters: (*left*) $\sigma = 10^{-1}$, $\kappa = 10^{-2}$, $c_1 = -c_2 = 0.14$, (*right*) $\sigma = 10^{-2}$, $\kappa = 10^{-4}$, $c_1 = -c_2 = 14$

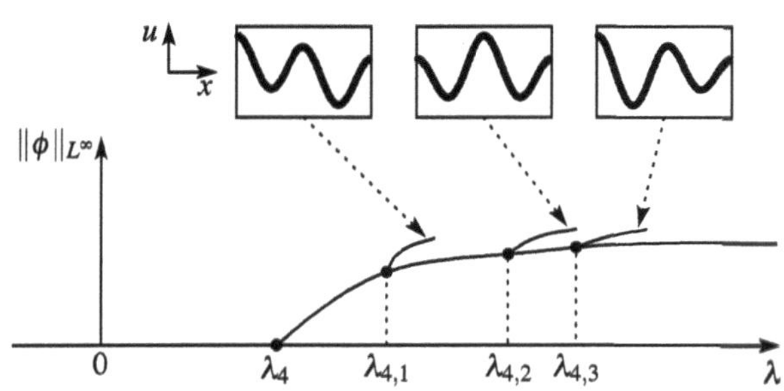

Fig. 4 Bifurcation diagram in the (λ, ϕ)-space in the case $b > 0$, $m = 0$, $k = 4$. The primary branch $\mathscr{C}_{4,0}$ of 4-mode solutions and the secondary branches $\mathscr{D}_{(4,1),0}$, $\mathscr{D}_{(4,2),0}$, $\mathscr{D}_{(4,3),0}$ of mixed mode solutions are depicted

no secondary bifurcations. Then the bifurcation diagrams are similar to those of the standard Cahn-Hilliard equation and stable steady states are constant or monotone, and so there are no stable microdomain structures. In the presence of the elastic interaction, $b \neq 0$, the primary bifurcations emerged from a trivial branch are essentially the same as those of the standard Cahn-Hilliard equation, but, on the primary branch

Fig. 5 Asymmetric bilayers in the presence of the trans-bilayer coupling energy. The gradient energy and trans-bilayer coupling coefficients are $\gamma_i = 6 \times 10^{-4}$, $a_0 = 2 \times 10^{-2}$, $a_1 = 10^{-4}$. The other parameters: (*left*) $\phi_1^0 \approx 0.6$, $\phi_2^0 \approx -0.2$, $\sigma = 10^{-1}$, $\kappa = 10^{-2}$, $c_1 = -c_2 = 1.4 \times 10^{-1}$, $s_0 = s_1 = -1$ with the Neumann boundary condition (4), (*right*) $\phi_1^0 \approx 0.5$, $\phi_2^0 \approx 0.1$, $\sigma = 10^{-1}$, $\kappa = 10^{-3}$, $c_1 = -c_2 = 1.4 \times 10^{-3}$, $s_0 = s_1 = +1$ with the periodic boundary condition

$\mathscr{C}_{k,m}$ of k-mode solutions, there exist secondary bifurcation points $(\lambda_{k,k'}, \phi_{k,k'})$ which produce the secondary branch $\mathscr{D}_{(k,k'),m}$ consisting of (k, k')-mixed mode solutions where $k' = 1, \dots, k-1$ (Fig. 4). It may be predicted that k-mode solutions become linearized stable through the $k-1$ secondary bifurcations as $\lambda \to \infty$. In fact, we can prove that there exists a stable k-mode solution if λ is sufficiently large. By the numerical bifurcation analysis, we can also observe that each secondary bifurcation point $\lambda_{k,k'}$ is monotone decreasing at $b > 0$, $\lambda_{k,k'} \downarrow 0$ as $b \uparrow \infty$, and $\lambda_{k,k'} \uparrow \infty$ as $b \downarrow 0$. This observation implies that the intensity of the stabilization is corresponding to the size $|b|$ of the elastic interaction and the system converges to the standard Cahn-Hilliard equation as the elastic interaction becomes smaller, $b \to 0$.

3.2 Trans-Bilayer Interaction

Biomembranes are generally asymmetric with respect to the bilayer membrane midsurface, but there may be some interrelation between the molecule distribution in one monolayer and that of the opposing monolayer. Simulations imply that there are various mechanisms for synchronized lipid sorting and deformation of the bilayer [1, 17, 27, 29, 49, 56]. Even in the simple model described in Sect. 2, there exist at least three trans-bilayer interaction terms: the trans-bilayer coupling energy (3), two terms, and curvature energy in (2). One of the other important trans-bilayer interactions is a bilayer thickness energy, i.e., an energy penalty for deviations of the intermonolayer distance from an optimal bilayer thickness [24, 29, 35]. Each of them acts to make the bilayer symmetric or anti-symmetric depending on the mechanisms of the interaction and properties of the constituent molecules, e.g., indirect mechanisms through the elastic energy (molecular stiffness, macroscopic elastic moduli), spontaneous curvature difference (molecular shape), bilayer thickness energy (molecular length), and a direct mechanism by the trans-bilayer coupling energy (molecular conformational and electrostatic structures [2, 27, 57], cholesterol flip-flop mobility [13, 27, 37]), and so on. In particular, there exists rigidity moduli heterogeneity in mixed lipid bilayers because the L_o phase domains are more tightly packed and rigid than the L_d domains due to the saturated hydrocarbon chains in sphingolipids and phospholipids belonging to the L_o phase. Consequently, by balancing such cooperative or competitive tran-bilayer interaction terms, the two order parameters, denoting the local lipid composition of the two monolayers, tend to be in-phase or anti-phase (Figs. 3 and 5). Thus, subdomain structures in one monolayer may serve as an infrastructure for the microdomain formation in the opposing monolayer.

4 Conclusion

In this article, fundamental interacting effects in mixed lipid bilayers have been reviewed and discussed. It is especially noteworthy that the stabilization effect of the elastic interaction on the microdomain formation can be proved and that many different mechanisms of synchronized lipid sorting and deformation of the bilayer are suggested by simulations. Furthermore, differences in the properties of the constituent molecules (e.g., shape, stiffness, length, conformational and electrostatic structures, and so on) can be reflected in the model interactions, as well as macroscopic modulus differences. Such differences often intensify a specific interaction, and so they can play an important role in chemical and mechanical processes in mixed molecule membrane systems. Actual biomembranes are quite complex interacting systems composed of many different species of molecules, but nevertheless the fundamental interactions, together with further protein-protein and protein-lipid interactions, may be utilized depending on the situation to organize appropriate morphological structures in the membranes.

References

1. Allender, D.W., Schick, M.: Phase separation in bilayer lipid membranes: effects on the inner leaf due to coupling to the outer leaf. Biophys. J. **91**, 2928–2935 (2006)
2. Baciu, C.L., May, S.: Stability of charged, mixed lipid bilayers: effect of electrostatic coupling between the monolayers. J. Phys. Condens. Matter **16**, S2455 (2004)
3. Baumgart, T., Hess, S.T., Webb, W.W.: Imaging coexisting fluid domains in biomembrane models coupling curvature and line tension. Nature **425**, 821–824 (2003)
4. Bozic, B., Kralj-Iglic, V., Svetina, S.: Coupling between vesicle shape and lateral distribution of mobile membrane inclusions. Phys. Rev. E Stat. Nonlin. Soft Matter Phys. **73**, 041915 (2006)
5. Brown, D.A., London, E.: Structure and function of sphingolipid- and cholesterol-rich membrane rafts. J. Biol. Chem. **275**, 17221–17224 (2000)
6. Cahn, J., Hilliard, J.: Free energy of a nonuniform system: interfacial free energy. J. Chem. Phys. **28**, 258–267 (1958)
7. Chazal, N., Gerlier, D.: Virus entry, assembly, budding, and membrane rafts. Microbiol. Mol. Biol. Rev. **67**, 226–237 (2003)

8. Collins, M.D., Keller, S.L.: Tuning lipid mixtures to induce or suppress domain formation across leaflets of unsupported asymmetric bilayers. Proc. Natl. Acad. Sci. USA **105**, 124–128 (2008)
9. Cooke, I.R., Deserno, M.: Coupling between lipid shape and membrane curvature. Biophys. J. **91**, 487–495 (2006)
10. Derganc, J.: Curvature-driven lateral segregation of membrane constituents in Golgi cisternae. Phys. Biol. **4**, 317–324 (2007)
11. Dietrich, C., Bagatolli, L.A., Volovyk, Z.N., Thompson, N.L., Levi, M., Jacobson, K., Gratton, E.: Lipid rafts reconstituted in model membranes. Biophys. J. **80**, 1417–1428 (2001)
12. Dietrich, C., Volovyk, Z.N., Levi, M., Thompson, N.L., Jacobson, K.: Partitioning of Thy-1, GM1, and cross-linked phospholipid analogs into lipid rafts reconstituted in supported model membrane monolayers. Proc. Natl. Acad. Sci. USA **98**, 10642–10647 (2001)
13. Hamilton, J.A.: Fast flip-flop of cholesterol and fatty acids in membranes: implications for membrane transport proteins. Curr. Opin. Lipidol. **14**, 263–271 (2003)
14. Hanzal-Bayer, M.F., Hancock, J.F.: Lipid rafts and membrane traffic. FEBS Lett. **581**, 2098–2104 (2007)
15. Heinrich, M., Tian, A., Esposito, C., Baumgart, T.: Dynamic sorting of lipids and proteins in membrane tubes with a moving phase boundary. Proc. Natl. Acad. Sci. USA **107**, 7208–7213 (2010)
16. Helfrich, W.: Elastic properties of lipid bilayers: theory and possible experiments. Z. Naturforsch. (C) **28**, 693–703 (1973)
17. Hirose, Y., Komura, S., Andelman, D.: Coupled modulated bilayers: a phenomenological model. Chem. Phys. Chem. **10**, 2839–2846 (2009)
18. Huttner, W.B., Zimmerberg, J.: Implications of lipid microdomains for membrane curvature, budding and fission. Curr. Opin. Cell Biol. **13**, 478–484 (2001)
19. Kamal, M., Millis, D., Grzybek, M., Howard, J.: Measurement of the membrane curvature preference of phospholipids reveals only weak coupling between lipid shape and leaflet curvature. Proc. Natl. Acad. Sci. USA **106**, 22245–22250 (2009)
20. Kodama, H., Komura, S.: Frustration-induced ripple phase in bilayer membranes. J. Phys. II Fr. **3**, 1305–1311 (1993)
21. Komura, S., Shimokawa, N., Andelman, D.: Tension-induced morphological transition in mixed lipid bilayers. Langmuir **22**, 6771–6774 (2006)
22. Leibler, S.: Curvature instability in membranes. J. Phys. **47**, 507–516 (1986)
23. Leibler, S., Andelman, D.: Ordered and curved meso-structures in membranes and amphiphilic films. J. Phys. (Paris) **48**, 2013–2018 (1987)
24. Lewis, B.A., Engelman, D.M.: Lipid bilayer thickness varies linearly with acyl chain length in fluid phosphatidylcholine vesicles. J. Mol. Biol. **166**, 211–217 (1983)
25. Liang, Q., Ma, Y.Q.: Curvature-induced lateral organization in mixed lipid bilayers supported on a corrugated substrate. J. Phys. Chem. B **113**, 8048–8055 (2009)
26. MacKintosh, F.C., Safran, S.A.: Phase separation and curvature of bilayer membranes. Phys. Rev. E **47**, 1180–1183 (1993)
27. May, S.: Trans-monolayer coupling of fluid domains in lipid bilayers. Soft Matter **5**, 3148–3156 (2009)
28. Mercker, M., Ptashnyk, M., Kühnle, J., Hartmann, D., Weiss, M., Jäger, W.: A multiscale approach to curvature modulated sorting in biological membranes. J. Theoret. Biol. **301**, 67–82 (2012)
29. Mercker, M., Richter, T., Hartmann, D.: Sorting mechanisms and communication in phase-separating coupled monolayers. J. Phys. Chem. B **115**, 11739–11745 (2011)
30. Minami, A., Yamada, K.: Domain-induced budding in buckling membranes. Eur. Phys. J. E **23**, 367–374 (2007)
31. Pande, G.: The role of membrane lipids in regulation of integrin functions. Curr. Opin. Cell Biol. **12**, 569–574 (2000)
32. Parthasarathy, R., Yu, C., Groves, J.T.: Curvature-modulated phase separation in lipid bilayer membranes. Langmuir **22**, 5095–5099 (2006)

33. Pencer, J., Jackson, A., Kučerka, N., Nieh, M.P., Katsaras, J.: The influence of curvature on membrane domains. Eur. Biophys. J. **37**, 665–671 (2008)
34. Ramaswamy, S., Toner, J., Prost, J.: Nonequilibrium fluctuations, traveling waves, and instabilities in active membranes. Phys. Rev. Lett. **84**, 3494–3497 (2000)
35. Rawicz, W., Olbrich, K.C., McIntosh, T., Needham, D., Evans, E.: Effect of chain length and unsaturation on elasticity of lipid bilayers. Biophys. J. **79**, 328–339 (2000)
36. Reigada, R., Buceta, J., Lindenberg, K.: Nonequilibrium patterns and shape fluctuations in reactive membranes. Phys. Rev. E **71**, 051906 (2005)
37. Risselada, H.J., Marrink, S.J.: The molecular face of lipid rafts in model membranes. Proc. Natl. Acad. Sci. USA **105**, 17367–17372 (2008)
38. Risselada, H.J., Marrink, S.J.: Curvature effects on lipid packing and dynamics in liposomes revealed by coarse grained molecular dynamics simulations. Phys. Chem. Chem. Phys. **11**, 2056–2067 (2009)
39. Roux, A., Cuvelier, D., Nassoy, P., Prost, J., Bassereau, P., Goud, B.: Role of curvature and phase transition in lipid sorting and fission of membrane tubules. EMBO J. **24**, 1537–1545 (2005)
40. Rozovsky, S., Kaizuka, Y., Groves, T.: Formation and spatio-temporal evolution of periodic structures in lipid bilayers. J. Am. Chem. Soc. **127**, 36–37 (2005)
41. Rózycki, B., Weikl, T.R., Lipowsky, R.: Stable patterns of membrane domains at corrugated substrates. Phys. Rev. Lett. **100**, 098103 (2008)
42. Safran, S.A., Pincus, P., Andelman, D.: Theory of spontaneous vesicle formation in surfactant mixtures. Science **248**, 354–356 (1990)
43. Safran, S.A., Pincus, P., Andelman, D., MacKintosh, F.C.: Stability and phase behavior of mixed surfactant vesicles. Phys. Rev. A **43**, 1071–1078 (1991)
44. Seifert, U.: Curvature-induced lateral phase segregation in two-component vesicles. Phys. Rev. Lett. **70**, 1335–1338 (1993)
45. Semrau, S., Idema, T., Holtzer, L., Schmidt, T., Storm, C.: Accurate determination of elastic parameters for multicomponent membranes. Phys. Rev. Lett. **100**, 088101 (2008)
46. Simons, K., Ikonen, E.: Functional rafts in cell membranes. Nature **387**, 569–572 (1997)
47. Simons, K., Toomre, D.: Lipids rafts and signal transduction. Nat. Rev. Mol. Cell Biol. **1**, 31–41 (2000)
48. Taniguchi, T.: Shape deformation and phase separation dynamics of two-component vesicles. Phys. Rev. Lett. **76**, 4444–4447 (1996)
49. Tasaki, S.: Phase-separating elastic system of mixed lipid bilayers. Physica D **246**, 23–38 (2013)
50. Tian, A., Baumgart, T.: Sorting lipids and proteins in membrane curvature gradients. Biophys. J. **96**, 2676–2688 (2009)
51. Veatch, S.L., Keller, S.: Organization in lipid membranes containing cholesterol. Phys. Rev. Lett. **89**, 268101 (2002)
52. van der Goot, F.G., Harder, T.: Raft membrane domains: from a liquid-ordered membrane phase to a site of pathogen attack. Semin. Immunol. **13**, 89–97 (2001)
53. Veatch, S.L., Keller, S.: Separation of liquid phases in giant vesicles of ternary mixtures of phospholipids and cholesterol. Biophys. J. **85**, 3074 (2003)
54. Veatch, S.L., Keller, S.: Miscibility phase diagrams of giant vesicles containing sphingomyelin. Phys. Rev. Lett. **94**, 148101 (2005)
55. Veatch, S.L., Keller, S.: Seeing spots: complex phase behavior in simple membranes. Biochim. Biophys. J. **1746**, 172–185 (2005)
56. Wagner, A.J., Loew, S., May, S.: Influence of monolayer-monolayer coupling on the phase behavior of a fluid lipid bilayer. Biophys. J. **93**, 4268–4277 (2007)
57. Wagner, A.J., May, S.: Electrostatic interactions across a charged lipid bilayer. Eur. Biophys. J. **36**, 293–303 (2007)
58. Yanagisawa, M., Imai, M., Masui, T., Komura, S., Ohta, T.: Growth dynamics of domains in ternary fluid vesicles. Biophys. J. **92**, 115–125 (2007)

A Note on Reconstructing the Conductivity in Impedance Tomography by Elastic Perturbation

Eric Bonnetier and Faouzi Triki

Abstract We give a short review on the hybrid inverse problem of reconstructing the conductivity in a medium in $\mathbf{R}^n$, $n = 2, 3$, from the knowledge of the pointwise values of the energy densities associated with imposed boundary voltages. We show that given n boudary voltages, the associated voltage potentials solve an elliptic system of PDE's in the subregions where they define a diffeomorphism, from which stability estimates can be obtained.

Keywords Inverse conductivity · Calderón's problem · Hybrid methods · Stability

1 Introduction

Over the last decade, there has been considerable activity around multiphysics or hybrid inverse problems, where one tries to determine the coefficients inside a medium from the knowledge of some boundary and internal data. The term multiphysics refers to the fact that the medium to be imaged is probed using two types of waves: one type is sensitive to the contrast in the material coefficients one wants to image, the other type can carry the information revealed by the first waves to the boundary of the domain, where an observer can make measurements. Several modalities are proposed, that hopefully will allow for considerable progress in imaging. In photoacoustic imaging for instance, one illuminates the interior of an object with a short laser pulse which, by the photoacoustic effect, triggers accoustic waves that are measured on the surface [13]. Electroseismic imaging combines electromagnetic

E. Bonnetier (✉) · F. Triki
Laboratoire Jean Kuntzmann, Université Grenoble-Alpes and CNRS,
BP 53, 38041 Grenoble, France
e-mail: Eric.Bonnetier@imag.fr

F. Triki
e-mail: Faouzi.Triki@imag.fr

M. Wakayama et al. (eds.), *The Impact of Applications on Mathematics*,
Mathematics for Industry 1, DOI: 10.1007/978-4-431-54907-9_21,

waves and elastic waves to probe the subsoil, for oil prospection. Indeed, an electromagnetic field that propagates through a porous medium saturated with an electrolyte moves the electric charges sitting at the solid/fluid interfaces in the medium, creating hereby a mechanical wave, that can be measured on the soil surface [14].

Typically, the inversion procedure proceeds in two steps. Firstly, one seeks to retrieve information from the waves that are measured on the boundary of the object under study. This usually takes the form of an inverse source problem for a wave equation, where one tries to recover the initial value of the field (pressure, elastic dispacement) inside the object [15]. This provides internal data for the equations that govern the propagation of the waves that are sensitive to the contrast in material coefficients. The second step consists in recovering the values of these coefficients from the internal data. See e.g. [7, 9] in the case of photoacoustics.

Here, we consider the particular case of electrical impedance tomography (EIT) under elastic perturbation, where one probes a medium with acoustic waves (ultrasounds) while making electrical measurements on its boundary [3, 12]. The associated internal data consists in the pointwise values of the electrostatic energy density. In Sect. 2 we recall the set up and the main results concerning this multiphysics problem. In particular, a Lipschitz stability estimate holds, that shows that the difference of two conductivity maps is bounded by the difference of the corresponding internal data, contrarily to the Calderón problem of determining the conductivity from knowledge of the Dirichlet to Neumann map, where the dependence is logarithmic [1].

In Sect. 3, we consider the problem of reconstructing the conductivity γ in a bounded domain $\Omega \subset \mathbf{R}^n$ from n measurements of the electrostatic energy density, under the assumption that the matrix of measurements H is invertible. We revisit a strategy designed in [6], who obtained a Lipschitz stability estimate by showing that the voltage potentials u_i associated to the measurements solve a system of elliptic PDE's, the coefficients of which only depend on H and not on γ. If this system is solvable, it follows from elliptic regularity that $||u_{1,i} - u_{2,i}||_{H^1} \le C||H_1 - H_2||_{W^{1,\infty}}$, for two matrices of measurements H_1, H_2 corresponding to 2 conductivities γ_1, γ_2 and the associated voltage potentials $u_{1,i}, u_{2,i}$, that take the same boundary values. This leads to a stability estimate for $||\gamma_1 - \gamma_2||$, as one can derive a reconstruction formula for the conductivity in terms of the u_i's. The purpose of this note is to give a simpler derivation of the system of elliptic PDE's satisfied by the voltage potentials u_i associated to the measurements, using essentially a change of variables.

2 A Short Review of EIT Under Elastic Perturbation

Let Ω denote a bounded domain in $\mathbf{R}^n$ that contains a medium with conductivity γ, that satisfies $0 < \lambda < \gamma(x) < \lambda^{-1}$ a.e. $x \in \Omega$. In electrical impedance tomography, a boundary voltage potential g is applied to the boundary $\partial\Omega$ and the resulting flux of electric current $\gamma \dfrac{\partial u}{\partial n}_{|\partial\Omega}$ is measured, where u is constrained to solve the PDE

$$\begin{cases} \operatorname{div}(\gamma(x)\nabla u(x)) = 0 & \text{in } \Omega, \\ u(x) \qquad\qquad\quad = g(x) & \text{on } \partial\Omega, \end{cases} \tag{1}$$

where n denotes the outer normal on $\partial\Omega$. The Calderón problem consists in trying to determine γ from knowledge, or partial knowledge, of the Dirichlet-to-Neumann map $\Lambda_\gamma : g \to \gamma \frac{\partial u}{\partial n}_{|\partial\Omega}$ [16].

This problem is notoriously ill-posed: it can be shown that the difference of 2 smooth conductivities can be controlled in terms of the difference of the norms of the associated Dirichlet-to-Neumann map only in a logarithmic way [1, 16]

$$||\gamma_1 - \gamma_2||_{L^\infty(\Omega)} \leq C \ln(||\Lambda_{\gamma_1} - \Lambda_{\gamma_2}||)^\sigma,$$

for some constants $C > 0$ and $0 < \sigma < 1$. This is what really restricts the use of electrical impedance tomography for diagnosis purposes, albeit its many other advantages such as being non-invasive, portable and low-cost.

It has been suggested that coupling the electrical measurements with elastic perturbations may improve the quality of reconstructions. In [3], a model is analyzed in which focalized ultrasound perturbations are sent into the medium while making the electrical measurements. The ultrasound perturbations affect a small sphere $z + \omega$ around a point $z \in \Omega$. Assuming that the associated elastic deformation changes the density of charges in the medium, the perturbed conductivity in the neighborhood of z takes the form $\gamma_{z,\omega}(x) = \gamma(x)\nu(x)$, where ν is the ratio of the volume of the sphere dilated by the ultrasound waves to its original volume. The function ν, is assumed to be known (in principle, it could be determined experimentally by varying the amplitude of the ultrasound waves). One can compare the solution of (1) to the voltage potential $u_{z,\omega}$ associated with the perturbed conductivity $\gamma_{z,\omega}$, that satisfies

$$\begin{cases} \operatorname{div}(\gamma_{z,\omega}(x)\nabla u_{z,\omega}(x)) = 0 & \text{in } \Omega, \\ u(x) \qquad\qquad\qquad\quad = g(x) & \text{on } \partial\Omega, \end{cases}$$

to obtain the asymptotic expansion of $u - u_{z,\omega}$ and in particular

$$\begin{aligned} \int_{\partial\Omega} (u_{z,\omega} - u) g \, d\sigma &= \int_{z+\omega} \gamma(x) \frac{(\nu(x)-1)^2}{\nu(x)+1} \nabla u(x) \cdot \nabla u(x) \, dx + o(|\omega|) \\ &= \gamma(z) |\nabla u(z)|^2 \int_{z+\omega} \frac{(\nu(x)-1)^2}{\nu(x)+1} \, dx + o(|\omega|), \end{aligned}$$

provided γ is sufficiently smooth, so that

$$\gamma(z)|\nabla u(z)|^2 = \left(\int_{z+\omega} \frac{(\nu(x)-1)^2}{\nu(x)+1} \, dx \right)^{-1} \int_{\partial\Omega} (u_{z,\omega} - u) g \, d\sigma + o(1).$$

We note that the above right-hand side consists in known information, if one makes electrical measurements when the medium is perturbed by ultrasounds focalized around z, and in the absence of such perturbation. Moving the focus z of the ultrasound beam throughout Ω yields thus knowledge of the electrostatic energy density $\gamma(z)|\nabla u(z)|^2$ throughout the whole or a part of Ω. Alternatively, one could probe the medium with ultrasounds that are plane waves, as described in [8], and recover similarly the pointwise value of the electrostatic energy density.

If u does not have critical points inside Ω, the knowledge of $H(x) = \gamma\nabla|u(x)|^2$ for $x \in \Omega$, allows us to rewrite (1) as

$$\operatorname{div}\left(\frac{H(x)}{|\nabla u(x)|^2}\nabla u(x)\right) = 0 \quad \text{in } \Omega,$$

thus transforming the inverse problem into that of solving a nonlinear PDE, since one could then recover the conductivity in the form $\gamma(x) = \frac{H(x)}{|\nabla u(x)|^2}$.

This PDE is degenerate and turns out to be hyperbolic [3, 4] and thus, difficult to analyze. Instead, one may use several applied boundary voltages g_i to collect internal data of the form $\gamma|\nabla u_i|^2$, and by polarization $\gamma\nabla u_i \cdot \nabla u_j$. Uniqueness of the reconstruction using several applied boundary voltages has been shown (see [11] in 2D and for instance [5] in 3D) as well as Lipschitz stability estimates. If γ_1, γ_2 are two sufficiently smooth conductivities and if H_1, H_2 are the associated matrices of measurements $\gamma\nabla u_i \cdot \nabla u_j$, then

$$||\ln(\gamma_1) - \ln(\gamma_2)||_{W^{1,\infty}} \leq C||H_1 - H_2||_{W^{1,\infty}},$$

for some constant $C > 0$ [5]. These stability results may explain the quality of the numerical reconstructions reported in [3, 11]. At their root is a reconstruction formula [5, 6, 11], such as that discussed in the next section.

3 Reconstructing the Conductivity via the Solution of an Elliptic PDE

Let $\Omega \subset \mathbf{R}^n$ be a smooth bounded domain, and let $\gamma \in \mathscr{C}^\infty(\Omega)$ denote a conductivity that satisfies

$$0 < \lambda < \gamma(x) < \lambda^{-1} \qquad \text{a.e. } x \in \Omega,$$

for some constant λ. Let $g_1, \ldots, g_n \in \mathscr{C}^2(\partial\Omega)$ and let u_i, $1 \leq i \leq n$ denote the solutions to

$$\begin{cases} \operatorname{div}(\gamma\nabla u_i) = 0 & \text{in } \Omega \\ u_i = g_i & \text{on } \partial\Omega. \end{cases} \tag{2}$$

The internal associated measurements are the pointwise values of the electrostatic energy densities

$$(H_{ij}(x))_{1\leq i,j\leq n} = (\gamma(x)\nabla u_i(x)\cdot\nabla u_j(x))_{1\leq i,j\leq n}.$$

We assume in this note that H is invertible in a subdomain $\omega\subset\Omega$. In 2 dimensions, one can choose $g_1 = x$, $g_2 = y$ to garantee that this assumption is actually verified in the whole of Ω [2], whatever the values of γ. In higher dimensions, this is no longer true: see for instance [10] for a conductivity which induces critical points. However, one can always find $n+1$ boundary voltages so that n of them will satisfy the condition $\det(H) > 0$ locally [5]. It is established in [6], that the u_i's satisfy a system of PDE's, the coefficients of which only depend on H. If solvable, one can infer from this system a reconstruction formula for the conductivity and stability estimates. In this section, we show a simple derivation of this system.

To this end, we consider the change of variables $\xi_i = F_i(x) = u_i(x)$. In what follows, and when the context is clear, we use the same notation for a function $v(x)$ and the transformed function $v\circ F^{-1}(\xi)$. We also denote $DF(x) = \left(\frac{\partial F_k}{\partial x_l}\right)_{1\leq k,l\leq n}$ the Jacobian matrix of F. Since $\det(H)\neq 0$, F maps ω diffeomorphically into a smooth domain $F(\omega)$. We can further assume that $\det(DF) > 0$ in ω. Let $\tilde{H}$ be defined by

$$\tilde{H} = \frac{H}{\sqrt{\det(H)}} = \frac{\gamma DF DF^T}{\gamma^{n/2}\det(DF)},$$

so that

$$\gamma^{n/2-1}\det(DF)DF^{-1} = DF^T\tilde{H}^{-1}. \tag{3}$$

One easily checks that the equilibrium condition for u_j in (2) transforms into

$$\begin{aligned} 0 &= \mathrm{div}_\xi\left(\frac{\gamma DF DF^T}{\det(DF)}\circ F^{-1}(\xi)\nabla_\xi\xi_j\right), \\ &= \mathrm{div}_\xi\left(\gamma^{n/2}\tilde{H}\nabla_\xi\xi_j\right) \\ &= \frac{n}{2}\gamma^{n/2-1}\nabla_\xi\gamma\cdot\tilde{H}\nabla_\xi\xi_j + \gamma^{n/2}\mathrm{div}_\xi\left(\tilde{H}\nabla_\xi\xi_j\right), \end{aligned} \tag{4}$$

for $\xi\in F(\omega)$. Hence, denoting $\tilde{H}_j = \tilde{H}\nabla_\xi\xi_j$, it follows that

$$\frac{n}{2}\frac{\nabla_\xi\gamma}{\gamma}\cdot\tilde{H}_j = -\mathrm{div}_\xi\left(\tilde{H}_j\right). \tag{5}$$

Moreover, transforming the relation $\Delta u_j = \mathrm{div}(\nabla u_j)$ in the ξ-variables we obtain

$$\Delta u_j\circ F^{-1}(\xi)(\det(DF))^{-1} = \mathrm{div}_\xi\left(\frac{DF DF^T}{\det(DF)}\circ F^{-1}(\xi)\nabla_\xi\xi_j\right)$$

$$= \mathrm{div}_\xi \left(\gamma^{n/2-1} \frac{\gamma DFDF^T}{\gamma^{n/2}\det(DF)} \circ F^{-1}(\xi) \nabla_\xi \xi_j \right)$$
$$= \mathrm{div}_\xi \left(\gamma^{n/2-1} \tilde{H}_j \right).$$

Expanding the divergence of the right-hand side of the above equation and combining with (5), we arrive at

$$\Delta u_j \circ F^{-1}(\xi)(\det(DF))^{-1} = \frac{2}{n} \gamma^{n/2-1} \mathrm{div}_\xi \left(\tilde{H}_j \right). \quad (6)$$

Further, we note that

$$\mathrm{div}_\xi(\tilde{H}_j) = \sum_{k=1}^{n} \frac{\partial(\tilde{H}_j)_k}{\partial \xi_k} = \sum_{k=1}^{n} \sum_{l=1}^{n} \frac{\partial(\tilde{H}_j)_k}{\partial x_l} \frac{\partial x_l}{\partial \xi_k} = \mathrm{trace}(DF^{-1} D\tilde{H}_j).$$

Using the above relation and (3) to transform (6) back into the x-variables, we obtain

$$\Delta u_j(x) = \frac{2}{n} \gamma^{n/2-1} \det(DF)\, \mathrm{trace}(DF^{-1} D\tilde{H}_j)$$
$$= \frac{2}{n}\, \mathrm{trace}(DF^T \tilde{H}^{-1} D\tilde{H}_j),$$

which leads to

Proposition 1 *Assume that $det(H) > 0$ in an open set $\omega \subset \Omega$. Then, the vector field $U = (u_j)_{1 \le j \le n}$ satisties the strongly elliptic system of PDE's*

$$\sum_{k=1}^{n} \frac{\partial^2 U}{\partial x_k^2} - \frac{2}{n} \frac{\partial \tilde{H}}{\partial x_k} \tilde{H}^{-1} \frac{\partial U}{\partial x_k} = 0 \qquad x \in \omega. \quad (7)$$

The system (7) is exactly that derived in Sect.5 of [6], In dimension $n = 2$, it can be cast in divergence form

$$\sum_{k=1}^{n} \frac{\partial}{\partial x_k} \left(\tilde{H}^{-1} \frac{\partial U}{\partial x_k} \right) = 0 \qquad x \in \omega,$$

and one obtains a coercive system. Let H_1, H_2 denote the measurement matrices associated to 2 smooth conductivity maps γ_1 and γ_2

$$(H_k)_{ij} = \gamma_k \nabla u_{k,i} \cdot \nabla u_{k,j},$$

where $u_{k,i}$ is the solution to (2) with $\gamma = \gamma_k$. Let $U_k = (u_{k,i})_{1 \le i \le n}$ for $k = 1, 2$. It follows from the Lax-Milgram Lemma that

$$||U_1 - U_2||_{H_0^1} \leq C||H_1 - H_2||_\infty.$$

In dimension 3, the system (7) cannot be written in divergence form, however it is Fredholm. As was noted in [6], one may thus be able to derive similar stability estimates, provided $n/2$ is not an eigenvalue of the associated operator. What are the conditions that ensure this is still an open question.

Finally, we show how one can recover the conductivity from the solutions to (7). Recalling (5), we see that

$$\frac{\nabla_\xi \gamma}{\gamma} \circ F^{-1}(\xi) = -\frac{2}{n} \tilde{H}^{-1} \operatorname{trace}(DF^{-1} D\tilde{H}),$$

where $\operatorname{trace}(DF^{-1}D\tilde{H})$ is the vector with components $\operatorname{trace}(DF^{-1}D\tilde{H}_j)$, $1 \leq j \leq n$. Changing to the x-variables yields

$$DF^{-T} \frac{\nabla \gamma}{\gamma}(x) = -\frac{2}{n} \tilde{H}^{-1} \operatorname{trace}(DF^{-1} D\tilde{H}).$$

Multiplying both sides by $\gamma^{n/2-1}\det(DF)$ and using (3), it follows that

$$\tilde{H}^{-1} DF \nabla_x (\ln(\gamma))(x) = -\frac{2}{n} \tilde{H}^{-1} \operatorname{trace}(DF^T \tilde{H}^{-1} D\tilde{H}).$$

In other words, we obtain

$$\nabla_x (\ln(\gamma))(x) = -\frac{2}{n} DU^{-1} \operatorname{trace}(DU^T \tilde{H}^{-1} D\tilde{H}),$$

from which one can deduce a stability estimate on γ as in [6].

References

1. Alessandrini, G.: Stable determination of conductivity by boundary measurements. App. Anal. **27**, 153172 (1988)
2. Alessandrini, G., Nesi, V.: Univalent σ-harmonic mappings. Arch. Rational Mech. Anal. **158**, 155–171 (2001)
3. Ammari, H., Bonnetier, E., Capdeboscq, Y., Tanter, M., Fink, M.: Electrical impedance tomography by elastic deformation. SIAM J. Appl. Math. **68**(6), 1557–1573 (2008)
4. Bal, G.: Cauchy problem for ultrasound modulated EIT. To appear in analysis and PDE (2013).
5. Bal, G., Bonnetier, E., Monard, F., Triki, F.: Inverse diffusion from knowledge of power densities. Inverse Probl. Imag. **7**(2), 353–375 (2013)
6. Bal, G., Monard, F.: Inverse diffusion problem with redundant internal information. Inverse Probl. **29**(8), 084001 (2012)
7. Bal, G., Ren, K., Uhlmann, G., Zhou, T.: Quantitative thermoacoustics and related problems. Inverse Probl. **27**, 055007 (2011)

8. Bal, G., Schotland, J.C.: Inverse scattering and acousto-optics imaging. Phys. Rev. Lett. **104**, 043902 (2010)
9. Bal, G., Uhlmann, G.: Reconstructions for some coupled-physics inverse problems. Appl. Math. Lett. **25–7**, 1030–1033 (2012)
10. Briane, M., Milton, G.W., Nesi, V.: Change of sign of the correctors determinant for homogenization in three-dimensional conductivity. Arch. Rational Mech. Anal. **173**, 133–150 (2004)
11. Capdeboscq, Y., Fehrenbach, J., de Gournay, F., Kavian, O.: Imaging by modification: numerical reconstruction of local conductivities from corresponding power density measurements. SIAM J. Imag. Sci. **2**, 1003–1030 (2009)
12. Gebauer, B., Scherzer, O.: Impedance-acoustic tomography. SIAM J. Appl. Math. **69–2**, 565–576 (2008)
13. Li, C., Wang, L.V.: Photoacoustic tomogtaphy and sensing in biomedicine. Phys. Med. Biol. **54**, R59–R97 (2009)
14. Pride, S.R.: Governing equations for the coupled electro-magnetics and acoustics of porous media. Phys. Rev. B **50**, 15678–15696 (1994)
15. Stefanov, p., Uhlmann, G.: Multi-wave methods via ultrasound, vol. 60, pp. 271–324. In Inside Out II, MSRI Publications (2012).
16. Uhlmann, G.: Electrical impedance tomography and Calderón's problem. Inverse Probl. **25**, 123011 (2009)

Applicability of Bayesian Methods for Loss Ratio Estimation

Hiroki Kondo and Shingo Saito

Abstract In an earlier paper, we proposed a Bayesian approach towards estimating the Value-at-Risk of an insurance loss ratio, taking into account both the parameter risk and the model risk. In this paper, we apply the approach to real data and evaluate the plausibility of the estimators.

Keywords Bayesian inference · Parameter risk · Model risk · Loss ratio · Value-at-Risk

1 Introduction

It is a significant component of risk management for an insurance company to estimate the future loss ratio (the total losses divided by the total premiums) of a line of business, given the data for previous years. Such estimation problems require us to take into account not only the *process risk* (caused by the stochastic nature of the model) but also the *parameter risk* (caused by the parameter estimation error) and the *model risk* (caused by using a wrong model). In [1], the authors presented a Bayesian approach to estimate the *Value-at-Risk* (VaR) of the future annual loss ratio, assuming that the annual loss ratios are independent and identically distributed. Here, VaR means the same as high quantile: for a real number α slightly smaller

H. Kondo
Graduate School of Mathematics, Kyushu University, 744, Motooka, Nishi-ku, Fukuoka 819-0395, Japan
e-mail: h-kondo@math.kyushu-u.ac.jp

S. Saito (✉)
Faculty of Arts and Science, Kyushu University, 744, Motooka, Nishi-ku, Fukuoka 819-0395, Japan
e-mail: ssaito@artsci.kyushu-u.ac.jp

M. Wakayama et al. (eds.), *The Impact of Applications on Mathematics*, Mathematics for Industry 1, DOI: 10.1007/978-4-431-54907-9_22,

than 1, the 100α % VaR of a random variable y is defined as the least y_0 for which $P(y \le y_0) \ge \alpha$; in this paper, we shall be concerned only with continuous distributions, so that the VaR y_0 always satisfies $P(y \le y_0) = \alpha$.

This paper aims to evaluate the plausibility of the estimators given in Kondo and Saito [1], by applying them to the real data of the loss ratios of fire insurance provided by the General Insurance Association of Japan [2]. Owing to the Tohoku earthquake and other natural disasters, we can think of the data as containing realisations of the VaR. Thus we shall compare the estimators and the realisations of the VaR, and investigate the extent to which our formulae can be used for risk management.

2 Description of the Formulae for the VaR Estimators

This section describes the formulae to be used to estimate the VaR of the future loss ratio, basically following [1]. The next paragraph explains the difference between our formulae and those given in Kondo and Saito [1], and a reader unfamiliar with [1] may skip the paragraph and the remark below without loss of continuity.

Section 2 of [1] gives five estimators, ranging from the estimator (i) incorporating only the process risk to the estimator (v) incorporating the parameter risk and the process risk as well. The estimators (i)–(iv) are scale invariant in the sense that if the given data $\mathbf{x} = (x_1, \dots, x_n)$ is multiplied by a constant k, then the estimators are also multiplied by k. However, the estimator (v) is not scale invariant because the posterior probability p that the model is normal is not scale invariant. This problem can be remedied by giving the parameter space $\Theta = \mathbb{R} \times \mathbb{R}_{>0}$ the improper prior $f(\mu, \tau) \propto \tau^{-1/2}$, instead of $f(\mu, \tau) \propto \tau^{-1}$ used in $(*)$ on page 87 of [1]. We shall describe below the formulae for the VaR estimators based on the prior $f(\mu, \tau) \propto \tau^{-1/2}$, without explaining in detail how to derive them, because the derivation proceeds in much the same way as in Kondo and Saito [1]. The formulae to be used in this article and those given in Kondo and Saito [1] turn out to show little difference in the numerical example in Sect. 3.

Remark 1 In the derivation of the estimator (v) in Kondo and Saito [1], it was crucial that we placed the same improper prior $f(\mu, \tau)$ on the parameter space Θ both in the normal model N and the log-normal model LN. This can be rephrased as employing the Bayesian inference on the expanded parameter space $\{\mathrm{N}, \mathrm{LN}\} \times \Theta$ with an improper prior $f(M, \mu, \tau)$ that does not depend on the model $M \in \{\mathrm{N}, \mathrm{LN}\}$.

2.1 Description of the Formulae

Let $\mathbf{x} = (x_1, \dots, x_n)$ be the given data of the annual loss ratios in the past n years, where $x_1, \dots, x_n$ are positive and not all equal. We shall estimate the 100α % VaR of the future loss ratio, where α is a real number slightly less than 1. Set

$$m_x = \frac{1}{n}\sum_{i=1}^{n} x_i, \qquad s_x = \left(\frac{1}{n}\sum_{i=1}^{n}(x_i - m_x)^2\right)^{\frac{1}{2}},$$

$$m_{\log x} = \frac{1}{n}\sum_{i=1}^{n} \log x_i, \qquad s_{\log x} = \left(\frac{1}{n}\sum_{i=1}^{n}(\log x_i - m_{\log x})^2\right)^{\frac{1}{2}}.$$

(i) Normal without parameter risk
We assume that the loss ratios have the normal distribution, and do not incorporate the parameter risk or the model risk. Then the estimator is

$$m_x + z_\alpha s_x,$$

where z_α is the α-quantile of the standard normal distribution.

(ii) Log-normal without parameter risk
We assume that the loss ratios have the log-normal distribution, and do not incorporate the parameter risk or the model risk. Then the estimator is

$$\exp(m_{\log x} + z_\alpha s_{\log x}).$$

(iii) Normal with parameter risk
We assume that the loss ratios have the normal distribution, and incorporate the parameter risk but not the model risk. Then the estimator is

$$m_x + \sqrt{\frac{n+1}{n}}\, t_\alpha(n) s_x,$$

where $t_\alpha(n)$ is the α-quantile of Student's t-distribution with n degrees of freedom.

(iv) Log-normal with parameter risk
We assume that the loss ratios have the log-normal distribution, and incorporate the parameter risk but not the model risk. Then the estimator is

$$\exp\left(m_{\log x} + \sqrt{\frac{n+1}{n}}\, t_\alpha(n) s_{\log x}\right).$$

(v) Normal and log-normal with parameter and model risk
We assume that the loss ratios have either the normal distribution or the log-normal distribution, and incorporate both the parameter risk and the model risk. If we set

$$p = \frac{s_{\log x}^n \prod_{i=1}^n x_i}{s_{\log x}^n \prod_{i=1}^n x_i + s_x^n},$$

then the estimator is the solution $q > 0$ to the equation

Table 1 Loss ratios of fire insurance, provided by the General Insurance Association of Japan [2]

Fiscal year	1995	1996	1997	1998	1999	2000	2001	2002	2003
Loss ratio (%)	30.5	32.7	31.1	42.3	44.9	38.3	40.6	35.2	35.3
Fiscal year	2004	2005	2006	2007	2008	2009	2010	2011	2012
Loss ratio (%)	73.9	45.6	47.2	41.0	40.5	39.9	38.3	155.1	79.2

$$pF_{t(n)}\left(\frac{q - m_x}{\sqrt{(n+1)/n}\, s_x}\right) + (1-p)F_{t(n)}\left(\frac{\log q - m_{\log x}}{\sqrt{(n+1)/n}\, s_{\log x}}\right) = \alpha,$$

where $F_{t(n)}$ is the cumulative distribution function of Student's t-distribution with n degrees of freedom.

Remark 2 We think of the normal distribution as the primary distribution for the loss ratios, and the log-normal distribution as the alternative distribution, so that we can view the difference between (i) and (iii) as the parameter risk, and the difference between (iii) and (v) as the model risk. If we would like to think of the log-normal distribution as the primary distribution, then we need to find a more heavy-tailed distribution to be used as the alternative distribution, but in such a case an estimator containing the model risk appears to be difficult to obtain.

3 Data Analysis

3.1 Description of the Data

We shall use the real data, shown in Table 1, of the loss ratios of fire insurance between fiscal years 1995–2012, calculated in total for the member companies of the General Insurance Association of Japan. Strictly speaking, since each value is "the ratio of claims paid plus loss adjustment expenses to net premiums written" according to [3], the assumption that the annual loss ratios are independent and identically distributed may not be perfectly appropriate; we should also investigate how the data has been affected by changes in insurance rating plans of the companies. Nevertheless, the data seems to be sufficiently suitable for our purposes, and for simplicity we shall use it without any modifications.

The loss ratio was approximately doubled in 2004 due to the damage in Japan caused by several strong typhoons, including Typhoon Songda. The gigantic value in 2011 was, needless to say, mainly due to the Tohoku earthquake, but the year also saw a large number of typhoons and the Thailand floods, which affected the companies via reinsurance. The influence of the earthquake remained in 2012.

Table 2 Comparison of actual loss ratios and VaR estimators

Year	Actual value	Estimated 99.9 % VaR					VaR estimated by (v)			
		(i)	(ii)	(iii)	(iv)	(v)	99 %	99.9 %	99.99 %	99.999 %
2004	73.9	51.6	54.4	58.5	65.5	62.7	52.4	62.7	77.7	101.3
2011	155.1	71.2	74.9	78.1	86.5	86.1	68.7	86.1	108.3	138.0
2012	79.2	135.7	137.9	154.5	176.5	176.5	116.6	176.5	266.0	408.7

3.2 Analysis

We shall compare each of the large values in 2004, 2011, and 2012 with the corresponding estimated VaR obtained by using the ratios in the previous years as the given data. For instance, we compare the actual value 73.9 in 2004 with the VaR estimators obtained by taking $\mathbf{x} = (30.5, \ldots, 35.3)$, of length $2003 - 1995 + 1 = 9$, as the given data. The results are shown in Table 2.

Recall that we interpret the difference between the estimators (i) and (iii) as the parameter risk, and the difference between (iii) and (v) as the model risk. The estimators (ii) and (iv) are shown for comparison purposes.

The loss ratios before 2004 can all be thought of as normal; it was in 2004 that we first had an unusual year (since 1995). The 99.9 % VaR estimated by (v) is somewhat smaller than the actual loss ratio.

The year 2011 was so devastating that the actual loss ratio was larger than even the estimated 99.999 % VaR (it was smaller than the estimated 99.9999 % VaR, which was 179.7). This may suggest the inadequacy of our formula, but it may also reflect the inappropriateness of the use of the VaR itself in the prediction of such a calamity. In contrast, the catastrophic year 2011 has caused such an upsurge in VaR estimators in subsequent years that the VaR estimators will be rather too large for the purpose of risk management within the near future.

4 Concluding Remarks

The data analysis indicates that our formulae often underestimates the VaR. It means that the normal distribution, used as the primary distribution, and the log-normal distribution, used as the alternative distribution, are not heavy-tailed enough to capture the behaviour of the loss ratios properly. We can artificially use the 99.999 % VaR, say, for risk management on a practical level, but such approach is neither objective nor theoretically satisfactory; it is certainly a subject of further research. In cases where we can reasonably assume that the given data contains a disastrous year such as 2011, our formulae risks overestimating the VaR because the year makes the estimators too large. It should also be noted that as far as the data used in this paper is concerned, the parameter risk and the model risk, illustrated as the difference between the estimators (i)–(v), is not as significant as the effect of the Tohoku earthquake.

References

1. Kondo, H., Saito, S.: Bayesian approach to measuring parameter and model risk in loss ratio estimation. J. Math. Ind. **4**(B), 85–89 (2012)
2. The General Insurance Association of Japan: Hoken Shumoku Betsu Songai Ritsu (loss ratios by line). http://www.sonpo.or.jp/archive/statistics/syumoku/pdf/index/syumoku_songairitu.pdf (2013). Accessed 22 Feb 2014
3. The General Insurance Association of Japan: General Insurance in Japan fact book 2012–2013. http://www.sonpo.or.jp/en/publication/pdf/fb2013e.pdf (2013). Accessed 22 Feb 2014

Simple Mathematical Models for Complex Industrial Processes

Frank R. de Hoog and Robert S. Anderssen

Abstract For the successful solution of an industrial problem a crucial, but often difficult first step, is to very clearly define the question that needs to be resolved as this will determine the nature of the modelling that is most appropriate. For example, if the aim is to determine an integrated quantity such as the total volumetric flow of paint between two rolls rather than point estimates, such as the velocity distribution of the paint between the rolls, then one would use a simpler model as integrated quantities are usually more robust with respect to model simplification. This and the fact that successful historical manufacturing processes worked because of their inherently robust, and have been made more so through process optimization, explains why simple models are often successful in capturing the essence required for decision making. Nevertheless, though the final model that resolves the matter is often disarmingly simple, the path to its identification is built on the availability of sophisticated mathematical results, knowledge and expertise. The relevance of this is exemplified in this paper using the reverse roll coating of steel and aluminium strip.

Keywords Industrial processes · Mathematical modelling · Roll coating · Lubrication theory

1 Introduction

Throughout history, the development of mathematics has made substantial contributions to technology. For example, mathematics was important in the develop-

F. R. de Hoog (✉) · R. S. Anderssen
CSIRO (Commonwealth Scientific and Industrial Research Organisation), GPO Box 664, Canberra, ACT 2601, Australia
e-mail: frank.dehoog@csiro.au

R. S. Anderssen
e-mail: bob.anderssen@csiro.au

M. Wakayama et al. (eds.), *The Impact of Applications on Mathematics*, Mathematics for Industry 1, DOI: 10.1007/978-4-431-54907-9_23,

ment of weapons such as the Greek catapult, the development of Galileo's telescope and the design of clocks. However, it is also the case that technology has been an important driver for the development of mathematics. The heat produced during the boring of gun barrels, for example, was the motivation for Fourier's work on the use of trigonometric series to solve differential equations and this is by no means an isolated example. Problems in surveying were a key driver for the subsequent development of geometry building on the results of Euclid; experimentation in agriculture drove the development of statistics; the construction of the first computers saw the birth of numerical analysis. The list goes on and on. Clearly, applications have shaped the development of the mathematical sciences.

What is less clear is why developments of mathematics driven by one application should be useful in a myriad of other applications, a phenomenon that Wigner [1] refers to as "the unreasonable effectiveness of mathematics". Indeed, why is it the case that mathematics devised for simple applications should even be applicable for situations that are much more complicated? It is this latter question that we seek to address in this paper and we do so based on models that we have worked on in the area of industrial process modelling.

We begin by making a number of key observations:

- Simple qualitative solutions can be effective. For example, since increasing (decreasing) the pressure between two paint rolls is known to decrease (increase) the paint film thickness, a measurement of the paint thickness that has been applied can be utilized, as a simple process control, to decide how to change the pressure to achieve the required paint thickness.
- Many industrial processes are robust in the sense that they work effectively under a wide range of operating conditions. This is particularly true for the processes that have a long history. For example, the rolling of metal sheets dates back to sketches (in about 1495) of a rolling mill by Leonardo da Vinci (see, for example, Roberts [2]). Processes that were not robust were simply not adopted, although this has, to a certain extent, now changed with the introduction of computer based control. Often robustness results because key parts of the process tend to dominate. From a mathematical point of view, such processes are only weakly coupled to their environment, and a reductionist approach is often effective. As a result, it is often possible to develop simple models that focus on the dominant aspect even when the process appears to be very complicated.
- It is frequently the case that an analysis is required to scale up (or down) a piece of equipment or place it in a different environment such as a centrifuge. If the process is robust, only a few key non-dimensional parameters dominate and a simple scaling analysis provides most of the information required.
- Usually, the geometry associated with industrial processes is relatively simple. Often aspect ratios such as the thickness to width ratios in metal rolling and lubrication films are small. Other non-dimension parameters, such as the ratios of material stiffnesses or reaction rates, are also often small, making industrial modelling a fertile area for dimensional reduction and perturbation analysis.

In this paper, we illustrate these points by examining simple models for reverse roll coating. As we shall see, although the models are simple, the path to deriving these simple models is less simple and involves a number of steps that are still not fully understood. This in turn provides new research opportunities for mathematicians and illustrates that industrial mathematics is a "win–win" activity for both industry and the mathematical sciences.

2 Reverse Roll Coating

Reverse roll coating is used to add a relatively thick (compared to forward roll coating) protective and/or cosmetic layer on strip steel and aluminium. We use the term reverse roll coating here to mean that the applicator and backup rolls move in opposite directions at their point of contact (i.e. They both move clockwise or both move anticlockwise). In some applications, the applicator and pickup rolls also move in opposite directions at their point of contact but here they move in the same direction (Fig. 1).

Paint from the paint tray is delivered to a gap between the pick-up and applicator rolls. The purpose of this gap is to meter the amount of paint flowing between the rolls. Once through the gap, the paint splits into two films, one of which is returned to the paint tray by the pick-up roll. The other film is transported to the junction between the applicator roll and the back-up roll where the paint is wiped onto the metal strip. The painted strip is then transferred to an oven where the paint is cured at high temperatures for 15–30 s. Although this residence time is quite short, the strip speed is typically 3 m/s and, consequently, the length of the oven is substantial with a long length of strip in the oven at any one time. In addition, the dry film thickness of the paint, as it exits the oven, can be measured accurately, while that for the wet film thickness, before it enters the oven, is problematical and unreliable. For this reason, model based control is required. However, for that, an appropriate model of the process is required.

Once the film thickness on the applicator roll is known, a simple consideration of mass balances will determine the film thickness on the strip once the peripheral speeds of the applicator and backup rolls are known. Consequently, the key questions that require resolution are:

- What is the flow between the pick-up and applicator rolls?
- How does the paint split between the pick-up and applicator rolls?

2.1 Film Splitting

Before considering film splitting, we highlight some relevant facts about coating flows. Specifically, we consider the stationary slot coater in Fig. 2, which is delivering

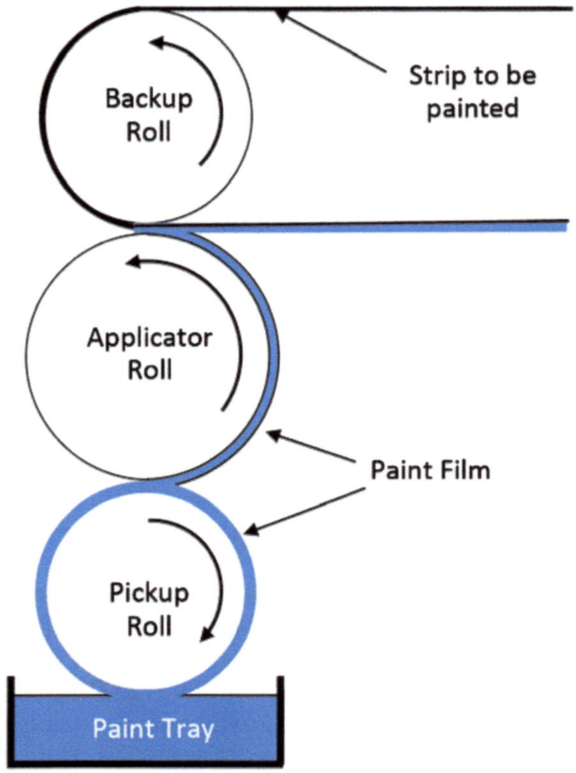

Fig. 1 Schematic of reverse roll coating

a metered flow onto a substrate which is moving to the left with velocity S. Our aim is to find a relationship between the asymptotic thickness and radius of curvature R_c of the meniscus.

This problem was solved when the capillary number is small by Landau and Levich [3], and was subsequently derived independently by a number of authors including Bretherton [4]. The capillary number is the non-dimensional number $\mathrm{Ca} = \mu S/\sigma$, where μ is the viscosity and σ is the surface tension. It is a measure of the ratio of stresses due to fluid viscosity to stresses due to surface tension. For small capillary numbers, surface tension dominates and the meniscus region near the slot can be well approximated by an arc of a circle. For typical operating parameters for a slot coater, the capillary number is indeed small.

The lubrication equation for a free surface (see, for example, Bretherton [4]) takes the form

$$\frac{d^3h}{dx^3} = 3\mathrm{Ca}\left(\frac{h - h(\infty)}{h^3}\right),$$

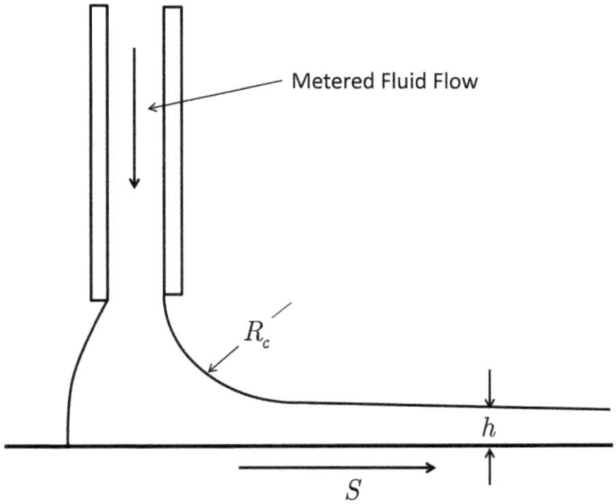

Fig. 2 Schematic of slot coating

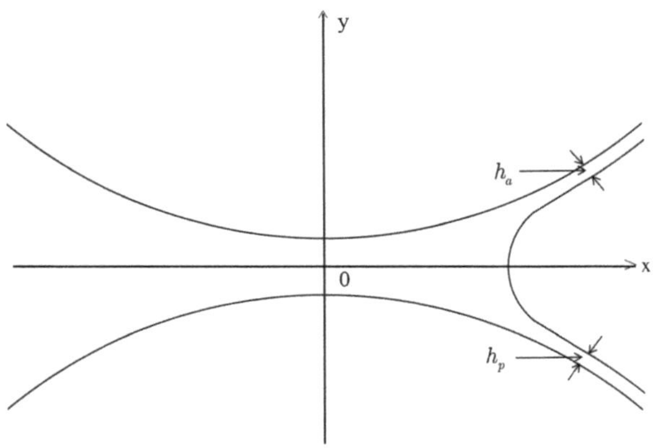

Fig. 3 Schematic of film splitting

where h is the film thickness, and we require that the linearized curvature, $d^2h/dx^2 \to 1/R_c$ as $x \to -\infty$, in order that the paint film connects smoothly with the region where the geometry of the meniscus can be well approximated by an arc of a circle. It now follows from a simple scaling argument that

$$\frac{h(\infty)}{R_c} = C\,(\mathrm{Ca})^{\frac{2}{3}}, \quad C = \text{constant}.$$

We now apply the above analysis to the film splitting problem as shown in Fig. 3. If the surface tension forces continue to dominate, then the meniscus will again be well approximated by an arc of a circle. Furthermore, we can also treat the film on the pick-up and applicator rolls independently to obtain

$$\frac{h_p}{R_c} = C\left(\frac{\mu S_p}{\sigma}\right)^{\frac{2}{3}} \quad \text{and} \quad \frac{h_a}{R_c} = C\left(\frac{\mu S_a}{\sigma}\right)^{\frac{2}{3}},$$

from which it follows that

$$\frac{h_a}{h_p} = \left(\frac{S_a}{S_p}\right)^{\frac{2}{3}}. \tag{1}$$

This is precisely the sort of simple formula that is required for process control purposes. There is a difficulty, however, with justifing its use for reverse roll coating. Although the assumption of small capillary number is valid for slot coating, it is not usually valid for reverse roll coating due to the much greater line speed. Indeed, a typical capillary number for reverse roll coating is about 10–15. However, on performing numerical simulations for film splitting for higher capillary numbers, one observes that the flow patterns actually become much simpler (There are, for example, no recirculation regions.) and the simple formula given by (1) still provides an excellent approximation. This has also been observed and validated by others. For example, experimental results of Benkreira [5] and finite element calculations reported by Cole et al. [6] demonstrate that (1) holds with an exponent of approximately 0.65 for capillary numbers greater than 1.

It is fortuitous that the simple equation for film splitting provides a good approximation even though it was derived as an asymptotic result for small capillary numbers, an assumption that is invalid for reverse roll coating. However, this is not unusual in our experience and can be attributed to the robustness of many industrial processes. It illustrates that, even though a simple analysis is performed by using quite specific assumptions, the resulting formula often gives an accurate approximation for more general situations.

2.2 Flow Between the Pick-up and Applicator Rolls—Simplified Model

We now examine the flow of paint between the pick-up and applicator rolls. For rigid rolls, the flow through the gap is well understood, but the theory is not applicable as the applicator roll has a soft polyurethane cover that varies in thickness between approximately 2 and 5 cm. However, for clarity of presentation, we will assume, in this section, that the pickup roll is rigid and the applicator roll is completely elastic. The modification required for an elastic layer of finite thickness on the applicator roll is addressed in the next section.

Our starting point is the integrated form of the Reynolds equation

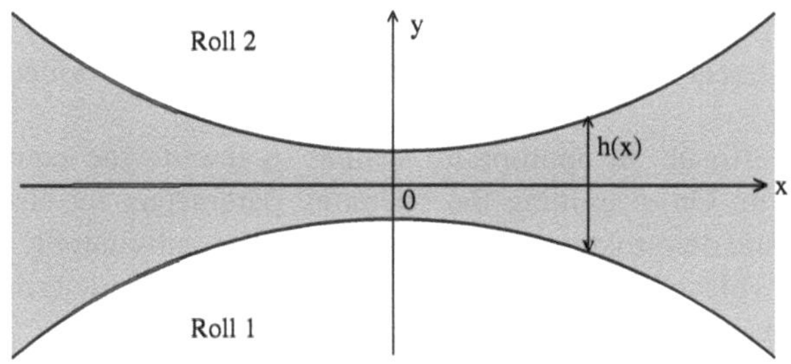

Fig. 4 Schematic of a flooded roll gap

$$\frac{dp}{dx} = 12\mu\left(\frac{Sh - Q}{h^3}\right),$$

where p is the pressure, μ is the viscosity, S is the average peripheral speed of the pick-up and applicator rolls, Q is the total flow between the rolls and $h\,(x)$ is the gap between the rolls, as shown in Fig. 4. Because numerical results indicate that the position of the meniscus has virtually no effect on the flow between the rolls, the simplest boundary conditions are assumed to hold. Specifically, we have used the Sommerfield condition which corresponds to both the inlet and outlet being flooded, as shown in Fig. 4. For the flooded situation, it is easy to verify that $Q = Sh^*$ where

$$h^* = \int_{-\infty}^{\infty} h^{-2}\,(x)\,dx \Big/ \int_{-\infty}^{\infty} h^{-3}\,(x)\,dx.$$

The elastic deformation takes the form

$$v\,(x) = -\frac{1}{\pi E}\int_{-\infty}^{\infty} \log\,(x - s)^2 p\,(s)\,ds + C, \tag{2}$$

where v is the normal displacement, E is the Young's modulus of the applicator roll and C is a constant. For notational convenience, it is assumed that the Poisson ratio is zero. The fact that the deformation can only be determined up to an arbitrary constant is a feature of contact theory in two dimensions. On combining the elastic deformation with the undeformed roll geometry, it follows that

$$h\,(x) = h\,(0) + \frac{x^2}{2R} - \frac{1}{\pi E}\int_{-\infty}^{\infty} \left\{(x - s)\log\,(x - s)^2 + s\log s^2\right\} p'\,(s)\,ds,$$

where R is the harmonic mean of the radii of the pick-up and applicator rolls. It is also convenient to introduce the force per unit width $F = \int_{-\infty}^{\infty} p\,(s)ds$ as this is a measured operating parameter. Note that on substituting for p' from Reynolds equation, a non-linear integral equation that determines the deformed roll geometry is obtained.

We now investigate an appropriate scaling to rewrite the equations in non-dimensional form. On examining the operating parameters for strip coating, we find that the elastic deformation of the applicator roll is substantially larger than the paint thickness. This indicates that the gap geometry is largely determined by elastic deformation and suggests the scaling

$$\lambda := \frac{24\mu SR}{\pi E h^2(0)};\ p \to \left(\pi E \sqrt{\frac{h(0)\lambda^3}{2R}}\right) p;\ F \to \pi E h(0)\lambda^3 F;$$
$$x \to \left(\sqrt{2Rh(0)\lambda^3}\right) x;\ h \to h(0)h;\ h^* \to h(0)h^*.$$

This yields

$$h(x) = 1 + \lambda^3 \left[x^2 - \int_{-\infty}^{\infty} \left\{ (x-s)\log(x-s)^2 + s\log s^2 \right\} p'(s)\,ds \right],$$
$$p'(x) = \lambda \left(\frac{h - h^*}{h^3} \right). \tag{3}$$

Since λ is quite large for strip painting, it follows that the term within the square brackets will be small relative to $h(x)$. If this term was zero, it would yield the pressure distribution due to pure elastic contact.

We have solved the system of Eq. (3) numerically for various values of λ and the results are plotted in Fig. 5. It appears that both the non-dimensional gap geometry and the pressure distribution are converging to a limit as $\lambda \to \infty$. In addition, for large λ, $h^* \approx 1$ and $F \approx 0.138$. On converting back to dimensional quantities, we obtain

$$Q \approx 0.645SR \left(\frac{12\mu S}{ER}\right)^{\frac{3}{5}} \left(\frac{ER}{F}\right)^{\frac{1}{5}}. \tag{4}$$

This simple equation for the flow through the gap is, when $\lambda \geq 5$, accurate to within a few percent of the full numerical solution. A more rigorous derivation of (4) can be obtained using the methods developed in Bissett [7].

2.3 Flow Between the Pick-up and Applicator Rolls—Modified Model

The model examined in the previous section was a simplification, as it assumed that the entire applicator roll was elastic. In reality, the applicator roll has an elastic

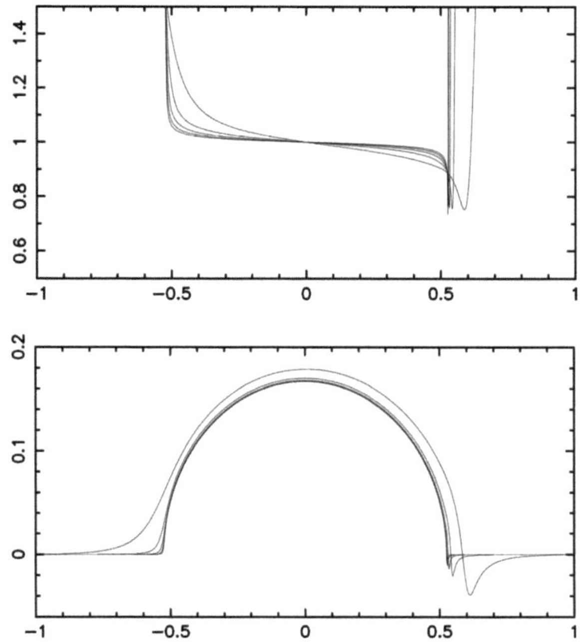

Fig. 5 Non-dimensional gap and pressure distribution for $\lambda = 4, 8, 12, 16$ and 20

outer polyurethane layer of between 2 and 5 cm thickness on a rigid cylindrical core. The thickness varies due to the fact that the roll needs constant refurbishing to accommodate and correct for surface damage resulting from wear, especially at the contact region between the edges of the roll and strip.

The fact that only a relatively thin part of the applicator roll is elastic makes a substantial difference to the flow through the gap. Consequently, it is necessary to modify the elastic deformation formula given in (2) to take this into account. Specifically, for frictionless contact, it follows that (see, for example, Gladwell [8])

$$v(x) = \frac{1}{\pi E} \int_{-\infty}^{\infty} k\left(\frac{x-s}{2b}\right) p(s)\, ds, \tag{5}$$

where

$$k(x) = 4 \int_{0}^{\infty} \frac{(3\sinh\omega - \omega)\cos\omega x}{\omega\left(6\cosh\omega + \omega^2 + 10\right)} d\omega. \tag{6}$$

Again, it is assumed for simplicity that the Poison ratio is zero. Unlike (3), Eq. (5) determines the displacement uniquely. As in the previous section, it is necessary to find the pressure distribution when the pick-up roll and applicator roll are in contact

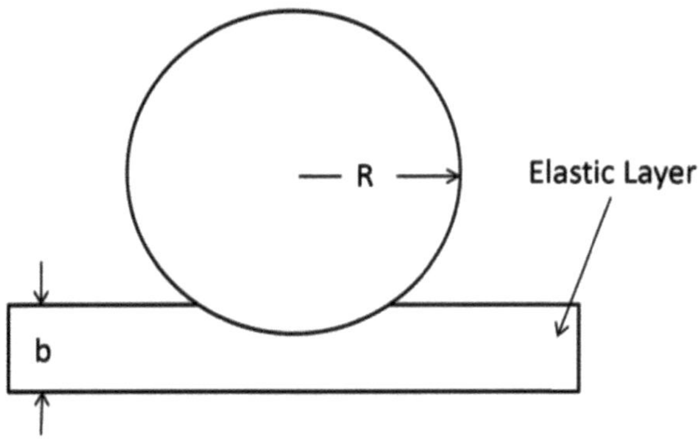

Fig. 6 Equivalent contact problems

with no fluid present. To leading order, this is equivalent to the contact of a rigid cylinder with an elastic layer that is attached to a rigid base as shown in Fig. 6.

Specifically, we need information about the solution of

$$v(0) - \frac{x^2}{2R} = \frac{1}{\pi E}\int_{-c}^{c} k\left(\frac{x-s}{2b}\right) p(s)\,ds,$$

where c is the half length of contact. In particular, we require information about the solution of

$$\int_{-\varepsilon}^{\varepsilon} [k(\eta - \xi) - k(\xi)]\hat{p}(\xi)\,d\xi + \eta^2 = 0 \tag{7}$$

in the neighborhood of $\eta = \varepsilon$. This is important, as it is required to match the pressure distribution developed at the inlet with the pressure distribution due to the elastic contact.

On replacing (2) with (5), it follows that

$$h(x) = h(0) + \frac{x^2}{2R} + \frac{b}{\pi E}\int_{-\infty}^{\infty} \left\{K\left(\frac{x-s}{b}\right) - K\left(\frac{s}{b}\right)\right\} p'(s)\,ds,$$

where $K(x) = \int_0^x k(s)ds$. Repeating the steps in Sect. 2.3, we obtain the approximation

$$Q \approx SR\phi(\varepsilon)\left(\frac{12\mu S}{ER}\right)^{\frac{3}{5}}\left(\frac{ER}{F}\right)^{\frac{1}{5}},$$

where ϕ is a smooth function, $\varepsilon = c/b$ and c is the half width of the contact between the pick-up and applicator rolls. The term $\phi(\varepsilon)$ is closely related to $\hat{p}$ [the solution of Eq. (7)]. Because ε is not a measured operating parameter, the approximation

$\phi(\varepsilon) \approx \Phi\left(Eb^2/RF\right)$ has been used in our implementation, where the function Φ is calculated via a lookup table. The agreement with experimental results is well within 10 % of the wet film thickness and a substantial fraction of this can be attributed to experimental error. This is well within the tolerances required for process based control.

3 Concluding Remarks

In this paper, the modelling and analysis of reverse roll coating has been used to highlight the opportunites which arise when exploring how to formulate simple models which explain and exploit the essence of the robust and stable nature of successful industrial processes. In particular, the opportunities include:

- For many modern industrial situations, simple formulas relating measureable properties to quantities that must be controlled are required for the effective implementation of model based process control procedures.
- Even though a simple analysis is performed using quite specific assumptions, the resulting formulas often gives a useful characterization of the more general situations.
- The importance of first identifying the question to be answered, as it determines the nature of the modelling that is most appropriate.
- Industrial decision making is more often than not based of knowledge about integrated properties (e.g. volume, total energy, thickness) rather than on point estimates (e.g. specific position, velocity, acceleration).

Acknowledgments The numerical results shown in this paper were obtained with the assistance of Huu-Nhon Huynh.

References

1. Wigner, E.P.: The unreasonable effectiveness of mathematics in the natural sciences. Commun. Pure Appl. Math. **13**, 1–14 (1960)
2. Roberts, W.: Cold rolling of steel. Marcel Dekker, New York (1987)
3. Landau, L., Levich, B.: Dragging of a liquid by a moving plate. Acta Physiochim. USSR **17**, 42–54 (1942)
4. Bretherton, F.P.: The motion of long bubbles in tubes. J. Fluid Mech. **10**, 166–188 (1961)
5. Benkreira H., Edwards M.F., Wilkinson M.F.: Roll coating of purely viscous liquids. Chem. Eng. Sci. **36**, 429–434 (1981)
6. Cole, D.J., Macosko, C.W., Scriven, L.E.: Film splitting in forward roll coating. J. Fluid Mech. **171**, 183–207 (1985)
7. Bissett E.J.: The line contact problem of elastohydrodynamic lubrication i. Asymptotic structure for low speeds. Proc.R. Soc. Lond. **424**, 393–407 (1989)
8. Gladwell, G.M.L.: Contact problems in the classical theory of elasticity. Sijthoff and Noordhoof, Groningen (1980)

Principal Component Analysis and Laplacian Splines: Steps Toward a Unified Model

J. P. Lewis, Taehyun Rhee and Mengjie Zhang

Abstract Principal component analysis models are widely used to model shapes in medical image analysis, computer vision, and other fields. The "Laplacian" spline approaches including thin-plate splines are also used for this purpose. These alternative approaches have complementary advantages and weaknesses: a low-rank principal component analysis model has some "knowledge" of the data being modeled, but cannot exactly fit arbitrary data, whereas spline models can fit arbitrary data but have only a generic smoothness assumption about the character of the data. In this contribution we show that the data fitting problem for these two approaches can be put into a common form, by making use of a relation between the data covariance and the Laplacian. This suggests the possibility of a unified approach that combines the advantages of each.

Keywords Principal component analysis · Covariance · Polyharmonic splines · Image registration and tracking

1 Introduction

Principal component analysis (PCA) and Laplacian splines (LS) are popular approaches to modeling shapes in medical image analysis, computer vision, computer graphics, and other fields. PCA and LS are frequently employed as prior models

J. P. Lewis (✉) · T. Rhee · M. Zhang
Victoria University School of Engineering and Computer Science,
PO Box 600, Wellington 6140, New Zealand
e-mail: jplewis@ecs.vuw.ac.nz; noisebrain@gmail.com

T. Rhee
e-mail: taehyun.rhee@ecs.vuw.ac.nz

M. Zhang
e-mail: mengjie.zhang@ecs.vuw.ac.nz

M. Wakayama et al. (eds.), *The Impact of Applications on Mathematics*,
Mathematics for Industry 1, DOI: 10.1007/978-4-431-54907-9_24,

to regularize inverse problems. In the context of face fitting, for example, both PCA [14] and LS approaches [4] have been employed. These distinct approaches have complementary advantages and weaknesses. Here we show that PCA and LS can be represented in (nearly) a common form, thereby revealing a relationship between these models.

We denote by "Laplacian splines" the family of interpolation schemes that minimize the integral of a squared derivative operator on the function, e.g. membrane (Laplace) interpolation, biharmonic interpolation and thin-plate splines. These methods can interpolate arbitrary data, however they do not embody any statistics of the particular data, preferring only that the interpolation be as smooth (in the sense of minimising the integrated squared derivative) as possible. These splines are commonly used for shape registration in medical imaging and computer vision [2, 6, 9]. In computer graphics they have been used for surface deformation, implicit surface modeling, and other purposes [3, 18].

PCA has the advantage that it captures the Gaussian statistics of a class of shapes and thus has some "knowledge" of what is (and is not) a reasonable shape from that class. On the other hand, PCA is usually employed as a low rank model, and thus it cannot exactly represent all possible shapes. This is appropriate in cases where there is measurement noise in the shape data acquisition. However in some applications (for example in movie visual effects and computer graphics) we assume that the data is high-quality and has little or no noise. A particular example is facial motion capture for movie applications [10, 11]. The difference in facial position between two mental states (perhaps "calm" and "contempt") may be quite small, 1 mm or less. Thus the facial motion capture process requires that very accurate data be obtained. A PCA model obtained during an initial motion capture will not exactly span the data captured in subsequent capture sessions, due to differing placement of motion capture markers or other reasons, but accurate tracking of the face position is nevertheless required.

The complementary strengths and limitations of the PCA and LS approaches are illustrated in Fig. 1 and summarized in this table:

	Low rank PCA	Laplacian spline
"Knowledge" of the data	Yes	No
Can fit data exactly	No	Yes

2 PCA

A PCA model is of the form

$$\mathbf{f} = \mathbf{U}\mathbf{c} + \mathbf{m}$$

where $\mathbf{U}$ is a matrix whose columns are the eigenvectors of the data covariance matrix, $\mathbf{c}$ are coefficients, and $\mathbf{m}$ is the mean shape (mean data vector). In our discussion we

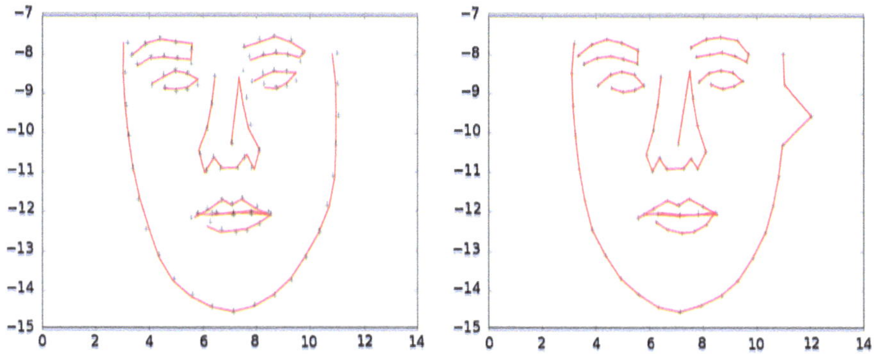

Fig. 1 *Left* A reduced-rank PCA model does not exactly fit the data. In this figure the model points are connected with lines for visualization. *Right* a spline (in this case a linear spline) fits the data exactly, but does not exclude unreasonable data such as the displaced point in this figure

take $\mathbf{f}$ as a discrete, uniformly sampled, and one-dimensional signal for simplicity. PCA is usually applied to multidimensional data by "vectorizing" the data into one-dimensional form by scanning the multiple dimensions in some arbitrary but consistent order, though intrinsically multidimensional variants of PCA have also been proposed [5]. We also take the mean to be zero without loss of generality.

Most commonly, the PCA model discards all eigenvectors corresponding to small eigenvalues, with the assumption that these variances are noise, or alternately that a simpler model that captures most but not all of the variation is preferable.

A form that will expose the relationship to LS is

$$\min_{\mathbf{c}} \|\mathbf{c}\|_{\Lambda}^{2} + \lambda^{T}\mathbf{S}(\mathbf{Uc} - \mathbf{d})$$

i.e. minimize a norm on the coefficients subject to interpolating some (sparse) data using the PCA model. Here $\mathbf{d}$ is a vector of data points, with zero (or any other value) where no data is available, and $\mathbf{S}$ is selects only the rows of $\mathbf{Uc} - \mathbf{d}$ corresponding to the available data (i.e. $\mathbf{S}_{r,c} = 1$ selects the cth row, zeros elsewhere). λ is a Lagrange multiplier vector (distinguish from Λ, the diagonal matrix of eigenvectors λ_k).

Because the expectation $\mathbb{E}[\mathbf{cc}^T] = \mathbf{U}^T\mathbb{E}[\mathbf{ff}^T]\mathbf{U} = \mathbf{U}^T\mathbf{CU} = \Lambda$ with $\mathbf{C}$ the covariance, the expectation squared of an individual coefficient is $c_k^2 \sim \lambda_k$. Thus the standard choice for the prior $\|\|_{\Lambda}$ is $\sum c_k^2/\lambda_k$. This prior is commonly used in applications, e.g. [1, 15] and many others. We rewrite the PCA fitting problem as

$$\min_{\mathbf{c}} \quad \mathbf{c}^T\Lambda^{-1}\mathbf{c} + \lambda^T\mathbf{S}(\mathbf{Uc} - \mathbf{d}) \tag{1}$$

3 Laplacian Splines

Laplacian splines minimize an energy $\int_\Omega \|\mathbf{Lf}\|^2 \, d\mathbf{x}$ (where $\mathbf{L}$ is a derivative operator such as the gradient or Laplacian) subject to interpolating specified constraints. Thin plate splines are related, minimizing an energy

$$\iint \left(f_{xx}^2 + 2 f_{xy}^2 + f_{yy}^2 \right) dx dy$$

that includes additional cross terms.

The constraint of interpolating the available data can be expressed as

$$\min_{\mathbf{f}} \|\mathbf{Lf}\|^2 + \lambda^T \mathbf{S}(\mathbf{f} - \mathbf{d}) \tag{2}$$

where $\mathbf{L}$ is a discrete approximation to a derivative operator. Again we choose a one-dimensional and uniformly sampled version of the problem to simplify the discussion. In this case the Laplacian has the form

$$\text{const} \cdot \begin{bmatrix} 2 & -1 & 0 & 0 \dots \\ -1 & 2 & -1 & 0 \dots \\ 0 & -1 & 2 & -1 \dots \\ & & \dots & \end{bmatrix} \tag{3}$$

4 Relating PCA and Laplacian Splines

The desired signal $\mathbf{f}$ can be represented with a PCA or other orthogonal basis as $\mathbf{f} = \mathbf{Vc}$. Substituting this into (2) gives

$$\min_{\mathbf{c}} \quad \mathbf{c}^T \mathbf{V}^T \mathbf{L}^2 \mathbf{V} \mathbf{c} + \lambda^T \mathbf{S}(\mathbf{Vc} - \mathbf{d}) \tag{4}$$

If $\mathbf{V}$ are the eigenvectors of $\mathbf{L}$, then $\mathbf{V}^T \mathbf{L}^2 \mathbf{V}$ is diagonal and (1), (4) are in the same form. This might initially suggest that $\mathbf{L}^2$ plays the role of the covariance matrix in PCA. This interpretation is not entirely satisfactory in that, while the second difference (Laplacian) matrix is positive semidefinite, it reflects a covariance that decays very quickly. However, the eigenvalues of a matrix are also the eigenvalues of its inverse. This suggest that $\mathbf{L}^2$ may play the role of the *inverse* covariance, i.e. the precision matrix.

Can $\mathbf{L}^2$ be interpreted as a (pseudo) inverse covariance? Several authors have noted a relationship of this form [8, 13] but without elaboration. To motivate this interpretation we note the following:

1. The matrix $\left[C_{r,c}\right] = \rho^{|r-c|}$ is known as the Kac-Murdock-Szego matrix, and for $\rho \neq 1$ has the inverse [7]

$$\frac{1}{1-\rho^2}\begin{bmatrix} 1 & -\rho & 0 & 0 & \dots \\ -\rho & 1+\rho^2 & -\rho & 0 & \dots \\ 0 & -\rho & 1+\rho^2 & -\rho & \dots \\ & & \dots & & \end{bmatrix}$$

For ρ near 1 this is an approximate Laplacian. $C_{r,c} = \rho^{|r-c|}$ also appears in the literature on the Discrete Cosine Transform (DCT), as a generic covariance matrix for images [12].

2. It is also known in the DCT literature that cosines are the eigenvectors of both $\left[C_{r,c}\right] = \rho^{|r-c|}$ [12] and of the discrete Laplacian approximation (3) [16].

5 Conclusion

PCA and spline models have complementary advantages and drawbacks. This observation raises the question of whether it is possible to create a unified model that incorporates the relative advantages of each. Such a unified model would be particularly suitable for computer vision tracking applications in which high accuracy is required. In these applications an exact fit to the data is needed (thus necessitating use of a spline), but a reasonable shape prior is also beneficial, for example as a robust low-dimensional parameterization of the search space.

We have shown that the data fitting problem for both PCA and LS can be put in the form $\mathbf{c}^T \Lambda^{-1}\mathbf{c} + \lambda^T\mathbf{S}(\mathbf{U}\mathbf{c}-\mathbf{d})$ with Λ, $\mathbf{U}$ obtained from the eigen-decomposition of the covariance and Laplacian operator respectively. The proposed common formulation of PCA and LS suggests a unified model that "inserts" a LS to account for dimensions discarded by the reduced rank PCA, for example by orthogonalizing the LS eigenvectors with respect to the retained PCA basis. The eigenvalues of the combined basis then provide a prior $\|\|_\Lambda = \sum c_k^2/\lambda_k$ on the data. Our proposal resembles Probabilistic PCA [17], which inserts an isotropic covariance in the discarded dimensions. It differs in that the LS has a decaying rather than constant spectrum, and prefers smooth deformation.

References

1. Anjyo, K., Todo, H., J.: A practical approach to direct manipulation blendshapes. J. Graph. Tools 16(3), 160–176 (2012)
2. Bookstein, F.L.: Principal warps: thin-plate splines and the decomposition of deformations. IEEE Trans. Pattern Anal. Mach. Intell. **11**, 567–585 (1989)
3. Botsch, M., Sorkine, O.: On linear variational surface deformation methods. IEEE Trans. Vis. Comp. Grap. **14**(1), 213–230 (2008)
4. Chui, H., Rangarajan, A.: A new point matching algorithm for non-rigid registration. Comp. Vis. Image Underst. **89**(2–3), 114–141 (2003)

5. Ding, C., Ye, J.: Two-dimensional singular value decomposition (2Dsvd) for 2D maps and images. In: Proceedings of SIAM international conference on data mining (SDM'05), pp. 32–43 (2005)
6. Donato, G., Belongie, S.: Approximate thin plate spline mappings. In: Proceedings of the 7th European conference on computer vision-Part III. ECCV'02, pp. 21–31. Springer, London, UK, UK (2002)
7. Dow, M.: Explicit inverses of Toeplitz and associated matrices. ANZIAM J. **44 E**, E185–E215 (2003)
8. Fieguth, P.: Statistical Image Processing and Multidimensional Modeling. Information science and statistics. Springer, New York (2011)
9. Lewis, J., Hwang, H.J., Neumann, U., Enciso, R.: Smart point landmark distribution for thin-plate splines. In: Proceedings of SPIE medical imaging, pp. 1236–1243. San Diego (2004)
10. Pighin, F., Lewis, J.: Digital face cloning. SIGGRAPH Course. http://portal.acm.org (2005)
11. Pighin, F., Lewis, J.: Performance-driven facial animation. SIGGRAPH Course. http://portal.acm.org (2006)
12. Rao, K., Yip, P.: Discrete Cosine Transform: Algorithms, Advantages, Applications. Academic Press (1990)
13. Rasmussen, C., Williams, C.: Gaussian Processes for Machine Learning. Mit Press, Adaptive Computation And Machine Learning (2006)
14. Schneider, D., Eisert, P.: Fast nonrigid mesh registration with a data-driven deformation prior. In: IEEE 12th international conference on computer vision workshops (ICCV Workshops), pp. 304–311 (2009)
15. Seol, Y., Lewis, J., Seo, J., Choi, B., Anjyo, K., Noh, J.: Spacetime expression cloning for blendshapes. ACM Trans. Graph. **31**(2), 14:1–14:12 (2012)
16. Strang, G.: The discrete cosine transform. SIAM Rev. **41**(1), 135–147 (1999)
17. Tipping, M.E., Bishop, C.M.: Probabilistic principal component analysis. J. Royal Stat. Soc. Ser. B **61**, 611–622 (1999)
18. Turk, G., O'Brien, J.F.: Shape transformation using variational implicit functions. In: Proceedings of the 26th annual conference on computer graphics and interactive techniques. SIGGRAPH'99, pp. 335–342. ACM Press/Addison-Wesley Publishing Co., New York, USA (1999)

Mathematics-in-Industry Study Group (MISG) Steel Projects from Australia and New Zealand

Winston L. Sweatman

Abstract Projects from the steel industry have made a valuable contribution to mathematics-in-industry study groups (MISG). In the eight-year period 2004–2011, there were eight such projects brought to the annual Australia and New Zealand MISG. Three particular projects are reviewed in more detail. The first relates to the extraction of vanadium and other metalloids from molten raw metal sourced from iron sand. The second considers the heating of cold-rolled steel coils during annealing. Finally, the third concerns the continuous hot-dipped galvanising process and coating deformations that may arise therein.

Keywords Steel industry · Mathematical modelling · Mathematics in industry study group

1 Introduction

Mathematics-in-Industry Study Groups (MISG) have been run in Australia and New Zealand since 1984. They ordinarily last a week and occur annually in Summer (January/February). Typically four to seven projects are brought by industry to be investigated by teams of mathematicians. Each project team is coordinated by two or sometimes three moderators who additionally present and report results. The author's participation began in 2004 and continued in subsequent meetings hosted by Massey University, Albany (2004–2006) in New Zealand, and University of Wollongong (2007–2009), RMIT University, Melbourne, (2010–2012) and Queensland University of Technology (QUT), Brisbane (2013) in Australia. There were fifty-seven projects in total during the decade.

W. L. Sweatman (✉)
Centre for Mathematics in Industry, Institute of Natural and Mathematical Sciences, Massey University, Albany, Auckland, New Zealand
e-mail: w.sweatman@massey.ac.nz

M. Wakayama et al. (eds.), *The Impact of Applications on Mathematics*,
Mathematics for Industry 1, DOI: 10.1007/978-4-431-54907-9_25,

The steel industry has provided strong support for the MISG. In the eight years 2004–2011, eight MISG projects were brought by representatives of New Zealand Steel and Bluescope Steel Research. The projects have been varied involving both different aspects of the steel production process and different kinds of mathematics.

MISG Steel Projects 2004–2011

1. MISG 2004, Strip temperature in a metal coating line annealing furnace [14].
2. MISG 2005, Development of empirical relationships for metallurgical design of hot-rolled steel products [12].
3. MISG 2006, Development of empirical relationships for the mechanical properties of cold-rolled steel products [16].
4. MISG 2007, Strip track-off and buckling between transport rolls [6].
5. MISG 2008, Cold point determination in heat-treated steel coils [4, 13, 19].
6. MISG 2009(a), Coil slumping [2].
7. MISG 2009(b), Metal coating deformation [8–10].
8. MISG 2011, Recovery of vanadium during steel production [20].

It should be noted that these investigations are team efforts. Several people work on each MISG project. A number were involved in the writing of reports, and subsequent publications, upon which this review is based. Sections 2, 3 and 4 relate to projects moderated by the author (projects 5, 7 and 8). Section 2 (project 8) considers vanadium removal from molten raw iron early in steel production from iron sand [20]. The chemical reactions are modelled using a system of differential equations. Section 3 (project 5) models the heating of cold-rolled steel coils in an annealing furnace [4, 13, 19]. Gaps between coil layers complicate heat transfer. Partial differential equations for the process are solved using Sturm-Liouville Theory. Section 4 (project 7) concerns defects arising in sheet steel galvanization [8–10]. Previous analyses of the fluid dynamics are extended by the inclusion of shear terms.

2 MISG 2011: Recovery of Vanadium During Steel Production

In New Zealand, steel is produced from iron sands (titanomagnetite). The process requires the removal of vanadium and other metalloids which, in addition, are valuable by-products. The goal of the project was to further the modelling and optimisation of this extraction. Section 2.1 describes the relevant processes at the steel mill, in particular the operation of the vanadium recovery unit (VRU). In Section 2.2, the processes are characterised by chemical reactions, and modelled with differential equations. The chemical variables and reaction rates proved awkward to quantify. Section 2.3 considers these, outlines an indicative model built at the MISG and concludes the discussion.

2.1 Vanadium Recovery Unit Operation

Metalloids are extracted at the VRU in the early stages of steel production. The molten raw metal, produced from iron sand by reduction with coal and heating, is contained in a ladle. This vessel contains on average 71 tonnes of the liquid and is used for transport as well as processing. The chemical composition is sampled. As well as iron and metalloids, including vanadium, manganese, silicon and titanium, the molten metal contains residual carbon from the coal. As far as possible metalloids need to be removed while retaining iron and carbon for making steel. A secondary VRU function is to melt off skulls (solidified iron) which sometimes form on ladle walls. A three-tonne skull can take three VRU visits to remove. When melting skulls, the process is prioritised over metalloid recovery.

Oxidation of the metalloids at the VRU produces slag which floats to the surface and is mechanically removed. The oxygen is from two sources: an oxygen lance blows continuously onto the liquid surface; and solid millscale (iron oxides) is added in discrete batches. A nitrogen lance ensures the liquid is well-mixed.

The VRU is constrained by the need to avoid 'carbon boil'. Carbon reactions increase with temperature, in contrast to the metalloid reactions which decrease. Apart from needing carbon for steel, its oxidation can become a hazardous and damaging runaway production of heat and carbon monoxide CO, with eruption and overflow of molten metal. This is avoided by retaining silicon concentration at sufficiently high levels, adding iron silicon alloy FeSi if required. It is routine to add about 105 kg of the alloy, 1–2 min before the end of the VRU process.

The ladle temperature is typically 1,350–1,450 °C. Lower temperatures enable metal solidification to form skulls, higher temperatures promote carbon boil.

2.2 The Oxidation Model

The model developed at the MISG, represents the complicated chemistry of the VRU with a number of simpler processes. Within the ladle, the composition, temperature, and reactions, are assumed homogenous. Key oxidation processes are

Reaction	*Reaction Rate*	
$2V + 3FeO \rightarrow V_2O_3 + 3Fe,$	$r_1^{V} = k_1^{V}[V]^{p_1^{V}}[FeO]^{p_2^{V}},$	(1)
$Si + 2FeO \rightarrow SiO_2 + 2Fe,$	$r_1^{Si} = k_1^{Si}[Si]^{p_1^{Si}}[FeO]^{p_2^{Si}},$	(2)
$Ti + 2FeO \rightarrow TiO_2 + 2Fe,$	$r_1^{Ti} = k_1^{Ti}[Ti]^{p_1^{Ti}}[FeO]^{p_2^{Ti}},$	(3)
$Mn + FeO \rightarrow MnO + Fe,$	$r_1^{Mn} = k_1^{Mn}[Mn]^{p_1^{Mn}}[FeO]^{p_2^{Mn}},$	(4)
$C + FeO \rightarrow CO + Fe,$	$r_1^{C} = k_1^{C}[C]^{p_1^{C}}[FeO]^{p_2^{C}},$	(5)
$2V + Fe_2O_3 \rightarrow 2Fe + V_2O_3,$	$r_2^{V} = k_2^{V}[V]^{p_3^{V}}[Fe_2O_3]^{p_4^{V}},$	(6)

$$3\text{Si} + 2\text{Fe}_2\text{O}_3 \rightarrow 4\text{Fe} + 3\text{SiO}_2, \qquad r_2^{\text{Si}} = k_2^{\text{Si}}[\text{Si}]^{p_3^{\text{Si}}}[\text{Fe}_2\text{O}_3]^{p_4^{\text{Si}}}, \tag{7}$$

$$3\text{Ti} + 2\text{Fe}_2\text{O}_3 \rightarrow 4\text{Fe} + 3\text{TiO}_2, \qquad r_2^{\text{Ti}} = k_2^{\text{Ti}}[\text{Ti}]^{p_3^{\text{Ti}}}[\text{Fe}_2\text{O}_3]^{p_4^{\text{Ti}}}, \tag{8}$$

$$3\text{Mn} + \text{Fe}_2\text{O}_3 \rightarrow 2\text{Fe} + 3\text{MnO}, \qquad r_2^{\text{Mn}} = k_2^{\text{Mn}}[\text{Mn}]^{p_3^{\text{Mn}}}[\text{Fe}_2\text{O}_3]^{p_4^{\text{Mn}}}, \tag{9}$$

$$3\text{C} + \text{Fe}_2\text{O}_3 \rightarrow 2\text{Fe} + 3\text{CO}, \qquad r_2^{\text{C}} = k_2^{\text{C}}[\text{C}]^{p_3^{\text{C}}}[\text{Fe}_2\text{O}_3]^{p_4^{\text{C}}}. \tag{10}$$

Although, in the real VRU, further compounds and reactions are present, (1)–(10) are taken to represent the main features. The reactions are taken to be oneway and characterised by typical rates. The oxide removal from reaction, as slag or vapour, reduces reversion. Reactions (1)–(5) represent the oxidation action of the oxygen lance. The intermediate product FeO arises in the rapid reaction at the liquid surface

$$2\text{Fe} + \text{O}_2 \rightarrow 2\text{FeO} \tag{11}$$

and spreads by mixing throughout the ladle. The net effect of the oxygen lance is to increase the liquid temperature. The flame exceeds 2000°C at the surface. Reactions (6)–(10) relate to millscale which is added at room temperature. This solid powder, mainly iron oxides, is represented by its primary component Fe_2O_3. Up to three 1,000–1,500 kg discrete batches of millscale are added, typically beginning with 1, 100 kg added at the start of the VRU process. In order to react with other species, millscale must melt and mix into the liquid. The process takes about 5 min to complete. Accordingly the millscale reaction is modelled as a continuous process lasting 5 min. The overall effect of millscale upon liquid temperature is minimal.

Section 2.3 discusses [P], the fractional concentration of constituent P, and r_i^P, its reaction rates. Exponents p_i^P relate to P's availability and k_i^P is constant.

Although mass is added through the action of the oxygen lance and incorporation of millscale, this is balanced by decreases in the effective mass as oxides form and leave the liquid as slag or vapour (such as CO). Accordingly, the total mass involved in the VRU process M_T is treated as constant, a typical value being 71,000 kg.

Based on (1)–(10), the VRU is governed by the differential Eqs. (12)–(17)

$$\frac{d[\text{V}]}{dt} = -2r_1^{\text{V}} - 2r_2^{V}, \quad \frac{d[\text{Si}]}{dt} = -r_1^{\text{Si}} - 3r_2^{\text{Si}}, \quad \frac{d[\text{Ti}]}{dt} = -r_1^{\text{Ti}} - 3r_2^{\text{Ti}}, \tag{12}$$

$$\frac{d[\text{Mn}]}{dt} = -r_1^{\text{Mn}} - 3r_2^{\text{Mn}}, \quad \frac{d[\text{C}]}{dt} = -r_1^{\text{C}} - 3r_2^{\text{C}}, \tag{13}$$

$$\frac{d[\text{FeO}]}{dt} = \frac{J_{FeO}}{M_T} - 3r_1^{\text{V}\dagger} - 2r_1^{\text{Si}\dagger} - 2r_1^{\text{Ti}\dagger} - r_1^{\text{Mn}\dagger} - r_1^{\text{C}\dagger}, \tag{14}$$

$$\frac{d[\text{Fe}_2\text{O}_3]}{dt} = \frac{J_{\text{Fe}_2\text{O}_3}}{\text{M}_\text{T}} - r_2^{V} - 2r_2^{\text{Si}\dagger} - 2r_2^{\text{Ti}\dagger} - r_2^{\text{Mn}\dagger} - r_2^{\text{C}\dagger}, \tag{15}$$

$$\frac{d[\text{Fe}]}{dt} = -\frac{J_{\text{FeO}}}{\text{M}_\text{T}}^{\ddagger} + 3r_1^{\text{V}\ddagger} + 2r_2^{\text{V}\ddagger} + 2r_1^{\text{Si}\ddagger} + 4r_2^{\text{Si}\ddagger} + 2r_1^{\text{Ti}\ddagger}$$

$$+ 4r_2^{\text{Ti}\ddagger} + r_1^{\text{Mn}\ddagger} + 2r_2^{\text{Mn}\ddagger} + r_1^{\text{C}\ddagger} + 2r_2^{\text{C}\ddagger}, \tag{16}$$

$$c\frac{dT}{dt} = \frac{H_{FeO}}{M_T} - \frac{H_{Fe_2O_3}}{M_T} - q_T + Q_1^V r_1^V + Q_2^V r_2^V + Q_1^{Si} r_1^{Si} + Q_2^{Si} r_2^{Si}$$
$$+ Q_1^{Ti} r_1^{Ti} + Q_2^{Ti} r_2^{Ti} + Q_1^{Mn} r_1^{Mn} + Q_2^{Mn} r_2^{Mn}$$
$$+ Q_1^{C} r_1^{C} + Q_2^{C} r_2^{C} + Q_2^{Fe} r_2^{Fe}. \tag{17}$$

Expressions (12)–(13) guide the decrease in metalloids within the ladle. For example, manganese [Mn] is oxidised by the combined effects of oxygen lance ($-r_1^{Mn}$) and millscale ($-3r_2^{Mn}$). The stochiometric coefficients, -1 and -3, reflect the numbers of manganese atoms involved, cf. (4) and (9). Similarly (14)–(15) govern the oxidising agents. As well as consumption by reactions, there are inward mass flows due to the oxygen lance (J_{FeO}) and millscale ($J_{Fe_2O_3}$). Using lance blow-rate (1,200 m^3/h), J_{FeO} is estimated at 2.2 kg/s [20]. Iron is converted into FeO, so the equation for iron (16) also contains J_{FeO}. Millscale must melt and mix in, so Fe_2O_3 input is spread over a 5 min period. For an 1,100 kg initial input, $J_{Fe_2O_3} = 1,100$ kg/300 s $= 3.67$ kg/s, falling to zero after 5 min.

The final Eq. (17) models the temperature within the ladle T. The quantity c is the specific heat capacity of the liquid (mainly molten iron). A typical value is 450 J/(kg K). The oxygen lance (H_{FeO}) and millscale ($H_{Fe_2O_3}$) transfer heat by their respective mass flows. Using surface flame temperature (2,000 °C) for the FeO mass flow produces $H_{FeO} = 9.2 \times 10^5$ W [20]. The millscale is added at room temperature (25 °C) and, like the incorporation of Fe_2O_3, the cooling effect is spread over 5 min before falling to zero. Neglecting any latent energy required for melting or incorporation, $H_{Fe_2O_3} = 3.3 \times 10^6$ W is produced for an 1,100 kg input [20]. Cooling of the ladle q_T is estimated at 3.75 W/kg, based on the cooling rate of a ladle with a lid. A similar behaviour is assumed as the surface is either insulated by slag or exposed to the flame whose effects are included in H_{FeO} [20]. The quantities of form $Q_i^P r_i^P$ are contributions due to exothermal energy of reactions. Exothermicities Q_i^P are listed in Table 1. The symbols † and ‡ warn that an extra constant factor may be required depending upon the definition of [P]. This is discussed in Sect. 2.3.

2.3 Chemical Variables, Parameters and Discussion

The model contains several parameters. Some quantities are readily available from the literature or industry data but others prove more challenging. To show typical features, some indicative values were used in simulations at the MISG.

Further, the interpretation of [P] is awkward. For industrial application the natural quantities are the mass fractions, i.e. [P] is the mass of P in a sample divided by the sample's total mass. These are measured at the VRU and were adopted by the MISG. However, chemical reactions involve molecules of substances, and from that perspective molar fractions are preferable. The difference may be remedied by rescaling reaction rates using atomic masses. To indicate that this may be required, the reaction rates (14)–(16), are annotated with † and ‡, although to date simulations treat these as the same as their unmarked counterparts.

Table 1 Reaction rates, exothermicities and initial conditions

Reaction rates (1/s)		Exothermicities (J/kg)		Initial conditions	
k_1^{V}	0.1552	Q_1^{V}	1.0627×10^5	[Fe]	0.9519
k_1^{Si}	0.2265	Q_1^{Si}	2.6029×10^5	[C]	0.034
k_1^{Ti}	0.4815	Q_1^{Ti}	1.7441×10^5	[V]	0.0049
k_1^{Mn}	0.2057	Q_1^{Mn}	0.6501×10^5	[Mn]	0.0044
k_1^{C}	0.0546	Q_1^{C}	1.1444×10^5	[Ti]	0.0028
k_2^{V}	0.1008	Q_2^{V}	0.5301×10^5	[Si]	0.002
k_2^{Si}	0.1367	Q_2^{Si}	0.3324×10^5	T	1,400 (°C)
k_2^{Ti}	0.3732	Q_2^{Ti}	0.4120×10^5	M_T	71,000 (kg)
k_2^{Mn}	0.1367	Q_2^{Mn}	0.1212×10^5		
k_2^{C}	0.0151	Q_2^{C}	-1.5142×10^5		

The reaction rates r_i^P (1)–(10), depend on fractional concentrations [P] of the species involved, rate constants k_i^P and exponents p_i^P. In reality, k_i^P vary with temperature, decreasing for all reactions apart from the carbon ones. These could be modelled with Arrhenius relations, however, a constant is a reasonable approximation with small temperature variations. Reactions vary, p_i^P may involve stochiometric coefficients or be unity. For illustrative simulations $p_i^P = 1$ and Table 1 lists appropriate rough values for k_i^P provided by industry representatives. The table also lists a typical set of initial conditions. These vary between batches of iron.

Numerical simulations were broadly consistent with expectation [20], however, the model should be interpreted as indicative of the process, further work being required for accurate representation. Chemical reactions parameters may be obtainable by laboratory work although, as the model reactions are representing a larger set, the parameters may differ somewhat. Another approach would be a statistical fitting of model constants with data on previous VRU operation. A different strategy could model reactions (1)–(5) in dynamic balance, with production calculated from Gibb's free energies [20].

Several effects and reactions are not explicitly included, such as FeO's production by iron reacting with Fe_2O_3 or its loss as vapour and slag. Although temperature is modelled its effects are not. The addition of FeSi is not included nor allowance made for skulls. These effects may need further consideration and inclusion, although there may still be scope to further simplify the model, perhaps combining different metalloids into a representative quantity. Asymptotic analysis could show dominant effects. As for all modelling with limited data, a balance must be struck to capture key features without extraneous detail. The process appears suitable for optimal control procedures, such as the Pontryagin maximum principle, to guide the most effective use of oxygen lance and millscale [15, 20].

3 Cold Point Determination in Heat Treated Steel Coils

Cold-rolled steel sheets are brittle. Annealing is required to reform crystalline structure. For the New Zealand Steel process, studied at the 2008 MISG, the coiled metal strips are heated to approximately 1,000 K and maintained at this temperature for several hours. Poorly annealed steel that has not been heated for long enough must be reprocessed. The MISG challenge was to model the temperature within a coil, find the cold point which takes the most time to reach the required temperature and determine how long this takes. Calculations are complicated by the gaps between layers of the coil. Also, the heating of coil surfaces needs to be modelled. It is of practical difficulty to find temperatures within the furnace. A full account from MISG is given in [13] with additional study of the problem in [4, 19].

3.1 Modelling the Steel Coil

A steel coil is a hollow cylinder with steel strip layers lying next to one another in the radial direction. This structure leads to differing heat conductivities in radial and axial directions. Axially, along the z-direction, the conductivity k_z (J/m/s/K) is essentially that of steel k_s. Radial conduction is hampered by the gaps between the layers and the effective conductivity k_r is lower.

Heat is transported in various ways between the radial layers including direct metal contact, gaseous diffusion, and radiation [17, 28, 29]. However, these effects may be collectively represented by conduction across alternating layers of steel and gas of respective thicknesses d_s and d_g [18, 27]. Further the layered structure may then be treated as homogeneous with effective radial conductivity $k_r \approx (d_s + d_g)/(d_s/k_s + d_g/k_g)$, with equality in the limiting case [7]. Examples suggest at 1,000 K, halfway along the coil, k_r is about $1/2$–$3/4$ of k_s [13].

A complication is that rolled steel strips are thinner at their edges than in their centre (crowning). The corresponding larger gaps at the flat ends of a coil will lead to a reduced radial conductivity. This was considered in numerical simulations at the MISG but did not have any great impact. This may be in part due to the effective heating of the ends of coils in our case. Coil tension and differential expansion in heating may also affect radial conductivity leading to variation in the radial direction but this is likely to be small. Allowance for changes in steel properties with rising temperature may also be accounted for by the concept of mean action time [11, 13].

Temperature T satisfies

$$\frac{\partial (c_p \rho T)}{\partial t} = \frac{1}{r}\frac{\partial}{\partial r}\left(k_r r \frac{\partial T}{\partial r}\right) + \frac{\partial}{\partial z}\left(k_z \frac{\partial T}{\partial z}\right), \tag{18}$$

with relevant properties and parameters listed in Table 2. Boundary conditions are discussed in Sect. 3.2. Assuming $c_p\rho$, k_r and k_z constant, solution is by separation

Table 2 Typical properties of steel, strip, coil and furnace

Steel density	ρ	7,854 kg/m^3 at 300 K
Steel thermal conductivity	k_s	60.5–30 W/m/K at 300–1,000 K
Steel thermal capacity	c_p	434–1,169 J/kg/K at 300–1,000 K
Strip thickness		0.4–3 mm
Strip width/coil length	L	700–1,500 mm
Coil inner radius	a	254 mm
Coil outer radius	b	750 mm
Coil mass		10–20 tonnes
Initial coil temperature	T_0	300 K
Gas thermal conductivity	k_g	0.06 W/m/K
Furnace temperature	T_g	1,000 K
Furnace dimensions		6.5 × 6.5 × 4 m^3
Furnace circulation		800 m^3/min
Platform mass		37 tonnes

and Sturm-Liouville Theory [3, 4, 13]. Introduce dimensionless variables, $r = b\bar{r}$, $z = L\bar{z}$, $t = (\rho c_p L^2/k_z)\bar{t}$, $u = (T - T_g)/(T_0 - T_g)$ and relative diffusivity $D = k_r L^2/k_z b^2$. Overbars will be omitted. The coil is now a hollow unit cylinder, inner radius $\alpha = a/b$. Its temperature cools from 1 towards furnace temperature 0:

$$\frac{\partial u}{\partial t} = D\frac{1}{r}\frac{\partial}{\partial r}\left(r\frac{\partial u}{\partial r}\right) + \frac{\partial^2 u}{\partial z^2}, \quad [u]_{t=0} = 1. \tag{19}$$

Separating variables, with $u(r, z, t) = R(r)Z(z)\Theta(t)$,

$$\frac{\partial^2 R}{\partial r^2} + \frac{1}{r}\frac{\partial R}{\partial r} = -\lambda_m^2 R, \quad \frac{\partial^2 Z}{\partial z^2} = -\eta_n^2 Z, \quad \frac{\partial \Theta}{\partial t} = -(D\lambda_m^2 + \eta_n^2)\Theta, \tag{20}$$

and we have individual solutions, involving Bessel functions, of the form

$$u = A_{mn}(\mathrm{J}_0(\lambda_m r) + B_m \mathrm{Y}_0(\lambda_m r))(\sin \eta_n z + L_n \cos \eta_n z)e^{-(D\lambda_m^2 + \eta_n^2)t} \tag{21}$$

with $n, m = 1, 2, \ldots$. The constants, A_{mn}, λ_n, η_n, B_m, and L_n, take values determined by boundary conditions associated with the differential Eq. 19. A linear combination forms the general solution with which to solve the full boundary problem.

3.2 Boundary Conditions and Solution

Steel coils are transported in and out of the furnace on a ventilated steel platform. A coil stands, flat end on the platform, so that the axial z-direction is vertical. Both platform and contacting coil end are presumed to reach furnace temperature rapidly.

Hence $[u]_{z=0} = 0$ and cosine terms vanish from (21). The general solution is now

$$u = \sum_{n=1}^{\infty}\sum_{m=1}^{\infty} A_{mn}\, e^{-(D\lambda_m^2+\eta_n^2)t}\left(\mathrm{J}_0(\lambda r) + B_m \mathrm{Y}_0(\lambda r)\right)\sin\eta_n z. \tag{22}$$

Once λ_n, η_n and B_m are known, using Sturm-Liouville orthogonality and results from Chaps. 9 and 11 of [1], the initial value in (19) gives

$$\begin{aligned} A_{mn} &= \frac{\int_0^1 \sin\eta_n z\;dz\;\; \int_a^1 r\,(\mathrm{J}_0(\lambda_m r) + B_m \mathrm{Y}_0(\lambda_m r))\;dr}{\int_0^1 \sin^2\eta_n z\;dz\;\; \int_a^1 r\left(\mathrm{J}_0(\lambda_m r) + B_m \mathrm{Y}_0(\lambda_m r)\right)^2 dr} \\ &= \frac{4\,(1-\cos\eta_n)\,[rC_{m1}]_\alpha^1}{(\eta_n - \cos\eta_n \sin\eta_n)\,\lambda_m\left[r^2\left(C_m^2 + C_{m1}^2\right)\right]_\alpha^1}, \end{aligned} \tag{23}$$

where $C_m \equiv \mathrm{J}_0(\lambda_m r) + B_m \mathrm{Y}_0(\lambda_m r)$ and $C_{m1} \equiv \mathrm{J}_1(\lambda_m r) + B_m \mathrm{Y}_1(\lambda_m r)$. Other directly heated coil surfaces are also presumed to adopt furnace temperature. The remaining surfaces are indirectly heated by gas circulation within the furnace and this convection is taken to satisfy Newton's Law of Cooling.

In the MISG project [13], radiant heaters on the furnace ceiling were assumed to heat the upper coil end directly. So $[u]_{z=0,1} = 0$ and $\eta_n = n\pi$. The curved surfaces are heated indirectly and hence have $k_r \partial T/\partial r = \pm H(T - T_g)$. Various approaches for estimating the heat transfer coefficient H all give values in the range 3–5 W/m^2/K [13]. Rescaling: $h = Hb/k_r$, these remaining boundary conditions become

$$\left[\frac{\partial u}{\partial r}\right]_{r=\alpha} = -\left[\frac{\partial u}{\partial r}\right]_{r=1} = hu. \tag{24}$$

The λ_m are found numerically from the conditions for consistency of B_m

$$\begin{vmatrix} h\mathrm{J}_0(\alpha\lambda_m) + \lambda_m \mathrm{J}_1(\alpha\lambda_m) & h\mathrm{Y}_0(\alpha\lambda_m) + \lambda_m \mathrm{Y}_1(\alpha\lambda_m) \\ h\mathrm{J}_0(\lambda_m) - \lambda_m \mathrm{J}_1(\lambda_m) & h\mathrm{Y}_0(\lambda_m) - \lambda_m \mathrm{Y}_1(\lambda_m) \end{vmatrix} = 0 \tag{25}$$

and then

$$B_m = \frac{\left(\frac{\lambda_m}{h}\mathrm{J}_1(\lambda_m) - \mathrm{J}_0(\lambda_m)\right)}{\left(\mathrm{Y}_0(\lambda_m) - \frac{\lambda_m}{h}\mathrm{Y}_1(\lambda_m)\right)}. \tag{26}$$

The leading term associated with λ_1 and η_1 tends to dominate, with the decay term $e^{-(D\lambda_1^2+\eta_1^2)t}$. The cold point, is half-way along the coil at $z = 1/2$, and $r = r_c$ the extremum of $C_1(r)$. This point is closer to the inner curved surface due to its smaller area, and therefore smaller heat flux, than the outer surface.

As well as the boundary conditions adopted for the MISG project [13], alternatives have been considered [4, 19]. The first subsequent investigation explored the effect of extra direct heating to reduce annealing time in the furnace [4]. If the outer curved surfaces of a coil are heated directly then the boundary condition becomes $[u]_{r=1} = 0$ and λ_m is obtained from the B_m consistency condition

$$\big(h\mathrm{J}_0(\alpha\lambda_m)+\lambda_m\mathrm{J}_1(\alpha\lambda_m)\big)\mathrm{Y}_0(\lambda_m)-\big(h\mathrm{Y}_0(\alpha\lambda_m)+\lambda_m\mathrm{Y}_1(\alpha\lambda_m)\big)\mathrm{J}_0(\lambda_m)=0. \quad (27)$$

If, further, the inner surfaces are also heated directly, $[u]_{r=\alpha,1} = 0$, and λ_m satisfies

$$\mathrm{J}_0(\alpha\lambda)\mathrm{Y}_0(\lambda_m) - \mathrm{Y}_0(\alpha\lambda_m)\mathrm{J}_0(\lambda_m) = 0. \quad (28)$$

In both cases $B_m = -\mathrm{J}_0(\lambda_m)/\mathrm{Y}_0(\lambda_m)$. There may be potential cost savings [4].

On the furnace ceiling there are spaces between heaters. Additionally roof structures partially obscure them. This motivates the study of a coil which is not directly heated from above, only being heated by contact with the platform and convection [19]. The heat transfer coefficient H_E may differ slightly due to the gaps between layers. It is rescaled using $h_E = H_E L/k_z$, and the boundary condition on top of the coil becomes $[\partial u/\partial z]_{z=1} = -h_E u$. The solution differs from the other cases in that $\eta_n \in [n\pi - \pi/2, n\pi]$ and satisfies

$$h_E \tan(\eta_n) + \eta_n = 0. \quad (29)$$

The cold point's vertical location is at the extremum in the leading factor $\sin(\eta_1 z)$ in the upper half of the coil. With perfect insulation on the top surface, the cold point is located there. The timescale for heating is a factor $(D\lambda_1^2 + \pi^2)/(D\lambda_1^2 + \eta_1^2)$ larger than with the original MISG assumptions [19].

Other alternative boundary condition scenarios may be implemented in the MISG model in the same fashion. The applicability of the linear model and dominance of leading terms aids calculation. Overall the indirect heating boundary conditions are the primary constriction on heat flow to the cool point, rather than differences in conductivity due to the radial gaps.

4 MISG 2009: Metal Coating Deformation

To prevent corrosion, sheet steel is normally coated. The 2009 MISG considered the continuous hot-dipped galvanising process used by Bluescope Steel. For this, a steel strip is first passed through a bath of molten alloy (e.g. zinc/aluminium). It is then drawn upwards, and the thickness is controlled by a pair of air knives, before coating solidification. The air knives, high velocity air jets, force surplus alloy downwards back to the bath. A model was built in earlier research [21–25], however, the development of new advanced coatings has prompted re-evaluation. Defects have arisen with high air-knife pressure, in the worst case pocks occurring, which substantially thin the local coating. The new MISG model agrees with the previous one but extends it by including air-knife shear terms. Full details are given in [8, 10], and a summary in [9].

Table 3 Parameters with typical values

Density of the coating	ρ	3×10^3 kg m^{-3}
Density of steel	ρ_s	7×10^3 kg m^{-3}
Dynamic viscosity of the coating	μ	10^{-3}kg m^{-1}s^{-1}
Gravitational acceleration	g	9.8 m s^{-2}
Scale of thickness of coating	h_0	5×10^{-6} m
vertical length scale-half-width of air jet	L	5×10^{-3} m
Half-width of the steel strip	d	10^{-3} m
Upward speed of the steel strip	U	2.5 m s^{-1}
Maximum centreline speed of the air jet	U_a	30 m s^{-1}
Reynolds number	$\mathrm{Re} = \rho UL/\mu$	37, 500
Stokes number	$S = \rho g h_0^2/\mu U$	0.0003
Length ratio	$\varepsilon = h_0/L$	10^{-3}
Pressure scaling	$\mu U/\varepsilon^2 L$	5×10^5 kg m^{-1}s^{-2}
Shear scaling	$\mu U/\varepsilon L$	500 kg m^{-1}s^{-2}

4.1 Model and Steady-State Solution

A first-order partial differential equation governs the system. This determines the steady-state coating shape and the evolution of any defects that may form. Model development [8, 10] parallels Tuck [22], however, with the addition of air-knife shear effects. Thin coating assumptions are made, the flow modelled is two-dimensional, incompressible, laminar and unsteady, with Navier-Stokes equations:

$$\rho(u_t + uu_x + wu_z) = -p_x + \mu(u_{xx} + u_{zz}) - \rho g, \tag{30}$$

$$\rho(w_t + uw_x + ww_z) = -p_z + \mu(w_{xx} + w_{zz}), \; u_x + w_z = 0, \tag{31}$$

where t is time, x and z are respectively vertical and horizontal coordinates, u and w are corresponding fluid velocities, and p is pressure. Subscripts indicate differentiation. Table 3 gives parameters and typical values. The boundary conditions are

$$u = U, \quad w = 0 \;\text{ at }\; z = 0 \;\text{ (substrate)}, \tag{32}$$

$$\mu u_z = \tau_a(x), \quad p - p_a(x) = -\gamma\kappa, \quad h_t + uh_x = w \;\text{ at}$$
$$z = h(x,t) \;\text{ (free surface)}. \tag{33}$$

The term $\gamma\kappa$ relating to surface tension proves insignificant [8, 10], and will be neglected. Air-knife pressure $p_a(x)$ and shear stress $\tau_a(x)$ are to be specified. Again we nondimensionalize, setting $t = (L/U)\bar{t}$, $x = L\bar{x}$, $z = \varepsilon L\bar{z}$, $u = U\bar{u}$, $w = \varepsilon U\bar{w}$, $p = (\mu U/\varepsilon^2 L)\bar{p}$, $h = \varepsilon L\bar{h}$, $p_a(x) = (\mu U/\varepsilon^2 L)P(x)$ and $\tau_a(x) = (\mu U/\varepsilon L)G(x)$. As before, overbars will be omitted. To leading order (30)–(33) are

$$p_x = u_{zz} - S, \quad p_z = 0, \quad u_x + w_z = 0, \tag{34}$$

$$u = 1, \quad w = 0 \quad \text{at} \quad z = 0, \tag{35}$$

$$u_z = G(x), \quad p = P(x), \quad h_t + uh_x = w \quad \text{at} \quad z = h. \tag{36}$$

Ignoring terms with factor $\varepsilon^2\text{Re}$, Eqs. (34)–(36) are completely solved by

$$u = (S + P'(x))\left(\frac{1}{2}z^2 - hz\right) + zG(x) + 1, \tag{37}$$

$$w = \frac{1}{2}z^2 h_x (S + P'(x)) - P''(x)\left(\frac{1}{6}z^3 - \frac{1}{2}hz^2\right) - \frac{1}{2}z^2 G'(x), \tag{38}$$

except that the last boundary condition becomes a partial differential equation

$$h_t + Q_x = 0, \tag{39}$$

which is to be satisfied. The flux of the coating liquid is

$$Q = h + \frac{1}{2}h^2 G(x) - \frac{1}{3}h^3 (S + P'(x)). \tag{40}$$

In normal operation the system is in steady state. Flux Q is constant and identical for all x. The coating thickness is maximal immediately above the bath ($h = h_+$). It thins as it approaches the region of air-knife action. Then, above this region, it thickens again but approaches a lesser value than the original ($h = h_-$). The model indicates that outside the region of influence of the air knife the thickness h is relatively constant and it is a solution to the cubic Eq. (40). However, despite the same coefficients, the thickness above the air knife (h_-) differs from that below (h_+), and they must correspond to different roots. As x is varied, to pass through the region where the air knife is active, the coefficients of the cubic expression change. The two roots for h, realised on either side of the knife, approach one another until a control point is reached (h_c, x_c) where the cubic has a repeated root. Then

$$\frac{\partial Q}{\partial h}(h_c, x_c) = 1 + h_c G(x_c) - h_c^2 [S + P'(x_c)] = 0, \tag{41}$$

and so, selecting the applicable negative sign,

$$h_c = \frac{1}{2}\left[\frac{G(x_c)}{S + P'(x_c)}\right]\left(1 - \sqrt{1 + 4\left[\frac{S + P'(x_c)}{G(x_c)^2}\right]}\right). \tag{42}$$

Also, in steady state,

$$\frac{dQ}{dx} = 0 = h'[1 + hG(x) - h^2(S + P'(x))] + \frac{h^2}{2}G'(x) - \frac{h^3}{3}P''(x), \tag{43}$$

and from (41–43), at $x = x_c$, $G'(x_c) = 2h_cP''(x_c)/3$. Using earlier experimental results [5, 26], realistic functional forms for $P(x)$ and $G(x)$ are given in [8]:

$$P(x) = P_{\text{MAX}}(1 + 0.6x^4)^{-3/2}, \tag{44}$$

$$G(x) = \begin{cases} \text{sign}(x)G_{\text{MAX}}\left[\text{erf}(0.41|x|) + 0.54|x|e^{-0.22|x|^3}\right] & \text{if } |x| < 1.73 \\ \text{sign}(x)G_{\text{MAX}}\left[1.115 - 0.24\log|x|\right] & \text{if} |x| \geq 1.73. \end{cases} \tag{45}$$

In [8], the system was explored with a range of parameter values. A typical set had:

$$S = 0.0015, \quad P_{\text{MAX}} = 0.01, \quad G_{\text{MAX}} = 0.1. \tag{46}$$

For these values $x_c \approx -1.107$, $h_c \approx 6.172$, $Q_c \approx 3.529$, $h_- \approx 43$ and $h_+ \approx 3.33$.

4.2 Non-steady Evolution of Coating Deformations

The steady state solution behaves resonably. Increasing air knife pressure forces more coating alloy downwards, with increased upstream thickness, decreased downstream (final) thickness and a corresponding decrease in flux Q. Now consider a perturbation representing a small surface deformation. It is not clear how this might occur although some possibilities are discussed in [8, 10]. In [8], a spatial disturbance is modelled by perturbation $\delta(x) = -0.3e^{-(x-0.5)^2}$ to steady-state thickness. Equation (39) is rewritten $h_t + c(h, x)h_x = A(h, x)$ where, respectively,

$$c(h, x) = 1 + G(x)h - h^2(S + P'(x)), \quad A(h, x) = \frac{h^3}{3}hP''(x) - \frac{h^2}{2}G'(x), \tag{47}$$

are disturbance propagation speed and amplitude. The significance of $A(h, x)$ is limited to a very narrow region close to (h_c, x_c). At (h_c, x_c), $c(h, x) = 0$ by (41), however, $c(h, x)$ is significant downstream approaching $1 - h^2S$. Disturbances above the control point propagate upwards and could potentially affect the final product, while those below are of less interest as they propagate downwards back to the bath. In either case, the propagation speed depends on the coating thickness (47).

Linear analysis and complementary numerical exploration [8, 10, 22], have been used to compare $c(h, x)$ and $A(h, x)$ for different strengths of air knife. Gravity is more significant for a weaker jet as the thickness h is larger and air-knife shear $G(x)$ smaller. With thinner coatings disturbances tend to be more persistent. For high pressure the evolution of a perturbation appears to depend upon the shear terms.

The behaviour of the coating near to the air knife is determined by the interaction of shear and pressure. Beyond this region perturbations appear to be marginally stable. In some cases, fluid near the steel substrate travels faster than that at the surface but in other cases the reverse is true [8, 10]. In particular, higher pressures tend to lead to thinner coatings, more persistent shear terms and potentially disturbances travelling upwards faster than the steel sheet and breaking forwards (upward). At lower values of pressure, the region of influence of the air knife is narrow. Beyond, the effects of gravity tends to dominate with the fluid surface slower than the substrate and disturbances break backwards (downward). Numerical simulations illustrated that at some intermediate cases, where disturbances neither break forward nor backward, they may persist a long way upward with only a minor change in shape.

The variables are interrelated. Dimensionless pressure effectively increases as strip speed decreases. Either a stronger air-knife jet or a reduced strip speed thins the coating, however, the latter also allows more time for disturbances to decay.

Surface tension and metallurgical effects of solidification have been ignored. They may have some effect but it seems insignificant [8, 24]. The parameter values here are indicative. Dimensional values depend on the actual parameters and these will need to be more precise to apply to a specific situation.

5 Concluding Comment

The three varied projects reviewed were contributed to the Australian and New Zealand MISG by the steel industry. In an eight-year period 2004–2011, New Zealand Steel and Bluescope Steel Research brought eight such projects. Inspired by the requirement to model these industrial processes, interesting, instructive and enjoyable mathematical challenges have arisen.

Acknowledgments I am very grateful to New Zealand Steel, Bluescope Steel Research and the industry representatives who brought these projects to the MISG. I also thank and acknowledge team members involved with these projects, and other contributors to the MISG: hosts, directors, industry partners and participants. Involvement with these projects have been instructive and fun. Finally I am grateful to the organisers of Forum “Math-for-Industry 2013Ó for the opportunity to report on this work.

References

1. Abramowitz, M., Stegun, I.: Handbook of Mathematical Functions. Dover, New York (1970)
2. Anderssen, R., Fowkes, N., Hickson, R., McGuinness, M.: Analysis of coil slumping. In: Merchant, T., Edwards, M., Mercer, G. (eds.) Proceedings of the 2009 Mathematics and Statistics in Industry Study Group, pp. 90–108. University of Wollongong, Australia (2010)
3. Budak, B.M., Samarskii, A.A., Tikhonov, A.N.: A collection of Problems in Mathematical Physics. Dover, New York (1964)
4. Barry, S.I., Sweatman, W.L.: Modelling heat transfer in steel coils. ANZIAM J. **50**, C668–C681 (2009)
5. Elsaadawy, E.A., Hanumanth, G.S., Balthazaar, A.K.S., McDermid, J.R., Hrymak, A.N., Forbes, J.F.: Coating weight model for the continuous hot-dip galvanizing process. Metall. Mater. Trans. B **38B**, 413–424 (2007)
6. Fraser, W.B., Macaskill, C., McGuinness, M., Thornton, A.: Strip track-off and buckling between transport rollers. In: Merchant, T., Edwards, M., Mercer, G. (eds.) Proceedings of

the 2007 Mathematics and Statistics in Industry Study Group, pp. 13–31. University of Wollongong, Australia (2008)
7. Hickson, R., Barry, S., Mercer, G.: Exact and numerical solutions for effective diffusivity and time lag through multiple layers. ANZIAM J. **50**, C682–C695 (2009)
8. Hocking, G.C., Sweatman, W.L., Fitt, A.D., Breward, C.: Deformations during jet-stripping in the galvanizing process. J. Eng. Math. **70**, 297–306 (2011)
9. Hocking, G.C., Sweatman, W.L., Fitt, A.D., Breward, C.: Deformations arising during air-knife stripping in the galvanisation of steel. In: Günther, M., Bartel, A., Brunk, M., Schöps, S., Striebel, M. (eds.) Progress in Industrial Mathematics at ECMI 2010, Mathematics in Industry 17, pp. 311–317. Springer, Berlin Heidelberg (2012)
10. Hocking, G.C., Sweatman, W.L., Roberts, M.E., Fitt, A.D.: Coating deformations in the continuous hot-dipped galvanizing process. In: Merchant, T., Edwards, M., Mercer, G. (eds.) Proceedings of the 2009 Mathematics and Statistics in Industry Study Group, pp. 75–89. University of Wollongong, Australia (2010)
11. Landman, K., McGuinness, M.: Mean action time for diffusive processes. J. Appl. Math. Decis. Sci. **4**(2), 125–141 (2000)
12. Marchant, T., Nickerson, A., Scott, D., Taylor, S.: Development of empirical relationships for metallurgical design of hot-rolled steel products. In: Wake, G.C. (ed.) Proceedings of the 2005 Mathematics in Industry Study Group, pp. 53–72. Massey University, New Zealand (2005)
13. McGuinness, M., Sweatman, W.L., Baowan, D., Barry, S.I.: Annealing Steel Coils. In: Merchant, T., Edwards, M., Mercer, G. (eds.) Proceedings of the 2008 Mathematics and Statistics in Industry Study Group, pp. 61–80. University of Wollongong, Australia (2009)
14. McGuinness, M., Taylor, S.: Strip temperature in a metal coating line annealing furnace. In: Wake, G.C. (ed.) Proceedings of the 2004 Mathematics in Industry Study Group, pp. 23–45. Massey University, New Zealand (2005)
15. Pontryagin, S., Boltyanskii, V.G., Gamkrelize, R.V., Mishchenko, E.F.: The Mathematical Theory of Optimal Processes. Wiley, New Jersey(1962)
16. Scott, D.J., Russell, K., Scheffer, J.: Multi-variable relationships in a batch annealing process. In: Wake, G.C. (ed.) Proceedings of the 2006 Mathematics in Industry Study Group, pp. 33–55. Massey University, New Zealand (2007)
17. Sridhar, M.R., Yovanovicht, M.M.: Review of elastic and plastic contact conductance models: Comparison with experiment. J. Thermophys. Heat Trans. 8, pp. 633–640 (1994)
18. Stikker, U.O.: Numerical simulation of the coil annealing process. Math. Models Metall. Process Dev, Iron and Steel Inst, Spec. Rep. **123**, 104–113 (1970)
19. Sweatman, W.L., McGuinness, M., Barry, S.I.: Heat transfer during annealing of steel coils. In: Günther, M., Bartel, A., Brunk, M., Schöps, S., Striebel, M. (eds.) Progress in Industrial Mathematics at ECMI 2010, Mathematics in Industry 17, pp. 303–309. Springer, Berlin Heidelberg (2012)
20. Sweatman, W.L., Wake, G.C., Fullard, L., Bruna, M.: Recovering vanadium during the production of steel from iron sand. ANZIAM J. **53**, M1–M21 (2012)
21. Thornton, J.A., Graff, H.F.: An analytical description of the jet-finishing process for hot-dip metallic coatings on strip. Metall. Mater. Trans. B **7B**, 607–618 (1976)
22. Tuck, E.O.: Continuous coating with gravity and jet stripping. Phys. Fluids **26**(9), 2352–2358 (1983)
23. Tuck, E.O., Bentwich, M., Van der Hoek, J.: The free-boundary problem for gravity-driven unidirectional viscous flows. IMA J. Appl. Maths **30**, 191–208 (1983)
24. Tuck, E.O., Vanden-Broeck, J.-M.: Influence of surface tension on jet-stripped continuous coating of sheet materials. Amer. Inst. of Chem. Eng. J. **30**, 808–811 (1984)
25. Tu, C.V.: Optimisation of lip gap for thin film coating in the jet stripping process. Proceedings of 5th International Conference Manufacturing Engineering, University of Wollongong, Australia (1990)
26. Tu, C.V., Wood, D.H.: Wall pressure and shear stress measurements beneath an impinging jet. Exp. Therm. Fluid Sci. **13**, 364–373 (1996)

27. Willms, A.R.: An exact solution of Stikker's nonlinear heat equation. SIAM J. Appl. Math. **55**(4), 1059–1073 (1995)
28. Zhang, X., Yu, F., Wu, W., Zuo, Y.: Application of radial effective thermal conductivity for heat transfer model of steel coils in HPH furnace. Int. J. Thermophys. **24**(5), 1395–1405 (2003)
29. Zuo, Y., Wu, W., Zhang, X., Lin, L., Xiang, S., Liu, T., Niu, L., Huang, X.: A study of heat transfer in high-performance hydrogen Bell-type annealing furnaces. Heat Trans.-Asian Res. 30 (8) 615–623 (2001)

Applications of Integrable Nonlinear Diffusion Equations in Industrial Modelling

P. Broadbridge

Abstract There are useful integrable nonlinear diffusion equations that can be transformed directly to linear partial differential equations. The possibility of linearisation allows us to incorporate a much broader class of boundary conditions than would be available under reduction by a one-parameter Lie symmetry. By this means we can solve nonlinear boundary value problems of practical significance. Examples are given in the solidification of multi-phase materials with nonlinear thermal transport coefficients, infiltration of water in unsaturated soil and evolution of a metal surface by fourth-order curvature-driven diffusion. From this approach, there arise some open mathematical problems.

Keywords Nonlinear diffusion · Stefan problems · Steel solidification · Unsaturated flow · Infiltration · Surface diffusion · Grain boundaries · Integrable models

1 Introduction

Many industrial processes involve the transfer of heat, matter and charge, depicted quantitatively as scalar densities. Heat and mass transport has long been modelled by scalar diffusion equations. These partial differential equation models have withstood the test of time (e.g. [13]). If we assign constant values to transport coefficients, then we recover the familiar classical linear heat diffusion equations. However, in applications wherein the dependent concentration or temperature variable has a wide range, the transport coefficients might vary enough to cause a change in character of the solutions. For example, over a range of increasing temperatures commonly found in metal foundries, the heat diffusivity of steel decreases by a factor of ten [23].

P. Broadbridge (✉)
Department of Mathematics and Statistics, La Trobe University,
Melbourne 3086, Australia
e-mail: p.broadbridge@latrobe.edu.au

M. Wakayama et al. (eds.), *The Impact of Applications on Mathematics*,
Mathematics for Industry 1, DOI: 10.1007/978-4-431-54907-9_26,

Over an increasing range of water concentrations found in soils, the soil-water diffusivity may increase by four orders of magnitude [19]. In these circumstances, in order to accurately predict heat and mass transport, one must use nonlinear diffusion models in which the diffusion coefficient is a function of the dependent concentration variable.

In solving practical boundary value problems involving nonlinear diffusion, one usually relies on numerical approximation methods, for which there are many existing schemes in use. However it is not widely known that there are useful integrable one-dimensional nonlinear diffusion equations that can be transformed to linear equations. Just as linear equations can be solved with a variety of initial-boundary conditions, so can linearisable equations. Furthermore, for some boundary value problems in which temperature contours or concentration contours can be traced in space and time, the parameters of these integrable models need not be constant but merely piecewise constant over subintervals of the range of the dependent variable. Just as in linear models, exact solutions to nonlinear boundary value problems may have a range of complexity, from closed-form solutions in terms of familiar elementary functions or transcendental functions, to open-form series of transcendental functions. Usually, exact solutions can be evaluated much more efficiently than numerical solutions. They offer a chance to deduce simple meaningful relationships between input data (e.g. initial conditions, boundary data and system parameters) and output data (e.g. net quantity of heat or mass transported and its rate of change). Special problems that may be solved exactly may be used as bench tests to validate numerical schemes whose main utility stems from their applicability to a much broader range of problems that could not be solved exactly by known techniques.

In subsequent sections, some of the main linearisable equations will be introduced through their practical applications. The development of exact solutions often leads to new mathematical problems that may open up new areas of fundamental mathematical research. Some of these problems, still not fully solved, will be mentioned along the way.

2 Solidification of Iron

The general equation for nonlinear one-dimensional heat transport is

$$\rho(\theta)C(\theta)\frac{\partial\theta}{\partial t} = \frac{\partial}{\partial x}\left[k(\theta)\frac{\partial\theta}{\partial x}\right], \tag{1}$$

where θ is temperature, ρ is density, C is specific heat capacity, k is thermal conductivity, t is time and x is distance from a specified planar boundary. For a multi-phase material, the volumetric heat capacity ρC may be a discontinuous function of temperature. However the heat energy density,

$$u = \int_0^\theta \rho(\eta) C(\eta) d\eta$$

is a continuous invertible function of temperature. As shown long ago by Kirchhoff [16], the heat equation simplifies when u is the dependent variable:

$$\frac{\partial u}{\partial t} = \frac{\partial}{\partial x}\left[D(u)\frac{\partial u}{\partial x}\right], \tag{2}$$

with nonlinear diffusivity $D(u) = k(\theta(u))/[\rho(\theta(u))C(\theta(u))]$. For a molten metal solidifying on a thick conductive base, we consider the free boundary problem to locate the solidification front $s(t)$ and solve for $u(x,t)$, given:

$$\frac{\partial u}{\partial t} = \frac{\partial}{\partial x}\left[D_j(u)\frac{\partial u}{\partial x}\right]; \quad j = -1 \text{ for } x < 0, \;\; j = 1 \text{ for } 0 \leq x < s(t), \tag{3}$$

$$u(x,0) = U_\infty \text{ (const.)}, \quad x < 0, \quad s(0) = 0. \tag{4}$$

$$u \to U_\infty\,, \; x \to -\infty, \tag{5}$$

$$\lim_{x\to 0^-} u(x,t) = \lim_{x\to 0^+} u(x,t), \tag{6}$$

$$D_{-1}(u)\frac{\partial u}{\partial x} = D_1(u)\frac{\partial u}{\partial x}, \quad x = 0, \tag{7}$$

$$D(u)\frac{\partial u}{\partial x} = \lambda\frac{ds_1}{dt}, \quad x = s(t). \tag{8}$$

Condition (8) is the Stefan free boundary condition, involving release of latent heat per unit volume λ at the solidification front. Closely related multi-phase Stefan problems (e.g. [7]) and nonlinear Stefan problems ([15]) have been proven to have unique classical solutions. There is no expectation that the problem (3–8) will have multiple solutions. Therefore one may make use of the well-known fact that if the free boundary takes the form $s_1(t) = \gamma_1\sqrt{t}$ with γ_1 constant, then the whole problem is invariant under the scaling transformation

$$\bar{x} = e^{\varepsilon}x, \;\; \bar{t} = e^{2\varepsilon}t, \;\; \bar{u} = u.$$

Invariant solutions under this Lie symmetry group take the form

$$\phi = f(u); \;\; \phi = xt^{-1/2}.$$

Then the second-order partial differential equation for $u(x,t)$ reduces to a second-order ordinary differential equation for $f(u)$;

$$f(u) = -2[D(u)/f'(u)]'. \tag{9}$$

2.1 Integrable Nonlinear Diffusion Models

Fujita [10, 11] showed that Eq. (9) could be fully integrated twice, in the special cases

$$D = \frac{a}{b-u}, \quad (i)$$
$$D = \frac{a}{(b-u)^2 + c} \quad (ii), \tag{10}$$

with a, b, and c constant. Bluman et al. [1] showed that these are the cases for which the nonlinear diffusion equation has two independent Lie potential symmetries. Most usefully, in the restricted Case (ii) with $c = 0$, the nonlinear diffusion equation happens to be transformable to the standard linear heat equation [12, 22]. This allows us to solve the nonlinear diffusion equation with a variety of boundary conditions. The known integrable diffusion equations are equivalent, under the group of contact transformations, to one of the canonical forms [17]:

$$\begin{aligned} u_t &= u_{xx} && (i),\\ u_t + uu_x &= u_{xx} && (ii),\\ u_t &= [u^{-2}u_x]_x && (iii),\\ u_t &= [u^{-2}u_x]_x \pm 1 && (iv). \end{aligned} \tag{11}$$

Each of these canonical forms may be transformed to the linear diffusion equation (11)(i) by little more than a combination of the hodograph transformation and introduction of a potential variable by integration of u [8]. In the case of (11)(iv), two integrations are needed [4].

The nonlinear diffusion equation with $D = a/(b-u)^2$, may be reduced to (11)(iii) by a linear change of the dependent variable. Then, up to an additive function of t, define a potential variable w by $u = w_x$. Then

$$w_t = \frac{w_{xx}}{w_x^2} + h'(t), \tag{12}$$

for some function h. If $\bar{w} = w - h$, then

$$\bar{w}_t = \frac{\bar{w}_{xx}}{\bar{w}_x^2}.$$

This is equivalent to

$$x_t = x_{\bar{w}\bar{w}}.$$

The same scaling symmetry, and consequently the same reduced ODE (9) applies as before when the parameters of the integrable model are specified *piecewise* over the range of temperatures for each phase of a multi-phase material,

$$D_k(u) = \frac{a_k}{(b_k - u)^2}, \quad U_{k-1} \le u \le U_k, \tag{13}$$

$$u = U_k, \quad x = s_k(t) = \gamma_k t^{1/2}, \tag{14}$$

$$D_k(u)\frac{\partial u}{\partial x} = D_{k+1}(u)\frac{\partial u}{\partial x} + \lambda_k s_k'(t), \quad x = \gamma_k t^{1/2}. \tag{15}$$

For example, there are four distinct solid phases and one liquid phase of iron. When latent heat is released at a phase boundary ($\lambda > 0$), heat flux will not be continuous across the boundary. When $\lambda = 0$ but $D(u)$ is discontinuous at $u = U_k$, $u(x, t)$ will be continuous but $u_x(x, t)$ will be discontinuous at $x = \gamma_k t^{1/2}$. Finally, for the purpose of fitting experimental data for heat diffusivity, we may choose the two-parameter model $D = a_k(b_k - u)^{-2}$ to exactly fit the data continuously at some selected values $u = U_k$ where λ will be set to zero. This may be regarded as a fictitious phase change. For each of the phases, the reduced second-order ODE (9) has a two-parameter solution. These parameters, along with the free boundary parameters γ_k are uniquely determined by the boundary conditions. In this way, Tritscher and Broadbridge [25] solved for the Stefan problem of iron solidifying on copper. The whole problem is reduced to a system of transcendental equations involving elsmentary functions and error functions. If the original solution of Stefan's problem for water freezing is considered to be exact, then so is the solution of the multi-phase Stefan problem of iron solidifying on copper. The same can be done for any steel alloy or indeed for any multi-phase material. In principle, the same boundary problem could be solved with the three-parameter model (10)(ii), allowing the diffusivity $D(u)$ to have continuous first derivative at the fictitious transition temperatures, and the solution $u(x, t)$ to be twice differentiable at the moving fictitious phase boundaries.

Note that with uniform initial temperature on a semi-infinite base material, the temperature at the platform $x = 0$ must be constant, determined from $0 = f(U_0)$.

2.1.1 Relation to Special Integrable Abel Equations

Abel equations are among the most commonly occurring first-order nonlinear ordinary differential equations that are encountered in applications. The exact parametric solution of the nonlinear heat equation follows from an exact parametric solution of a particular form of Abel equation.

Now following (12), we define the potential variable w more specifically as

$$w = \int_0^x u(y, t)dy, \tag{16}$$

so that $x = 0 \iff \phi = 0 \iff w = 0$. Assume that u has been rescaled so that $D = u^{-2}$. Then the heat flux at $x = 0$ is $u^{-2}u_x = Rt^{-1/2}$, where $R = U_0^{-2}/f'(U_0)$. In this case, in (12), $h'(t) = -Rt^{-1/2}$. After applying the hodograph transformation,

$$x_t = x_{ww} - Rt^{-1/2}x_w. \tag{17}$$

Now define $p = wt^{-1/2}$. From the scaling symmetry with invariant $\phi = xt^{-1/2}$,

$$w = p(\phi)t^{1/2}, \;\; u = w_x = p'(\phi) = 1/\phi'(p). \tag{18}$$

Then (17) reduces to the second-order ordinary differential equation

$$\frac{-1}{2}p\phi'(p) + \frac{1}{2}\phi = \phi''(p) - R\phi'(p), \tag{19}$$

for which the general solution is

$$\phi = A(p - 2R) + B\left[(\frac{p}{2} - R)erf(\frac{p}{2} - R) + \frac{e^{-(p-2R)^2/4}}{\sqrt{\pi}}\right], \tag{20}$$

from which,

$$u = \frac{1}{\phi'(p)} = \frac{1}{A + \frac{B}{2}erf(\frac{p}{2} - R)}. \tag{21}$$

The general parametric solution for the second-order ODE for $\phi(u)$ is $p \to (u(p), \phi(p))$. Note that the three real parameters A, B and R are not independent. The boundary condition $\phi = 0 \iff w = 0 \iff p = 0$ implies

$$A = \frac{B}{2}\left[erf(R) + \frac{e^{-R^2}}{R\sqrt{\pi}}\right]. \tag{22}$$

This condition guarantees $\phi''(p)/\phi'(p) = -R$ at $p = 0$, which gives the correct heat flux, equal to $Rt^{-1/2}$ at $x = 0$.
Exchanging the dependent and independent variables in (9), we have

$$u''(\phi) = 2u^{-1}(u')^2 - \frac{1}{2}\phi u^2 u'(\phi). \tag{23}$$

This equation has a one-parameter group of scaling symmetries

$$\bar{u} = e^{\varepsilon}u, \;\; \bar{\phi} = e^{-\varepsilon}\phi. \tag{24}$$

For every one-parameter Lie symmetry group, there exists a set of canonical coordinates, consisting of an invariant Φ satisfying $d\Phi/d\varepsilon = 0$, and a translating variable ψ satisfying $d\psi/d\varepsilon = 1$ (e.g. [14]). In this case, we may choose $\Phi = \phi u$ and $\psi = log\phi$. Expressed in terms of the canonical variables, the ODE (23) is

$$\Phi''(\psi) = 2\Phi^{-1}(\Phi')^2 - \Phi' - \frac{1}{2}\Phi^2\Phi' - \frac{1}{2}\Phi^3. \tag{25}$$

By design, this equation is autonomous in ψ, so we can reduce to first order by defining $M = \Phi'(\psi)$ and $\Phi''(\psi) = M\frac{dM}{d\Phi}$. Then

$$M\frac{dM}{d\Phi} = 2\Phi^{-1}M^2 + [-1 - \frac{1}{2}\Phi^2]M - \frac{1}{2}\Phi^3. \tag{26}$$

This is an Abel equation of the second kind. In order to simplify it to canonical form, we define $\omega = \Phi^{-2}M$ and $z = \Phi^{-1} - \frac{1}{2}\Phi$, resulting in

$$\omega\omega'(z) = \omega + \frac{1}{2\sqrt{z^2+2}}. \tag{27}$$

This is one of the list of Abel equations for which the exact solution is known [20] (Sect. 1.3). By an analogous procedure, the similarity reduction of nonlinear diffusion equation with $D = u^{-1}$ leads to an ODE whose solution is equivalent to solving the separable Abel equation

$$\omega\omega'(z) = \omega - 2. \tag{28}$$

The three-parameter model (10) allows us to connect any diffusivity data points by a spline $D(u)$ that is differentiable. With that nonlinear diffusivity, the heat flow problem with constant-temperature boundary conditions on a semi-infinite domain, or with Stefan conditions on a finite domain, could be solved exactly after similarity reduction. On each spline segment, after linear change of variable, the diffusivity is represented by $D = (1+u^2)^{-1}$. With that model, Eq. (19) is replaced by

$$p''(\phi) = [1 + \frac{1}{4}p'(\phi)^2][p - \phi p' + 2R]. \tag{29}$$

This equation has a one-parameter symmetry group

$$\begin{aligned} \bar{\phi} &= \phi \cos \varepsilon - (\frac{p}{2} + R) \sin \varepsilon, \\ \bar{p} &= 2\phi \sin \varepsilon + (p + 2R) \cos \varepsilon - 2R. \end{aligned} \tag{30}$$

This is recognisable as the group of rotations in the plane with Cartesian coordinates $(2\phi, p+2R)$. The canonical coordinates are the polar coordinates (α, r); $(2\phi, p+2R) = (r \cos \alpha, r \sin \alpha)$, in terms of which, (29) is an autonomous equation

$$r''(\alpha) = (\frac{2}{r} + \frac{1}{4}r)(r')^2 + r + \frac{1}{4}r^3. \tag{31}$$

Following the standard reduction to Abel equation in canonical form, we choose $p = r'(\alpha)$, $\omega = pr^{-2}e^{-r^2/8}$ to obtain the separable equation

$$\omega\omega'(r) = r^{-3}e^{-r^2/4} + \frac{1}{4}r^{-1}e^{-r^2/4}. \tag{32}$$

3 Water Flow in Unsaturated Soil

Following the Darcy–Buckingham modelling approach, water flow in the downward direction of increasing depth z, is governed by the Richards equation, a nonlinear diffusion-convection equation

$$\frac{\partial\theta}{\partial t} = \frac{\partial}{\partial z}[D(\theta)\frac{\partial\theta}{\partial z}] - K'(\theta)\frac{\partial\theta}{\partial z}. \tag{33}$$

Here, θ is volumetric water content, D is the soil-water diffusivity with $D > 0$, $D'(\theta) > 0$, and K is the hydraulic conductivity with $K > 0$, $K'(\theta) > 0$ and $K''(\theta) > 0$. Within the equivalence class of (11)(ii), there is a solvable model

$$D = \frac{a}{(b-\theta)^2}, \quad K = \frac{\Lambda}{2(b-\theta)} + \gamma(b-\theta) + \beta, \tag{34}$$

with a, b, Λ, γ and β constant. This model is useful in approximating experimental data [27] but unlike other standard models that are better at matching transport coefficients, it admits useful exact solutions for water flow. After rescaling variables $(t, z, \theta) \to (t^*, z^*, \Theta)$ (as in [24]),

$$\frac{\partial\Theta}{\partial t^*} = \frac{\partial}{\partial z^*}\left[\frac{C(C-1)}{(C-\Theta.)^2}\frac{\partial\Theta.}{\partial z^*}\right] - \zeta\frac{C(C-1)}{(C-\Theta.)^2}\frac{\partial\Theta.}{\partial z^*} - (\zeta-1)\frac{\partial\Theta.}{\partial z^*}, \tag{35}$$

with $C \in (1, \infty)$, $\Theta. \in [0, 1]$. The final linear convection term may be removed by the standard change of reference frame, $\bar{z} = z^* - (\zeta - 1)t*$. A change of variables to

$$u = \frac{C - \Theta.}{\zeta(C[C-1])^{1/2}}e^{\zeta\bar{z}}, \quad \text{and } y = 1 - e^{-\zeta\bar{z}}, \tag{36}$$

effectively transforms the nonlinear convection term to zero, to arrive at the same equation that has been solved in the previous section,

$$\frac{\partial u}{\partial t^*} = \frac{\partial}{\partial\bar{z}}\left[u^{-2}\frac{\partial u}{\partial\bar{z}}\right]. \tag{37}$$

In this way, the integrable form of Richards' equation (35) has been solved subject to constant-flux boundary conditions [6, 21]. After some transformations, this involved solving a linear diffusion-covection equation with a constant coefficient replacing the factor $Rt^{-1/2}$ in (17). With much greater difficulty, Richards' equation (35) has more recently been formally solved subject to constant-concentration boundary conditions [24]. This required an assumption that for this boundary condition, the flux at $z^* = 0$ is a power series in $t^{1/2}$, with leading term $Rt^{-1/2}$. In the case of zero convection dealt with in the previous section, this is exactly the form of the flux at the boundary. With convection terms present, due to gravity in the case of soil-water flow, the flux at the boundary is modified by a series of terms of order $t^{(j-1)/2}$. This assumption was originally due to Philip [19]. There is experimental and computational evidence that it works in practice but after more than 40 years, it has still not been proven whether or not the Philip infiltration series converges. The solution of Richards' equation, subject to Dirichlet boundary conditions, involves a formal series of generalised hypergeometric functions. We have summed it to 160 terms, the partial sums are quite well behaved, and they appear to converge even at large-t but this has still not been proved. If it could be proved, then this would also settle the open question on the radius of convergence of the Philip infiltration series.

4 Surface Diffusion on Stable Metals

The integrable second-order nonlinear diffusion equations sit in a hierarchy of higher-order nonlinear equations that may be linearised by the same transformations. The nth-order integrable equations may be constructed as follows: (i) Begin with the nth-order constant-coefficient evolution equation

$$\frac{\partial x}{\partial t} = \sum_{j=0}^{n} c_j \frac{\partial^j x}{\partial w^j}.$$

(ii) Apply the hodograph transformation to obtain an equation for $w(x,t)$.
(iii) Differentiate throughout that equation for $w(x,t)$, with respect to x.
(iv) Write $\frac{\partial w}{\partial x}$ as u and $\frac{\partial^{j+1} w}{\partial x^{j+1}}$ as $\frac{\partial^j u}{\partial x^j}$.

For example, one of the fourth-order evolution equations that arises is

$$\frac{\partial u}{\partial t} = -B\frac{\partial^2}{\partial x^2}\left[\frac{1}{b+u}\frac{\partial}{\partial x}\left(\frac{1}{(b+u)^3}\frac{\partial u}{\partial x}\right)\right]. \tag{38}$$

This is the equation for the slope of a solid surface subject to curvature-driven surface diffusion on an anisotropic material that is close to isotropic [5]. This is the predominant mechanism for surface evolution of stable metals such as gold [18]. This equation has been solved with the boundary conditions of a microscopic symmetric grain boundary having constant slope at $x = 0$ [26]. This solution takes the form of a similarity solution $u = f(xt^{-1/4})$. As in the case of nonlinear heat conduction, such a similarity solution can still be obtained when parameters of model (38) are piecewise-constant as a function of temperature. The boundary groove problem has also been solved when the groove slope is time-dependent, controlled for example by temperature modifying surface tension [3]. This breaks the scale invariance but the solution may be expanded as a series of generalised hypergeometric functions, with the similarity solution as the leading term at small-t.

5 Conclusion and Outlook

The boundary value problem of solidifying iron that was discussed in Sect. 2, is reminiscent of the casting methods of the early Iron Age, when molten metal was poured into a passive mould. Production rates in modern continuous steel slab casters are twice as high as those predicted by that idealised boundary value problem. In fact, heat is removed from the copper slab by a water spray that converts sensible heat to latent heat at an approximately constant rate. This would prevent boundary heat flux decreasing in proportion to $t^{-1/2}$ as in the similarity solution. The linearisable models do not rely on having boundary conditions that are compatible with similarity solutions. This gives some hope that more realistic models of steel casting may yet be solved exactly. As shown in Refs. [3, 24], other nonlinear boundary value problems, without similarity invariance, may be solved by a series of generalised hypergeometric functions, with the coefficients determined uniquely from recurrence relations that are dictated by the boundary conditions. The series has so far not been proven to converge or diverge. However it is promising that the coefficients of Philip's infiltration series for water infiltration, predicted by such series, agree well with field measurements [24]. Unlike in one spatial dimension, in two or three spatial dimensions there are no nonlinear diffusion equations that may be transformed to a linear equation [2]. However, in some special cases of nonlinear diffusivity, there are Lie symmetries that lead to variable reductions and exact solutions (e.g. [9]). It is a major problem of Lie symmetry reductions that these apply to a limited class of initial-boundary data.

References

1. Bluman, G.W., Kumei, S., Reid, G.J.: New classes of symmetries for partial differential equations. J. Math. Phys. **29**, 806–811 (1988)
2. Broadbridge, P.: Nonintegrability of nonlinear diffusion-convection equations in two spatial dimensions. J. Phys. A Math. Gen. **19**, 1245–1257 (1986)
3. Broadbridge, P., Goard, J.M.: Grain boundary evolution with time dependent material properties. J. Eng. Math. **66**(1–3), 87–102 (2010)

4. Broadbridge, P., Rogers, C.: On a nonlinear reaction-diffusion boundary-value problem: application of a Lie-BŁcklund symmetry. J. Austral. Math. Soc. (b) **34**, 318–332 (1993)
5. Broadbridge, P., Vassiliou, P.J.: The role of symmetry and separation in surface evolution and curve shortening. Symmetry, integrability and geometry: methods and applications (SIGMA), 7, 052, 19 pages (2011). http://dx.doi.org/10.3842/SIGMA.2011.052
6. Broadbridge, P., White, I.: Constant rate rainfall infiltration: a versatile non-linear model. 1. Analytic solution. Water Resour. Res. **24**, 145–154 (1988)
7. Cannon, J.R., Douglas, J., Hill, C.D.: A multi-boundary Stefan problem and the disappearance of phases. J. Math. Mech. **17**, 21–33 (1967)
8. Clarkson, P.A., Fokas, A.S., Ablowitz, M.J.: Hodograph transformations on linearizable partial differential equations. SIAM J. Appl. Math. **49**, 1188–1209 (1989)
9. Edwards, M.P.: Classical symmetry reductions of nonlinear diffusion convection equations. Phys. Lett. A. **190**, 149–154 (1994)
10. Fujita, H.: The exact pattern of a concentration-dependent diffusion in a semi-infinite medium. Part I. Textile Res. J. **22**(11), 757–760 (1952)
11. Fujita, H.: The exact pattern of a concentration-dependent diffusion in a semi-infinite medium. Part II. Textile Res. J. **22**(12), 823–827 (1952)
12. Fujita, H.: The exact pattern of a concentration-dependent diffusion in a semi-infinite medium. Part III. Textile Res. J. **24**, 234–240 (1954)
13. Fulford, G.R., Broadbridge, P.: Industrial Mathematics: Case Studies in the Diffusion of Heat and Matter. Cambridge University Press, Cambridge (2001)
14. Ibragimov, N.H.: CRC Handbook of Lie Group Analysis of Differential Equations, vol. 1. CRC Press, Boca Raton (1993)
15. Kenmochi, N.: A new proof of the uniqueness of solutions to two-phase Stefan problems for nonlinear parabolic equations. In: Hoffmann, K.-H., Sprekels, J. (eds.) Free Boundary Value Problems, pp. 101–126. Birkhauser, Basel (1990)
16. Kirchhoff. :Theorie der Wärme. Leipzig (1891)
17. Mikhailov, A.V., Shabat, A.B., Sokolov, V.V.: The symmetry approach to classification of integrable equations. In: Zakharov, V.E. (ed.) What is Integrability ?, pp. 115–184. Springer, Heidelberg (1991)
18. Mullins, W.W.: Theory of thermal grooving. J. Appl. Phys. **28**, 333–339 (1957)
19. Philip, J.R.: Theory of infiltration. Adv. Hydrosci. **5**, 215–296 (1969)
20. Polyanin, A.D., Zaitsev, V.F.: Handbook of Exact Solutions for Ordinary Differential Equations, 2nd edn. Chapman-Hall/CRC Press, Boca Raton (2003)
21. Sander, G.C., Parlange, J.-Y., Kuhnel, V., Hogarth, W.L., Lockington, D., O'Kane, J.P.J.: Exact nonlinear solution for constant flux infiltration. J. Hydrol. **97**(34), 341–346 (1988)
22. Storm, M.L.: Heat conduction in simple metals. J. Appl. Phys. **22**, 940–951 (1951)
23. Touloukian, Y.S. (ed.): Thermophysical Properties of Matter, vol. 1. Plenum, New York (1970)
24. Triadis, D., Broadbridge, P.: Analytical model of infiltration under constant-concentration boundary conditions. Water Resour. Res. **46**(3), W03526 (2010). doi:10.1029/2009WR008181
25. Tritscher, P., Broadbridge, P.: A similarity solution of a multiphase Stefan problem incorporating general nonlinear heat conduction. Int. J. Heat Mass Transfer **37**, 2113–2121 (1994)
26. Tritscher, P., Broadbridge, P.: Grain boundary grooving by surface diffusion: an analytic nonlinear model for a symmetric groove. Proc. Roy. Soc. A **450**, 569–587 (1995)
27. White, I., Broadbridge, P.: Constant rate rainfall infiltration: a versatile non-linear model. 2. Applications of solutions. Water Resour. Res. **24**, 155–162 (1988)

User Interfaces for Character Animation and Character Interaction

Takaaki Shiratori

Abstract The heart of visual storytelling is the performances of characters. Through their performances, they express the story, demonstrate their emotions, personality and motivation, and ultimately entertain people. However, design of appealing performances for characters and interaction with the animated characters are key for artists to enhance the experience of the storytelling. In this paper, we present user interfaces that allow efficient creation of animated content and the enhancement of the interactive experience. Specifically, we will describe recent progress on three main issues of character animation and interaction: motion capture, animation design, and interactive control.

Keywords Motion capture · Structure from motion · Animation design" and 3D modeling · Performance interface · Iterative design

1 Introduction

The heart of visual storytelling is the performances of characters [1, 2]. Via their performances, characters express the story, demonstrate their emotions and personality, motivation, and ultimately entertain people. Nowadays, thanks to free or inexpensive animation software, even novices can create animated content and share their creations on the internet. However, designing appealing and effective performances for characters is still a difficult and time-consuming task because of the complexity of the skeletal structure and the required physical plausibility of the performances. We are interested in user interfaces that allow efficient creation of such content and the enhancement of the experience for viewers or players by leveraging insights on how humans achieve/express/experience performances for animated content creation.

T. Shiratori (✉)
Visual Computing Group, Microsoft Research, Beijing, People's Republic of China
e-mail: takaakis@microsoft.com

M. Wakayama et al. (eds.), *The Impact of Applications on Mathematics*,
Mathematics for Industry 1, DOI: 10.1007/978-4-431-54907-9_27,

In this paper, we present our recent progress on user interfaces for animations of characters, involving three main areas: motion capture [3, 4], animation design [5, 6], and interactive control [7–10]. The first area, *motion capture*, is about techniques for capturing the motion of humans and/or animals in 3D. This technique is useful for not only 3D content creation but also rehabilitation, training, and biomechanics. However, conventional systems require expensive instrumentation of the environment. We describe a new motion capture system that uses body-mounted cameras to go beyond conventional motion capture systems [3] (Sect. 2). The second area, *animation design*, is about systems to design performances for characters. A typical way to design an animation is to specify keyframes or edit motion capture data for each degree of freedom (DOF) frame by frame (e.g., Autodesk Maya[1]). With this procedure, even professional artists need several days to create a 10-s animation that satisfies the director's requirements. To make the animation pipeline more efficient and intuitive, we describe a puppeteering interface that allows users to perform the motion for characters [5] (Sect. 3). The last area, *interactive control*, is about system designs that enable users to use performance interfaces to interact with characters. Now that various motion sensing devices are commercially available, the next issue is how to design intuitive, immersive, and effective interaction systems. We describe a performance interface that allows users to create and animate an arbitrary shape of characters [10] (Sect. 4). We conclude this paper by describing possible directions of these research fields.

2 Motion Capture from Body-Mounted Cameras

Motion capture has been used to provide much of the character motion in several recent live action movies. In *Avatar*, motion capture was used to animate characters riding on direhorses and flying on the back of mountain banshees. To capture realistic motion for such scenes, the actors rode horses and robotic mock-ups in an expansive motion capture studio requiring a large number of cameras [11]. Coverage and lighting problems often prevent directors from capturing motion in natural settings or in other large environments. Inertial systems, such as an Xsens motion capture[2] and the one described by Vlasic and colleagues [12], allow capture to occur in outdoor spaces but are designed to recover only the *relative* motion of the joints, not the global root motion.[3]

We introduced a wearable system of outward-looking cameras that allow the reconstruction of the relative and the global motion of an actor outside of a laboratory or closed stage [3]. The cameras can be mounted on casual clothing (Fig. 1a), are easily mounted and removed using Velcro attachments, and are lightweight enough to allow unimpeded movement. To estimate the pose of the cameras throughout the capture, we used structure-from-motion (SfM), a computer vision technique with

[1] http://www.autodesk.com/products/autodesk-maya.

[2] http://www.xsens.com/.

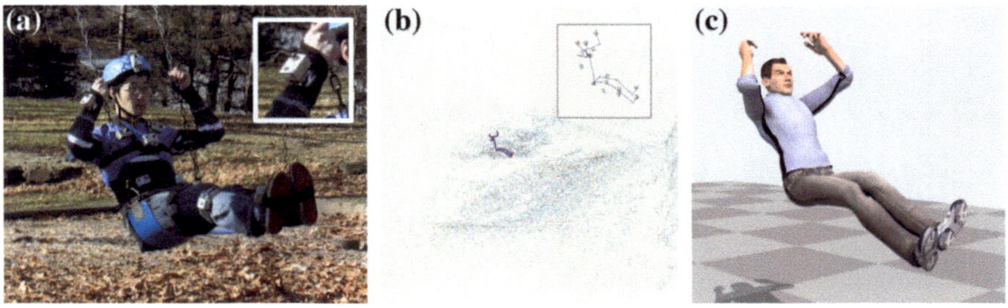

Fig. 1 Capturing both relative and global motion in natural environments using cameras mounted on the body. **a** Body-mounted cameras. **b** Skeletal motion and 3D structure. **c** Rendered actor

which, given geometrically consistent 2D feature correspondences among images, the 6D poses of the cameras and the 3D locations of the 2D features can be estimated [13, 14]. The estimated camera movements from a range-of-motion sequence are used to automatically build a skeleton using co-occurring transformations of the limbs connecting each joint [15]. The reconstructed cameras and skeleton (Fig. 1b) are used as an initialization for an overall optimization to compute the root position, orientation, and joint angles while minimizing the image matching error. Reference imagery of the capture area is leveraged to reduce drift. We render the motion of a skinned character by applying the recovered skeletal motion (Fig. 1c).

To compute appropriate estimates of human motion across time, our SfM-based solution considers the articulation of body-mounted cameras with the underlying skeleton of the actor and fits them to image measurements:

$$\{\mathbf{O}^*, \mathbf{A}^*\} = \underset{\mathbf{O},\mathbf{A}}{\operatorname{argmin}}\; E_r + \lambda_{\mathbf{O}} E_{\mathbf{O}} + \lambda_{\mathbf{A}} E_{\mathbf{A}}, \tag{1}$$

where $\mathbf{O}$ and $\mathbf{A}$ are the time-series data of the root position and the joint angles, respectively. E_r accounts for reprojection errors of the 3D reconstruction with measured image feature locations using the skeleton constraint for the cameras:

$$E_r = \sum_{j,t,p} \left\| P_j\left(\mathbf{X}_p, t, \mathbf{O}, \mathbf{A}\right) - \mathbf{x}_{j,t,p} \right\|_{\Sigma}^{2}, \tag{2}$$

where $P(\cdot)$ is a camera projection function, and j, t, and p are indices of cameras, time, 3D points, respectively. $\mathbf{X}$ is the location of the 3D structure point in the world coordinate system, and $\mathbf{x}$ is the corresponding 2D measurement. $E_{\mathbf{O}}$ and $E_{\mathbf{A}}$ can be also considered to obtain smooth motion. The differences of the root positions and joint angles between consecutive frames are minimized as

$$E_{\mathbf{O}} = \sum_t \|\mathbf{O}(t) - \mathbf{O}(t-1)\|_{\Sigma}^2, \tag{3}$$

$$E_{\mathbf{A}} = \sum_t \|\mathbf{A}(t) - \mathbf{A}(t-1)\|_{\Sigma}^2. \tag{4}$$

Fig. 2 Results of our motion capture system in natural environments. **a** Swinging on a swing. **b** Swinging on a monkey bar. **c** Running on a street

Fig. 3 Subject walking along a long winding path. *Left* a photo of the scene manually superimposed to the reconstructed scene (the *red curve* represents the trajectory of the subject), and *right two* the reconstructed walking motion and sparse 3D structure

These terms are effective, particularly when the camera poses estimated from the absolute and relative registration contain undesirable jitter.

By utilizing forward kinematics for the estimated camera poses, the global and relative motion of an actor can be captured outdoors under a wide variety of lighting conditions or in extended indoor regions without any additional equipment (Figs. 2 and 3). We also avoid some of the missing data problems introduced by occlusions between the markers and cameras in traditional optical motion capture, because, in our system, any visually distinctive feature in the world can serve as a marker in the traditional systems. A by-product of the capture process is a sparse 3D structure of the scene. This structure is useful as a guide for defining the ground geometry and as a first sketch of the scene for 3D animators and directors. We evaluate our approach against motion capture data generated by a Vicon optical motion capture system[4] and report a mean joint position error of 1.76 cm and a mean joint angle error of 3.01° on the full range-of-motion sequence used for skeleton estimation. Our results demonstrate that the system can reconstruct actions that are difficult to capture with traditional motion capture systems, including outdoor activities in direct sunlight, activities that are occluded by near by proximal structures, and extended indoor activities.

Our prototype is the first, to our knowledge, to employ camera sensors for motion capture by measuring the environment and to estimate the motion of a set of cameras that are related by an underlying articulated structure. Current cameras are inexpensive, have form factors that rival inertial measurement units (IMUs), and are already embedded in everyday handheld devices. Our approach will continue to benefit from consumer trends that are driving cameras to become cheaper, smaller, faster, and more pervasive. Given the expected continuation of these technological trends, we believe that systems such as the one proposed here, will become viable alternatives to traditional motion capture technologies.

3 Expressing Animated Performances Through Puppeteering

For performances of characters in animated movies, animators often use some technique called *keyframing*, with which animators specify a pose of each joint of the characters at several timings and create motion between the poses through interpolation. This performance is first designed in a process called *blocking-in*: working from audio and the storyboards, the animator creates a rough version of the motion.[5]

Animators often spend one or more days to create a 10-s blocked-in animation with keyframing in a mouse-based animation system. Particularly in commercial productions, the blocked-in animation serves as a form of visual communication between the animator and the director, and the animator refines and redoes the blocked-in motion until the director is satisfied that it conveys the essence of the performance. With several days required between iterations, this process can be slow to converge. On the other hand, for live action movies, a director can guide the performance of actors, and the actors can rehearse their performance in real time. A user interface that would allow for such interactive visual feedback [16] and a quick revision cycle for animation productions would not only reduce the cost of the productions but might also result in a higher quality product by allowing more iterations of refinement.

We introduced a 3D puppeteering interface that allows an animator to *perform* the motion of a character or a simulation effect for the blocking phase [5] (Fig. 4), rather than keyframing it in one of the commercially available animation software

[4] http://www.vicon.com/.

[5] Copyright © 2013, IEEE. Reprinted with permission from [5].

Fig. 4 Animator puppeteering a character for a dialog, and frames from the resulting animation

packages. We hypothesize that the animator can perform motion much more quickly through puppeteering than he/she can animate in traditional keyframing software. Performing the motion allows the animator to quickly get a feeling for the timing of the performance while maintaining approximate control of the poses. A number of animators have observed that the timing of actions is the key element in conveying the performance of the character.

In addition to keyframing, alternative approaches used to convey the proposed performance include motion capture [17], selecting motion from a database [6], or videotaping the animator's performance. Our preliminary interviews with animators indicated that puppeteering showed significant promise for several reasons. It allows motions that are not physically realistic and non-human—a requirement for many character animations as well as for secondary animations such as fluid effects. Puppeteering provides the animator with full control over the design and timing of the motion whereas selecting from even a large database is necessarily more restrictive. Creating the blocked-in motion for the actual character geometry and rig brings it far closer to the final animation than a video of a human performance or motion capture.

We chose to use an iterative refinement process with a set of animators to better understand what they expected and needed from a puppeteering interface for blocked-in motion. Iterative refinement is the seminal idea of iterating between an interface refinement phase and a user test phase [18], and has been shown to result in significantly better systems in many domains. Seven professionally trained animators and two experienced animators created a variety of animations with our system.

Through the iterative refinement process, we learned about a number of factors that are important to animators in the design of an interface for blocking in motion: (1) the motion should be puppeteered, not motion-captured. The physical constraints of motion capture are not a good match to the exaggerated motion required of most animated characters. (2) Blocking in motion is truly a performance and as such, the speed with which the animator can record the timing and poses of the performance is critical for capturing the performance that is in his/her head. Many of the design decisions of the user interface were driven by the need for speed of capture as the animator moved through the DOFs of the character. (3) Synergies or correlations between joints are essential for natural human motion and are similarly important for animated characters. Our animators wanted to tie joints together with offsets in angle and timing so that they could more quickly capture a full performance. The iterative refinement process also generated a list of necessary user interface features such as control over the editable range, controllable delay, enabling/disabling particular DOFs, multiple live views, and saving and restoring settings.

Because the number of DOFs that could be puppeetered in one take was so limited, we quickly realized that it was crucial to facilitate layering of multiple takes each of which animates a small number of DOFs. For example, our animators first specified the path of the pelvis as a position on the ground plane (two DOFs) and in a second layering pass, added the up/down trajectory for the pelvis motion (one DOF). Audio was played to facilitate synchronization for those sequences that included it. Our approach to this problem is similar to previous layering techniques [19, 20].

The sensitivities (i.e., scale) and home position (i.e., offset) of the mapping between the animator's motion and that of the character proved to be crucial for agile control of the character. For translation, our animators liked a view-dependent mapping that mapped a tracked object's motion to target DOFs with respect to the current view. Given the user-specified home position ${}^{M}\mathbf{p}_0$ and the current position ${}^{M}\mathbf{p}$ of the tracked object in a motion capture coordinate system M, the displacement in an animation coordinate system A at time t is represented as

$$\Delta^{A}\mathbf{p}(t) = \mathbf{C}^{\mathsf{T}}\mathbf{S}\left({}^{M}\mathbf{p}(t-\Delta t) - {}^{M}\mathbf{p}_0\right), \tag{5}$$

where $\mathbf{C}$ represents the orientation matrix of the viewing camera in the animation coordinate system, $\mathbf{S}$ is a diagonal matrix consisting of user-specified sensitivities, and Δt is a user-specified parameter for the controllable delay feature. The mapped position of the target DOFs in A is computed as

$${}^{A}\mathbf{p}_d\,(s_t t) = {}^{A}\mathbf{p}_d(0) + \Delta^{A}\mathbf{p}(t), \tag{6}$$

where s_t is a sensitivity for time. For rotation, our animators preferred a view-independent mapping. Therefore, the rotation for the puppeteered joint/rig, $\mathbf{R}_d$, is computed as

$$\mathbf{R}_d\,(s_t t) = S\left({}^{M}\mathbf{R}(t-\Delta t){}^{M}\mathbf{R}_0^{\mathsf{T}}\right)\mathbf{R}_d(0), \tag{7}$$

where ${}^{M}\mathbf{R}_0$ and ${}^{M}\mathbf{R}$ represent the user-specified home orientation and the current orientation of the tracked object, respectively, and $S(\cdot)$ is a function that converts an input rotation matrix in the motion capture coordinate system into Euler angles, multiplies the angles by sensitivities, and returns a scaled rotation matrix in the animation coordinate system. Which approach the animators chose was determined by the behavior being animated. For example, local orientation is useful for specifying joint angles and global orientation is used for root orientation. Therefore, the interface allowed the animators to choose global or local mapping, and the global orientation of the target DOFs in the animation coordinate system, ${}^{A}\mathbf{R}_d$, is computed as

$${}^{A}\mathbf{R}_d(t) = \begin{cases} \mathbf{R}_d(t) & \text{if global mapping} \\ \mathbf{R}_d(t){}^{A}\mathbf{R}_p(t) & \text{if local mapping} \end{cases}, \tag{8}$$

where ${}^{A}\mathbf{R}_p$ represents the global orientation of the parent DOFs in the animation coordinate system A.

Fig. 5 Resulting animations created by user study participants with the puppeteering system

Our test animators demonstrated that puppeteering could provide significant time savings by allowing them to rapidly explore the elements of a performance (Fig. 5). With the puppeteering interface, an animator is now able to produce a credible performance of a 15 s scene in less than an hour, rather than over a period of several hours or longer. A number of the animations were created twice, once in the puppeteering interface and once in a traditional keyframing interface. In all cases, the puppeteering interface was more efficient and the animators felt that the animations were of equal effectiveness in conveying the important elements of their intended performance.

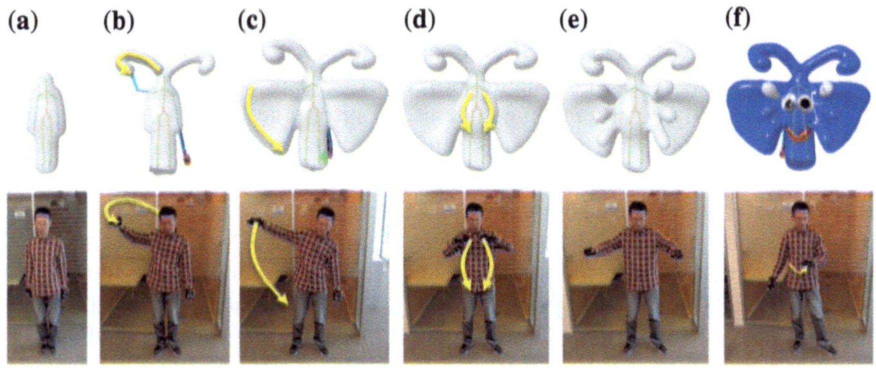

Fig. 6 Creating a 3D butterfly using the six operations of the BodyAvatar interface. **a** Scan. **b** Drag. **c** Sweep. **d** Sculpt. **e** Grow. **f** Paint

4 BodyAvatar: Creating Freeform 3D Avatars Using First-Person Body Gestures

While people watch animated content of movies, they can control and interact with animated characters in interactive content such as video games and online virtual worlds, where players are represented with 3D avatars. Particularly, with the popularity of Kinect,[6] these avatars can now directly mimic players' performance, making the experience ever more immersive. These 3D avatars are usually designed by professional game artists, with only limited player customization available such as by selecting from a collection of predesigned body parts, colors, and accessories. However, if one likes to be more creative and build an imaginary 3D character of non-predefined forms as their avatar, they usually have to master complex 3D modeling software, much beyond the skills of typical game players.[7]

Motivated to fill this gap and exemplified in a typical Kinect gaming setting, we present BodyAvatar, a system to allow the users to easily create their freeform 3D avatar out of imagination using full-body gestures as the input language like they do when playing Kinect games [10] (Fig. 6). These avatars may also be animated by the user's body similar to in Kinect games. Given that avatars typically take forms of living creatures, BodyAvatar focuses on creation of organic-looking 3D shapes but without structural constraints. BodyAvatar is unique from other gesture-based 3D modeling systems (e.g., [21, 22]), in that it centers around a first-person "you're the avatar" interaction metaphor. This metaphor is directly based on the fact that the

[6] http://www.xbox.com/en-US/kinect.

[7] Copyright © 2013, ACM, Inc. Reprinted by permission from [10].

users are creating a virtual representation of themselves. Instead of treating the 3D model as a separate passive object (third-person metaphor), the users consider their own body as a physical representation or proxy of the avatar being created. Based on an intuitive body-centric mapping, the users perform gestures to their own body as if wanting to modify it. The gestures in turn result in corresponding modifications to the 3D shape of avatar.

BodyAvatar was designed through a user-centered process. Our exploration started from an abstract concept: to let a user create the shape of a 3D avatar using their full body in ways most intuitive to them. To identify common patterns in how people intuitively express 3D shapes using their body, we adopted a Wizard-of-Oz method [23]. The "system" was simulated by a researcher drawing 2D sketches of 3D shapes on a whiteboard in front of the participant, representing the avatar being created. The participants used any body actions they felt appropriate to generate and modify the avatar until satisfied, while thinking aloud to explain the anticipated effect of each action. The researcher drew and modified the sketch according to the participants' actions and explanations.

The majority of the participants were dominated by a first-person mentality; they fantasized themselves as the avatar they were creating, and performed actions on or around their own body. One participant instead adopted a third-person style, i.e., imagining the avatar being in front of him, and performed actions in the air where he imagined the avatar to be. Another participant combined both styles. Almost all participants treated the avatar as a single entity to be continuously built upon, hence they naturally followed a general "generating basic shape" and then "adding features/details" workflow. For generating the basic shape, two strategies were observed: posing their full body to mimic the intended shape, and sketching the 2D silhouette in the air using their finger.

Several common actions were observed for adding features and details to the basic shape: *Growing* a new limb was fairly common. Many participants expressed this by extending their arms or legs outwards, mimicking the shape of the expected limb. For adding thinner features, such as an antenna on the head, many participants used a *pinching* gesture, starting from the respective part of their own body and moving away, as if pulling the feature out. Unlike growing limbs, the participants expected the shape of the new feature to follow the path of their hand movement. This also allowed the participants to create more complex geometric features than they can directly mimic using arms or legs. *Tracing* an imaginary shape was frequently observed. Several participants tended to use a finger to trace a curve, and use palms to trace a surface (either a free hanging surface, or the surface surrounding a volume traced using both hands, such as a big belly). Bimanual actions were commonplace, where participants simultaneously used both hands/arms to perform the same type of actions, mostly in a geometrically symmetric fashion. There were also occasional observations of asymmetric bimanual actions. For example, two participants used one hand to describe the shape of a feature, and the other hand to point to their body where they would like the feature to be added. Other than adding features, some participants also used their hand like a knife to *cut* unwanted parts. One participant used his hand like a brush on his body in order to paint color on the avatar. Although

Fig. 7 Avatar shapes created with BodyAvatar during user study

the participants did not see a real system, most of them already expressed much fondness in the concept of creating avatars using their body during brief interviews after the study. They thought the actions they came up with were quite intuitive.

Based on these observations, we implemented the operations for the BodyAvatar interface (Fig. 6a–f). The users start with the users posing their body to set the initial shape of the avatar such a simple stick or a four-legged animal (Fig. 6a). The avatar can then be *attached* to the users' body and continuously animated by body movement. Under the attached status, the users perform various gestures to their own body to edit the avatar progressively. For example, the users drag from their head to give the avatar an antenna (Fig. 6b), or gesture around their stomach to grow a fat belly for the avatar (Fig. 6d). In addition, two users may create an avatar collaboratively in a similar fashion. The static 3D shape of an avatar is modeled by an implicit surface constructed from a number of meta-balls in 3D space [24], and the kinematic structure is represented by a tree-structured skeleton. A triangle mesh is used to render the avatar. The mesh is generated from the meta-ball model using the marching-cube algorithm [25] and animated with the avatar skeleton. A texture map is used for coloring.

BodyAvatar enables free 3D avatar creation without requiring professional skills (Fig. 7). It aims at an intuitive, immersive, and playful experience for the user; the first-person metaphor provides both an intuitive frame of reference for gesture operations and an immersive "you're the avatar" feeling, and its game-like interaction style offers a playful atmosphere.

5 Conclusion

In this paper, we described our recent work on user interfaces for character animation and character interaction, ranging over three main issues: motion capture for live action movies, animation design for animated movies, and interactive control for interactive animated content.

Considering a wide variety of researches in this field being conducted, we believe that one potential direction of user interfaces for creating character animation is *suggestive interfaces* which suggests users how to create content and with which the users can decide whether they refer to or ignore them, instead of a system that automatically creates content based on domain knowledge such as databases, statistics or models learned with some machine learning methods. Such suggestions in turn enable the users to learn, for example, what kind of animations are very attractive and to create content by themselves with their own imagination and creativity in the end. We expect that various mathematical techniques combined with intuitive interfaces can grow novice people as animators and artists.

Another issue, mainly for interactive control, is haptic feedback to players. Despite the capability of motion sensing devices to directly map player's motion to a virtual character, typical feedback to players is binary vibration (on/off) regardless of intensity of their motion. However, humans obtain very rich information from sense of touch, and can manipulate objects in a hand without watching them, for example. Humans can also perceive the shape of an object only from sense of touch, which cannot be represented with the binary vibration. Much richer haptic feedback to players would open up a much wider range of possibilities for character interactions.

References

1. Thomas, F., Johnson, O.: The Illusion of Life: Disney Animation. Disney Editions, NY, USA (1995)
2. Block, B.: The Visual Story—Creating the Visual Structure of Film, TV and Digital Media. Focal Press, Elsevier Inc, MA, USA (2008)
3. Shiratori, T., Park, H.S., Sigal, L., Sheikh, Y., Hodgins, J.K.: Motion capture from body-mounted cameras. ACM Trans. Graph. **31**, 1–10 (2011) http://dx.doi.org/10.1145/1964921.1964926
4. Park, H.S., Shiratori, T., Matthews, I., Sheikh, Y.: 3D reconstruction of a moving point from a series of 2D projections. In: Proceedings of European Conference on Computer Vision (2010)
5. Shiratori, T., Mahler, M., Trezevant, W., Hodgins, J.K.: Expressing animated performances through puppeteering. In: Proceedings of IEEE Symposium on 3D User Interfaces. IEEE (2013). http://dx.doi.org/10.1109/3DUI.2013.6550198
6. Numaguchi, N., Nakazawa, A., Shiratori, T., Hodgins, J.K.: A puppet interface for retrieval of motion capture data. In: Proceedings of ACM SIGGRAPH/Eurographics Symposium on Computer Animation (2011)
7. Shiratori, T., Hodgins, J.K.: Accelerometer-based user interfaces for the control of a physically simulated character. ACM Trans. Graph. **123**, 1–9 (2008)
8. Willis, K.D.D., Poupyrev, I., Shiratori, T.: MotionBeam: A metaphor for character interaction with handheld projectors. In: Proceedings of ACM SIGCHI Conference on Human Factors in Computing Systems (2011)

9. Willis, K.D.D., Shiratori, T., Mahler, M.: HideOut: Mobile projector interaction with tangible objects and surfaces. In: Proceedings of International Conference on Tangible, Embedded and Embodied Interaction (2013)
10. Zhang, Y., Han, T., Ren, Z., Umetani, N., Tong, X., Liu, Y., Shiratori, T., Cao, X.: BodyAvatar: Creating freeform 3D Avatars using first-person body gestures. In: Proceedings of ACM Symposium on User Interface Software and Technology. ACM Inc (2013). http://dx.doi.org/10.1145/2501988.2502015
11. Welch, G., Foxlin, E.: Motion tracking: no silver bullet, but a respectable Arsenal. IEEE Comput. Graph. Appl. **22**(6), 24–38 (2002)
12. Vlasic, D., Adelsberger, R., Vannucci, G., Barnwell, J., Gross, M., Matusik, W., Popović, J.: Practical motion capture in everyday surroundings. ACM Trans. Graph. **35**, 1–9 (2007)
13. Hartley, R.I.: A Multiple View Geometry in Computer Vision. Cambridge University Press, Zisserman (2004)
14. Snavely, N., Seitz, S.M., Szeliski, R.: Photo tourism: exploring photo collections in 3D. ACM Trans. Graph. **25**(3), 835–846 (2006)
15. O'Brien, J.F., Bodenheimer, R.E., Brostow, G.J., Hodgins, J.K.: Automatic joint parameter estimation from magnetic motion capture data. In: Proceedings of Graphics Interface (2000)
16. Barnes, C., Jacobs, D.E., Sanders, J., Goldman, D.B., Rusinkiewicz, S., Finkelstein, A., Agrawala, M.: Video puppetry: a performative interface for cutout animation. ACM Trans. Graph. **124**, 1–9 (2008)
17. Shin, H.J., Lee, J., Shin, S.Y., Gleicher, M.: Computer puppetry: an importance-based approach. ACM Trans.Graph. **20**(2), 67–94 (2001)
18. Nielsen, J.: Iterative user-interface design. Computer **26**(11), 32–41 (1993)
19. Oore, S., Terzopoulos, D., Hinton, G.: A desktop input device and interface for interactive 3D character animation. In: Proceedings of Graphics Interface (2002)
20. Dontcheva, M., Yngve, G., Popović, Z.: Layered acting for character animation. ACM Trans. Graph. **22**(3), 409–416 (2003)
21. Schkolne, S., Pruett, M., Schroeder, P.: Surface drawing: Creating organic 3D shapes with the hand and tangible tools. In: Proceedings of ACM SIGCHI Conference on Human Factors in Computing Systems, pp. 261–268 (2001)
22. Sheng, J., Balakrishnan, R., Singh, K.: An interface for 3D sculpting via physical proxy. In: Proceedings of GRAPHITE (2006)
23. Kelley, J.F.: An iterative design methodology for user-friendly natural language office information applications. ACM Trans. Inf. Syst. **2**(1) (1984)
24. Nishimura, H., Hirai, M., Kawai, T., Kawata, T., Shirakawa, I., Omura, K.: Object modeling by distribution function and a method of image generation. Trans. Inst. Electron. Commun. Eng. Jpn **68**(4), 718–725 (1985)
25. Bloomenthal, J.: Polygonization of implicit surfaces. Comput. Aided Geom. Des. **5**(4), 341–355 (1988)

Thermodynamic Gibbs Formalism and Information Theory

Victor Ermolaev and Evgeny Verbitskiy

Abstract Links between Information Theory and Thermodynamics are well known. The concept of *entropy*, introduced by C. Shannon in 1948 in his groundbreaking work which gave birth to Information Theory, originates in Statistical Mechanics. In the past 60 years multiple links have been found, which led to new results. Information Theory has been incredibly successful in utilization of probabilistic methods in problems like data compression, prediction, classification, and coding of information sources. Most approaches and algorithms can be classified as *causal*, or omni-directional: the data are processed in a directed sequential fashion; distribution of the *present* with respect to the *past* is used for prediction or classification purposes. However, recently some novel approaches have been proposed in Information Theory. It turns out that the *non-causal* (bi-directional) approaches, i.e., when the influences of the *past* as well as of their *future* are taken into account, lead to very interesting and often superior solutions in problems like denoising and classification. The theory of Gibbs states in Statistical Mechanics—the so-called Gibbs formalism—provides the right framework for treatment of stochastic processes in a non-causal way. We will discuss specific information-theoretic algorithms based on the Gibbs formalism.

Keywords Information theory · Gibbs states · Thermodynamic formalism

V. Ermolaev · E. Verbitskiy (✉)
Johann Bernoulli Institute for Mathematics and Computer Science,
University of Groningen, Groningen, The Netherlands
e-mail: evgeny@math.leidenuniv.nl; e.verbitskiy@gmail.com

V. Ermolaev
Nedap N.V., Groenlo, The Netherlands
e-mail: victor.ermolaev@nedap.com

E. Verbitskiy
Mathematical Insititute, Leiden University, Leiden, The Netherlands

M. Wakayama et al. (eds.), *The Impact of Applications on Mathematics*,
Mathematics for Industry 1, DOI: 10.1007/978-4-431-54907-9_28,

1 Introduction

We view an *information source* as a stationary stochastic process $\{X_n\}_{n\in\mathbb{Z}}$. For simplicity, we assume that the process takes values in a finite *alphabet* S, e.g, $S = \{0, 1\}$, $\{A, C, G, T\}$, or $\{a, b, c, \dots, z\}$. Conditional distributions of the present ($n = 0$) given a complete past ($n < 0$)

$$\mathbb{P}\left(X_0 = a_0 | X_{-\infty}^{-1} = a_{-\infty}^{-1}\right) = \mathbb{P}\big(X_0 = a_0 | X_{-1} = a_{-1}, X_{-2} = a_{-2}, \dots\big), \quad a_i \in S, \tag{1}$$

are important characteristics of the process $\{X_n\}$. For example, Markov chains are defined by requiring that the distribution of the value at present depends only on the immediate past

$$\mathbb{P}\big(X_0 = a_0, | X_{-1} = a_{-1}, X_{-2} = a_{-2}, \dots\big) = \mathbb{P}\big(X_0 = a_0 | X_{-1} = a_{-1}\big).$$

The entropy rate of any stationary process $\{X_n\}$ can be naturally expressed in terms of its one-sided conditional probabilities (1)

Various classes of processes can be defined by requiring certain properties of the one-sided probabilities, or simply by providing a particular parametric form for (1): to mention just a few, chains with complete connections, countable mixtures of Markov chains, g-measures, absolutely regular processes.

One particular class of processes admitting a relatively "simple" description in terms of one-sided conditional probabilities is the so-called *variable-length Markov chains*.

The notion of *VLMC* first appeared in the context of Information theory in the seminal work of Rissanen [11], who introduced stochastic processes (chains) with memory of variable length: the distribution of the next symbol is dependent on a certain number of past values; this number is a function of the past itself:

$$\mathbb{P}\left(X_0 = a_0 | X_{-\infty}^{-1} = a_{-\infty}^{-1}\right) = \mathbb{P}\big(X_0 = a_0 | X_{-1} = a_{-1}, \dots, X_{-l} = a_{-l}\big), \tag{2}$$

where $l = l(\mathbf{a}) \in \mathbb{N}$ is a function of $\mathbf{a} = (a_{-1}, a_{-2}, \dots)$.

Allowing memory to be of variable length can capture long-range dependence, but does so in an economic fashion, by tracking "pasts" only to relevant depth. Note that a k-step Markov model for a process with values in a finite set S requires $\mathcal{O}(|S|^k)$ parameters. The algorithm "Context" developed in [11] is able to reconstruct (estimate) conditional distributions from sufficiently long samples of random processes. The algorithm outputs a tree, encoding the information on estimated conditional distributions. Each contains information about a context and a conditional distribution of a next symbol, given this context as an immediate past.

It is evident that the work of Rissanen has had a great impact because for information-theoretic problems like prediction of future values, or closely related compression and filtering problems, the knowledge of the conditional distributions $\mathbb{P}(X_0 = a_0, |X_{-1} = a_{-1}, X_{-2} = a_{-2}, \ldots)$ is crucial.

That paper drew a lot of attention and became a foundation for later work both from theoretical and applied points of view. A good example of applied research is the study of protein families in [1]. The idea of VLMC was applied to identify significant patterns in a set of related protein sequences. The developed method based on the "Context" singles out significant patterns of variable length surprisingly well without assuming any preliminary biological information. Another interesting example is in linguistic studies. In [6], application of VLMC was motivated by the linguistic challenge of retrieving rhythmic features from written texts (a set of daily newspapers). As a result an illustration compatible with the long standing conjecture that Brazilian Portuguese and European Portuguese belong to different rhythmic classes was provided.

What is less known that there are alternative approaches. For example, one can define and study random processes in terms of *two-sided* conditional distributions:

$$\mathbb{P}\left(X_0 = a_0 | X_{-\infty}^{1} = a_{-\infty}^{1}, X_1^{+\infty} = a_1^{+\infty}\right). \tag{3}$$

Here one is interested in conditional distribution of the present given the complete past and the complete future. It is seemingly strange and, possibly, even unnatural to approach random processes in this way. However, there are certain advantages in switching to a two-sided description.

In information theory a paradigm shift occurred recently after the publication of [13], where a *discrete universal denoising* algorithm (DUDE) was introduced. It was demonstrated that optimal denoising of processes corrupted by discrete memoryless channels can be achieved via estimation of two-sided conditional probabilities and a subsequent application of a simple Bayesian rule based on these probabilities. This development stimulated interest in two-sided approaches to information-theoretic problems.

On the other hand, the pioneering work of Dobrushin, Lanford, and Ruelle in the late 1960s on rigorous foundations of Statistical Mechanics provided the rigorous foundations for probabilistic description of interacting spins models, based on two-sided conditional probabilities like in (3). The probability law $\mathbb{P}$ is called Gibbs if the conditional probabilities are given by

$$\mathbb{P}\left(X_0 = a_0 | X_{-\infty}^{-1} = a_{-\infty}^{-1}, X_1^{+\infty} = a_1^{+\infty}\right) = \frac{1}{Z} \exp\left(-H\left(a_{-\infty}^{+\infty}\right)\right), \tag{4}$$

where H is the so-called *Hamiltonian*, and Z is a normalizing factor known as the *partition function*.

In the thermodynamic Gibbs formalism, information about conditional distributions is encoded using a potential—a family of functions describing the interactions between random variables. Rissanen's idea of variable memory can be interpreted directly in terms of *variable range* of the corresponding potentials. The question we address in the present paper is whether a direct application of the Gibbs formalism might improve solutions of some Information Theory problems.

2 Variable Length Memory Models of Stochastic Processes

Suppose $\{X_n \mid n \in \mathbb{Z}\}$ is a stationary process with values in a finite alphabet S. The space of realisations $\Omega := S^{\mathbb{Z}}$ is endowed with the product topology and the corresponding Borel σ-algebra $\mathscr{A}$. Denote by $\mathbb{P}$ the stationary law of $\{X_n\}$.

Definition 1 The stationary process $\{X_n\}$, equivalently the corresponding probability measure $\mathbb{P}$ on $(\Omega, \mathscr{A})$, is called a Variable Length Markov Chain (VLMC) if for any realisation of the *past* $a_{-\infty}^{-1} \in S^{\mathbb{Z}_{<0}}$, there exists a non-negative integer $\ell = \ell(a_{-\infty}^{-1})$ such that

$$\mathbb{P}\left(X_0 = a_0 | X_{-\infty}^{-1} = a_{-\infty}^{-1}\right) = \mathbb{P}\left(X_0 = a_0 \mid X_{-\ell}^{-1} = a_{-\ell}^{-1}\right). \tag{5}$$

The function $\ell : S^{\mathbb{Z}_{<0}} \to \mathbb{Z}_{\geq 0}$ is the length of what we referred to in the Introduction as the *relevant past* or *context* $a_{-\ell}^{-1}$. We explicitly assume that all contexts are *minimal*, i.e., $\ell = \ell(a_{-\infty}^{-1})$ is the minimal integer such that (5) holds.

The Variable Length Markov Chains can be conveniently represented with the so-called *probabilistic context trees*.

We say that a length-j sequence $a = a_{-j}^{-1} \in S^j$ is a *suffix* of a length-k sequence $b = b_{-k}^{-1} \in S^k$ if $j \leq k$ and $a_{-i} = b_{-i}$ for all $i = 1, \ldots, j$, notation $a_{-j}^{-1} \preceq b_{-k}^{-1}$ is appropriate here. If, moreover, $j < k$, then a is a proper suffix of b ($a \prec b$). Denote also by S^* the set of all finite words in the alphabet S, including the empty word $\mathbf{e}$:

$$S^* = \bigcup_{k=0}^{\infty} S^k.$$

Definition 2 Subset $\mathscr{T}$ of S^* is called a tree if it satisfies a *suffix property*: namely, for every $b = b_{-k}^{-1} \in \mathscr{T}$ and all $j = 1, \ldots, k-1$, $b_{-k+j}^{-1} \notin \mathscr{T}$. Equivalently, if $b \in \mathscr{T}$, then no proper suffix of b is in $\mathscr{T}$.

Let us explain the reference to $\mathscr{T}$ as a tree. The relation "$\prec$" gives a natural order on the sequence $\{b_{-k}^{-1}\}_{k\geq 1}$ in the following manner: $b_{-1} \prec b_{-2}^{-1} \prec b_{-3}^{-1} \cdots \prec b_{-k}^{-1}$. Therefore, given a set $\mathscr{T}$, satisfying the suffix property, all its elements can be represented as leaves of a certain tree rooted at the empty word $\mathbf{e}$. Nodes of such a tree at depth j are indexed by suffixes b_{-j}^{-1} of $b_{-k}^{-1} \in \mathscr{T}$, $j < k$. Each node has possibly as many offspring as the cardinality of the set S, all offspring differ from the ancestor by one extra symbol from S. We also say that a tree $\mathscr{T}$ with a suffix property is proper, if for every $(a_{-1}, a_{-2}, \ldots)$, there exists a unique word $b \in \mathscr{T}$ such that $b \prec (a_{-1}, a_{-2}, \ldots)$.

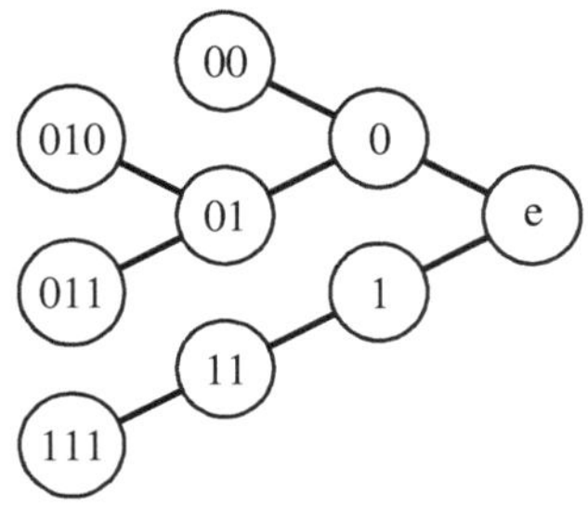

Fig. 1 Tree corresponding to a proper collection of binary words $\mathscr{T} = \{00, 010, 011, 111\}$

Definition 3 A probabilistic context tree over S is an ordered pair $(\mathscr{T}, \mathbf{p})$ where $\mathscr{T} \subset S^*$ is a proper tree, and $\mathbf{p} = \{\mathbf{p}(\cdot|b) : b \in \mathscr{T}\}$ is a family of probability measures on S indexed by elements of $\mathscr{T}$. Without loss of generality, we may assume that $\mathscr{T}$ is *irreducible*, i.e., no $b \in \mathscr{T}$ can be replaced by its proper suffix without violating the suffix property.

Clearly, the probabilistic suffix trees offer a convenient way to encode all the relevant information, necessary to represent a given VLMC: the tree $\mathscr{T}$ is simply the collection of all contexts, and the corresponding distributions $\mathbf{p}(\cdot|b)$ are given by

$$\mathbf{p}(a|b) = \mathbb{P}\left(X_0 = a|X_{-|b|}^{-1} = b\right).$$

The tree $\mathscr{T}$ will also be referred to as the *tree of uni-directional contexts*.

An important task is to reconstruct or estimate the probabilistic context tree based on a finite sample of the underlying VLMC. If somehow the tree $\mathscr{T}$ would be known, estimating the corresponding distributions $\mathbf{p}(\cdot|b)$ is straightforward. The main difficulty lies in the determination of the set of contexts. In the original paper [11], Rissanen proposed an algorithm *Context* for the solution of this problem. Since then a number of modifications have been proposed, e.g., [1, 14].

2.1 Bi-directional or Two-Sided Models

Let us now consider modelling random processes by means of two-sided contexts. Suppose $\mathscr{T}_1$ and $\mathscr{T}_2$ are finite subsets of S^*. Given a couple $(a_{-k}^{-1}, a_1^m) \in \mathscr{T}_1 \times \mathscr{T}_2$, we will refer to a_{-k}^{-1} as the *past* and to a_1^m as the *future*.

Definition 4 Let $\mathbb{P}$ be a stationary law of process $\{X_n\}$, then $\mathbb{P}$ (equivalently, $\{X_n\}$) is called a *variable length Markov field* if there exist two functions $\ell^-, \ell^+ : S^{\mathbb{N}} \to \mathbb{Z}_{\geq 0}$ such that for all $a_{-\infty}^{-1}, a_1^{+\infty}$, one has

$$\mathbb{P}\left(X_0 = \cdot | X_{-\infty}^{-1} = a_{-\infty}^{-1}, X_1^{+\infty} = a_1^{+\infty}\right)$$
$$= \mathbb{P}\left(X_0 = \cdot \middle| X_{-\ell^-(a_{-\infty}^{-1})}^{-1} = a_{-\ell^-(a_{-\infty}^{-1})}^{-1}, X_1^{\ell^+(a_1^{+\infty})} = a_1^{\ell^+(a_1^{+\infty})}\right).$$

This definition extends Definition 1 in a straightforward fashion. Here $\ell^-(a_{-\infty}^{-1})$ and $\ell^+(a_1^{+\infty})$ are lengths of relevant past and future contexts. In fact, one might require ℓ^-, ℓ^+ to depend both on $a_{-\infty}^{-1}, a_1^{+\infty}$, i.e., $\ell^- = \ell^-(a_{-\infty}^{-1}, a_1^{+\infty})$ and $\ell^+ = \ell^+(a_{-\infty}^{-1}, b_1^{+\infty})$. However, already the above choice: $\ell^- = \ell^-(a_{-\infty}^{-1})$ and $\ell^+ = \ell^+(a_1^{+\infty})$ provides sufficient flexibility for applications. Again, we assume explicitly the minimality of the context length functions ℓ^-, ℓ^+.

Definition 5 A product $\mathscr{T}_1 \times \mathscr{T}_2$ satisfies a suffix property if for every couple of strings $(a_{-k}^{-1}, a_1^m) \in \mathscr{T}_1 \times \mathscr{T}_2$, one has that $(a_{-k'}^{-1}, a_1^{m'}) \notin \mathscr{T}_1 \times \mathscr{T}_2$ for all $k' \le k$, $m' \le m$ such that $k' + m' < k + m$.

If $\mathscr{T}_1 \times \mathscr{T}_2$ satisfies the suffix property, we will refer to elements of $\mathscr{T}_1 \times \mathscr{T}_2$ as the set of *bi-directional contexts*. The set of bi-directional contexts must also satisfy the following property: for any infinite bi-directional context $(a_{-\infty}^{-1}, a_1^{+\infty})$, there must be a unique pair $(b, c) \in \mathscr{T}_1 \times \mathscr{T}_2$ such that $b \prec (a_{-1}, a_{-2}, \ldots)$ and $c \prec (a_1, a_2, \ldots)$, i.e., for every possible realisation of infinite past and infinite future, one is able to specify bi-directional context pair in $\mathscr{T}_1 \times \mathscr{T}_2$ uniquely. Under this condition, the set of bi-directional contexts can be represented as a tree. Similar to the one-sided case, if $\mathbf{p} = \{\mathbf{p}(\cdot|b, c)\}$ is a family of probability measures on S indexed by $(b, c) \in \mathscr{T}_1 \times \mathscr{T}_2$, then $(\mathscr{T}_1 \times \mathscr{T}_2, \mathbf{p})$ is called a bi-directional probabilistic context tree.

We say that the process $\{X_n\}$ is consistent with a bi-directional probabilistic context tree $(\mathscr{T}_1 \times \mathscr{T}_2, \mathbf{p})$ if for all $\mathbf{a} = (\ldots, a_{-1}, a_0, a_1, \ldots) \in S^{\mathbb{Z}}$, one has

$$\mathbb{P}(X_0 = a_0 | X_{-\infty}^{-1} = a_{-\infty}^{-1}, X_1^{+\infty} = a_1^{+\infty}) = \mathbf{p}(a_0 | a_{-k}^{-1}, a_1^m),$$

where k, m are such that $(a_{-k}^{-1}, a_1^m) \in \mathscr{T}_1 \times \mathscr{T}_2$.

As mentioned earlier, for a given process one can construct probabilistic one-sided context tree approximations rather easily and efficiently. Construction of two-sided or bi-directional probabilistic context trees is much more difficult. The main problem lies in the fact that it is not clear how to *grow* two-sided contexts, preserving the necessary uniqueness condition discussed above, in order to achieve progressively better approximations.

As a first step, one might consider whether the problem of (re-)construction of two-sided probabilistic context trees can be reduced to the corresponding one-sided problem. In particular, it is interesting to understand for which classes of stochastic processes such a reduction is possible, i.e., for which processes one is able to represent two-sided conditional probabilities in terms of one-sided conditional probabilities. Several models with this property were put forward by Yu and Verdú [15]. For example, in the *Backward-Forward product (BFP)* model, the conditional distribution of X_0 with given past and future is assumed to be proportional to the product of the two one-sided conditional distributions:

$$\mathbb{P}\left(X_0 = a_0 | X_{-\infty}^{-1} = a_{-\infty}^{-1}, X_1^{+\infty} = a_1^{+\infty}\right)$$
$$\propto \mathbb{P}\left(X_0 = a_0 | X_{-\infty}^{-1} = a_{-\infty}^{-1}\right) \times \mathbb{P}\left(X_0 = a_0 | X_1^{+\infty} = a_1^{+\infty}\right).$$

Furthermore, if $\{X_n\}$ is assumed to be Markov of some finite order, then one is able to express two-sided conditional probabilities in terms of one-sided ones.

It is also possible for a given process to estimate bi-directional probabilistic tree directly, without resorting to uni-directional constructions. However, the lack of natural order on the set of possible bi-directional contexts complicates the task significantly. Representation of bi-directional contexts as leaves of a tree is only possible under additional requirements on the way the tree is grown, i.e., preserving the fact that the set of contexts forms a partition.

Several methods have been proposed [3, 10, 15] and employ various ideas to achieve the desired goal. However, these constructions are too involved to be discussed in greater detail here.

2.2 Gibbs Formalism

In this section we argue that the Gibbs measures of Statistical Mechanics are a convenient alternative to context trees for purposes of bi-directional modelling of stochastic processes. Interactions, which define the Gibbs state, are suitable to encode finite, but variable memory. We also propose a method to estimate Gibbs interactions using Kozlov's theorem.

Our presentation of the Gibbs formalism is very limited, for a general treatment of Gibbs states see e.g., [7].

Definition 6 An interaction $\Phi = \{\phi_\Lambda : \Lambda \Subset \mathbb{Z}\}$ is a family of functions $\phi_\Lambda : \Omega \mapsto \mathbb{R}$, $\Omega = S^{\mathbb{Z}}$, indexed by finite subsets of $\mathbb{Z}$ (notation $\Lambda \Subset \mathbb{Z}$), such that for every $\Lambda \Subset \mathbb{Z}$, ϕ_Λ is a function of ω_Λ—the restriction of ω to the finite set Λ, i.e.,

$$\phi_\Lambda(\omega) = \phi_\Lambda(\omega_\Lambda).$$

Interaction $\Phi = \{\phi_\Lambda : \Lambda \Subset \mathbb{Z}\}$ is said

- to have *finite range* if
$$\phi_\Lambda \equiv 0,$$
for all $\Lambda \Subset \mathbb{Z}$ with sufficiently large diameter.
- to be *absolutely summable*, if for any $V \Subset \mathbb{Z}$,
$$\sum_{\Lambda \cap V \neq \varnothing} ||\phi_\Lambda|| < \infty,$$

where $||\phi_\Lambda|| = \sup_{\omega\in\Omega} |\phi_\Lambda(\omega)|$ is the sup-norm of ϕ_Λ.

For an absolutely summable interaction Φ, Hamiltonian for the set $V \Subset \mathbb{Z}$ with boundary conditions $\omega \in \Omega$ is defined as

$$H_V^\omega(\sigma) = \sum_{\Lambda\cap V\neq\emptyset} \phi_\Lambda(\sigma_V\,\omega_{V^c}), \tag{6}$$

where $\sigma_V\omega_{V^c}$ is defined as the configuration in Ω which coincides with σ on V and with ω on the complement of V in $\mathbb{Z}$.

For an absolutely summable interaction Φ, the Boltzmann weights are functions defined for all $\Lambda \Subset \mathbb{Z}$ and all boundary conditions ω as

$$\gamma_\Lambda^\Phi(\sigma_\Lambda|\omega_{\Lambda^c}) = \frac{e^{-H_\Lambda^\Phi(\sigma_\Lambda\omega_{\Lambda^c})}}{Z_\Lambda^\Phi(\omega)}, \tag{7}$$

where $Z_\Lambda^\Phi(\omega) = \sum_{\bar\sigma_\Lambda\in S^\Lambda} e^{-\beta H_\Lambda^\Phi(\bar\sigma_\Lambda\omega_{\Lambda^c})}$ is the normalising factor. Therefore, $\gamma_\Lambda^\Phi(\cdot|\omega_{\Lambda^c})$ is a probability distribution on S^Λ. In fact, the family of Boltzmann weights $\Gamma^\Phi = \{\gamma_\Lambda^\Phi(\cdot|\cdot) : \Lambda \Subset \mathbb{Z}\}$ is called a *specification*—a family of probability kernels with a number of useful properties [4]. Finally, we say that a Borel probability measure $\mathbb{P}$ is Gibbs with potential Φ, if $\mathbb{P}$ is consistent with the corresponding specification Γ^Φ: namely, Γ^Φ coincides with the conditional probabilities of $\mathbb{P}$ for all $\Lambda \Subset \mathbb{Z}$,

$$\mathbb{P}\left(X_\Lambda = \sigma_\Lambda|X_{\Lambda^c} = \omega_{\Lambda^c}\right) = \gamma_\Lambda^\Phi(\sigma_\Lambda|\omega_{\Lambda^c})$$

for $\mathbb{P}$-almost all $\omega \in \Omega$. In particular, the two-sided conditional probabilities are given by

$$\begin{aligned}\mathbb{P}\left(X_0 = \sigma_0|X_{-\infty}^{-1} = \omega_{-\infty}^{-1}, X_1^\infty = \omega_1^\infty\right) &= \gamma_\Lambda^\Phi\left(\sigma_0|\omega_{-\infty}^{-1}, \omega_1^\infty\right)\\ &= \frac{e^{-H_\Lambda^\Phi\left(\omega_{-\infty}^{-1}\sigma_0\omega_1^\infty\right)}}{\sum_{\xi_0\in S} e^{-H_\Lambda^\Phi(\omega_{-\infty}^{-1}\xi_0\omega_1^\infty)}},\end{aligned}$$

2.2.1 Telescoping Potentials

For a Gibbs measure $\mathbb{P}$ the corresponding interaction is not uniquely determined. Two interactions are called *physically equivalent* if the sets of corresponding Gibbs states coincide. Nevertheless, Kozlov [9] found a method to *reconstruct* potentials from Gibbs specifications.

Theorem 1 *Suppose $\mathbb{P}$ is a translation invariant Gibbs measure for some absolutely summable potential, and let $\Gamma = \{\gamma_\Lambda, \Lambda \Subset \mathcal{V}\}$ be the corresponding specification. Then $\mathbb{P}$ is also Gibbs for a potential $\Phi = \{\phi_\Lambda\}$ with a reference state*

$\theta = (\dots, s, s, s, \dots)$, where $\phi_\Lambda = 0$ for all $\Lambda \Subset \mathbb{Z}$ which are not intervals in $\mathbb{Z}$, and for an interval $\Lambda = [m,n]$, $m \le n$, the potential is given by

$$\phi_{[m,n]}\left(\sigma_{[m,n]}\right) = \ln \frac{\gamma_{[m,n]}\left(\theta_m\sigma_{(m,n]}|\theta_{[m,n]^c}\right)\gamma_{[m,n]}\left(\sigma_{[m,n)}\theta_n|\theta_{[m,n]^c}\right)}{\gamma_{[m,n]}\left(\sigma_{[m,n]}|\theta_{[m,n]^c}\right)\gamma_{[m,n]}\left(\theta_m\sigma_{(m,n)}\theta_n|\theta_{[m,n]^c}\right)} \tag{8}$$

Kozlov's potential ϕ_Λ measures the ratio of probabilities of words $\sigma_{[m,n]}, \theta_m\sigma_{(m,n]}$, $\sigma_{[m,n)}\theta_n$, and $\theta_m\sigma_{(m,n)}\theta_n$, conditioned on $\theta_{[m,n]^c}$. Suppose now we have a finite realisation $\{X_n\}_{n=1}^N$ of a random process, whose law we assume to be Gibbs for some unknown potential. The probability of a particular finite word b under $\mathbb{P}$ can be estimated by computing the frequency of occurrences of this word in the realization of our stochastic process, and hence, the one-sided conditional probabilities can be estimated as

$$\hat{\mathbb{P}}\left(X_0 = s_0|X_{-1} = s_{-1}, \dots X_{-j} = s_{-j}\right) = \frac{\#\left(s_{-j}^0\right)}{\sum_{\bar{s}_0}\#\left(s_{-j}^{-1}\bar{s}_0\right)},$$

where $\#(b)$ is the number of occurrences of word b in $\{X_n\}_{n=1}^N$ By analogy with the expression for Kozlov's potential, we propose to use the following "estimator" for the underlying potential: fix a symbol $s \in S$, and let

$$\hat{\phi}_{[m,n]}(\sigma_{[m,n]}) = \ln \frac{(\varepsilon + \#(s\sigma_{(m,n]}))(\varepsilon + \#(\sigma_{[m,n)}s))}{(\varepsilon + \#(\sigma_{[m,n]}))(\varepsilon + \#(s\sigma_{(m,n)}s))}, \tag{9}$$

where ε is a small positive constant, introduced to ensure that the faction is always non-negative.

Estimating a potential in accordance to (9) is relatively straightforward. The algorithm can be divided in two parts: (1) collecting statistics of word occurrences and (2) estimating the potential of certain sub-strings via (9).

Given a sufficiently long sample $\{X_n\}_{n=1}^N$, $N \gg 1$, we fix a number of parameters: $M \in \mathbb{N}$, $\delta, \varepsilon > 0$, and $s \in S$. The reference symbol $s \in S$ can be chosen arbitrarily, but the choice can have some effect on the performance of the algorithm. In practice, it is advisable to evaluate various reference symbols. By design, our Gibbs potential will have a finite range at most $2M + 1$. The natural choice for $M = \lfloor \frac{1}{c} \log_{|S|} N \rfloor$, where $\lfloor \cdot \rfloor$ is the integer part and $c > 1$. This choice ensures that every word b of length up to $2M + 1$ has a *reasonable chance* to appear in $\{X_n\}_{n=1}^N$, provided the probability of b is positive. Parameter δ is a threshold value: we will take into account contribution of word $b = (b_m, \dots, b_n)$ to $\hat{\phi}_{[m,n]}$ only in case the right-hand side of (9) is *significant*, i.e., has absolute value of at least δ: for any word $b = (b_m, \dots, b_n) \in S^{n-m+1}$ let

$$u_{[m,n]}(b) := \ln \frac{(\varepsilon + \#(sb_{(m,n]}))(\varepsilon + \#(b_{[m,n)}s))}{(\varepsilon + \#(b_{[m,n]}))(\varepsilon + \#(sb_{(m,n)}s))},$$

$$\text{and} \quad \hat{\phi}_{[m,n]}(b_{[m,n]}) = \begin{cases} u_{[m,n]}(b), & \text{if } |u_{[m,n]}(b)| > \delta, \\ 0, & \text{otherwise} \end{cases}$$

This approach ensures the *sparsity* of the potential. We note that since the process $\{X_n\}$ is stationary, our construction will yield a translation invariant potential $\hat{\phi}_{[m,n]}$: indeed, expression for the potential depends only the number of occurrences of a particular word, and not on the location of these occurrences. This allows us to consider intervals of the form $\Lambda = [1, n']$, $n' \leq 2M + 1$.

Finally, since we are interested in two-sided conditional probabilities, for every $b = (b_{-M} \dots b_M) \in S^{2M+1}$, we estimate the probability as

$$\hat{\mathbb{P}}\left(X_0 = b_0 | X_{-M}^{-1} = b_{-M}^{-1},\ X_1^M = b_1^M\right)$$
$$= \frac{\exp\left(-\sum_{[m,n]\ni 0} \hat{\phi}_{[m,n]}\left(b_{-M}, \dots, b_{-1}, b_0, b_1, \dots, b_M\right)\right)}{\sum_{\bar{b}_0 \in S} \exp\left(-\sum_{[m,n]\ni 0} \hat{\phi}_{[m,n]}\left(b_{-M}, \dots, b_{-1}, \bar{b}_0, b_1, \dots, b_M\right)\right)}.$$

It is very well possible, that despite the fact that formally our estimate of two-sided conditional probabilities depends on two-sided contexts b_{-M}^{-1} and b_1^M of length M, in practice however, conditional probabilities typically depend on shorter contexts b_{-k}^{-1}, b_1^m, with $k, m < M$. It happens automatically due to our threshold criterion involving δ. In the end, one could transfer information encoded by the potential into the bi-directional probabilistic graph. In the future, we plan to compare the trees produced by different algorithms.

3 Validation and Applications

We have validated the performance of our algorithm for estimation of a Gibbs potential on several sources with known potentials such as Markov chains (finite memory) and a particular class of Hidden Markov Chains with infinite memory which are outputs of binary symmetric memoryless channel applied to a binary symmetric Markov chain [12]. In all these cases, the algorithm provides an accurate approximation of the underlying Gibbs potential.

Novel methods for estimating *divergence* between two unknown sources have been proposed in [8]. Divergence appears naturally in the theory of Gibbs measures, where it plays a key role in variational principles, and in some sense measures the "distance" between Gibbs measures. To validate their methods, the authors of [8] used the divergence estimators in linguistic problems such as language classification and author recognition. Distance between two Gibbs measures can also be measured in a number of ways, for example, in terms of the underlying potentials, or in terms of the corresponding specifications. We will use specifications estimated by the algorithm proposed in the previous section.

Let us consider the following problem: suppose we are given two samples $\{x_n\}_{n=1}^{N_1}$ and $\{y_n\}_{n=1}^{N_2}$ of stationary stochastic processes with values in a common finite alphabet S, whose respective laws $\mathbb{P}$ and $\mathbb{Q}$ are Gibbs for some unknown potentials. Suppose also that we are presented with a third sample $\{z_n\}_{n=1}^{N_3}$ of a stochastic process, whose law is either $\mathbb{P}$ or $\mathbb{Q}$. The question is to classify the third sample, i.e., determine from which source, $\mathbb{P}$ or $\mathbb{Q}$, it originates.

A typical approach consists in evaluating the likelihood of the third sample $\{z_n\}_{n=1}^{N_3}$ with respect to $\mathbb{P}$ and $\mathbb{Q}$, respectively, and then choosing the most probable source as the classifier. Since the laws $\mathbb{P}$ and $\mathbb{Q}$ are unknown, they have to be estimated ($\hat{\mathbb{P}}$, $\hat{\mathbb{Q}}$) using the known samples $\{x_n\}_{n=1}^{N_1}$ and $\{y_n\}_{n=1}^{N_2}$. Equivalently, the source can be identified based on the sign of log-likelihoods:

$$\log \frac{\hat{\mathbb{P}}(z_1^{N_3})}{\hat{\mathbb{Q}}(z_1^{N_3})} = \begin{cases} >0 \;\Rightarrow\; \text{Answer } = \mathbb{P}, \\ <0 \;\Rightarrow\; \text{Answer } = \mathbb{Q}. \end{cases}$$

Equivalently, the decision can be based by evaluating sum of logarithms of one-sided conditional probabilities

$$\sum_{t=1}^{N_3} \log \frac{\hat{\mathbb{P}}(z_t|z_1^{t-1})}{\hat{\mathbb{Q}}(z_t|z_1^{t-1})} =: \sum_{t=1}^{N_3} R_n(z_1^{N_3}).$$

In the experiments described below, we will use the estimates for conditional probabilities using an algorithm for construction of probabilistic context trees developed by Bejerano and Yona [1].

Since our algorithm is geared towards estimation of two-sided conditional probabilities, we will also consider statistics based on ratios of those probabilities:

$$\sum_{n=1}^{N_3} \log \frac{\hat{\mathbb{P}}(z_n|z_1^{n-1}, z_{n+1}^{N_3})}{\hat{\mathbb{Q}}(z_n|z_1^{n-1}, z_{n+1}^{N_3})} =: \sum_{n=1}^{N_3} \tilde{R}_n(z_1^{N_3}).$$

In what follows we will compare behavior of local statistics $\{R_n\}$ and $\{\tilde{R}_n\}$ in a number of examples. Note that these statistics are *informative* only if the assumed values have large absolute value, i.e., when one model has a clear dominance in comparison with the second model. Hence, it is beneficial to disregard values R_n and $\tilde{R}_n$ with small absolute values. In the plots below, the horizontal broken lines indicate the threshold value. Finally, in the experiments we allowed the maximal memory to be equal to 4 in the one-sided model, and we allowed contexts of length up to 2 to the left and to the right in the two-sided model.

Example 1 (Order 1 Markov chains) We consider two binary Markov chains ($S = \{0, 1\}$), with transition matrices

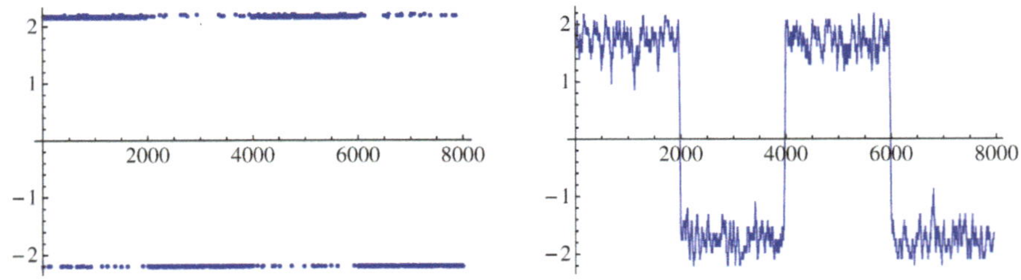

Fig. 2 Markov chains: graphs of R_n and running average of R_n

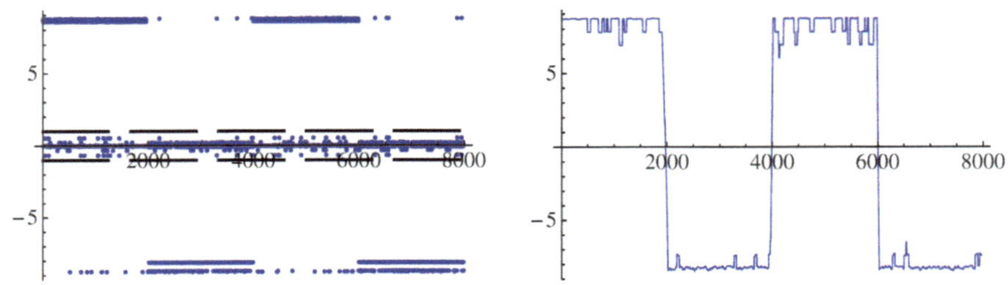

Fig. 3 Markov chains: graphs of $\tilde{R}_n$ and running average of $\tilde{R}_n$

$$P_i = \begin{pmatrix} p_i & 1-p_i \\ 1-p_i & p_i \end{pmatrix}, \quad i = 1, 2,$$

where the first chain jumps between different states with a small probability (p_1 is large), while the second chain prefers to switch states relatively often (p_2 is small). The sample $\{z_n\}$ is a mixture of 4 independent samples generated by models 1 and 2. Results of one-sided and two-sided scoring are depicted in Figs. 2 and 3, transitions between the samples are clearly visible.

Example 2 (Languages with common alphabet) We will consider the Latin alphabet $S = \{a, \ldots, z\}$ (only lower case letters). Our goal is to distinguish the Dutch and English languages. The training samples were: the Wikipedia article about the Netherlands in English, and the Wikipedia article about the Netherlands in Dutch. The test sample $\{z_n\}$ is a concatenation of two articles in English and in Dutch about Antwerpen. The result is presented in Fig. 4 for the one-sided approach and in Fig. 5 for the two-sided approach.

Example 3 (Author attribution) Let again $S = \{a, \ldots, z\}$. As test sample we used 4000 letter excerpts from "Mansfield Park" by Jane Austen and from the English translation of "Anna Karenina" by Leo Tolstoy. As test sample we used the following mixture:

1. 4000 letter excerpt from "Anna Karenina" used in training,
2. 4000 letter excerpt from "Mansfield Park" used in training,

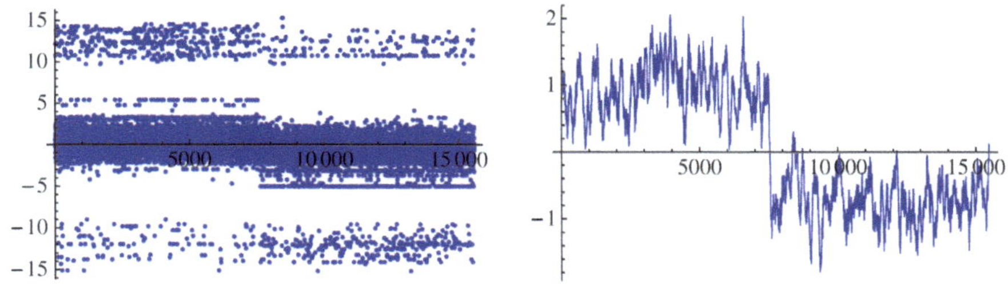

Fig. 4 Dutch versus English: graphs of R_n and running average of R_n

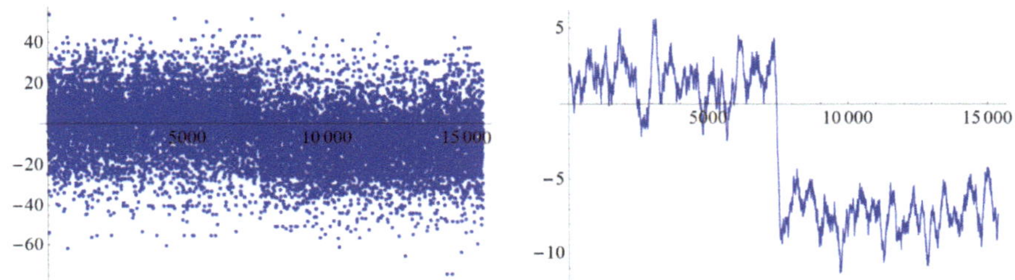

Fig. 5 Dutch versus English: graphs of $\tilde{R}_n$ and running average of $\tilde{R}_n$

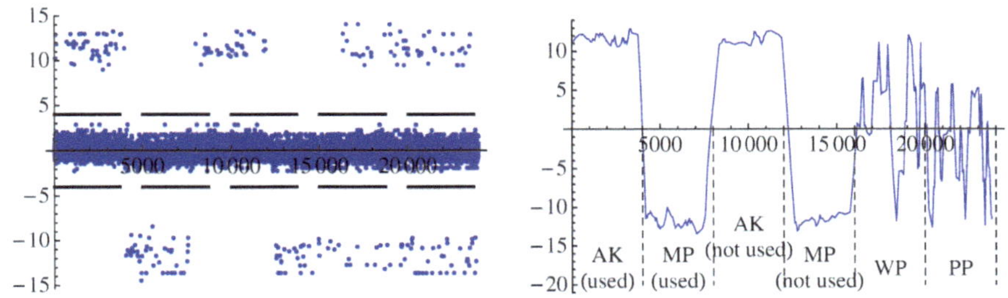

Fig. 6 Tolstoy versus Austen: graphs of R_n and running average of R_n

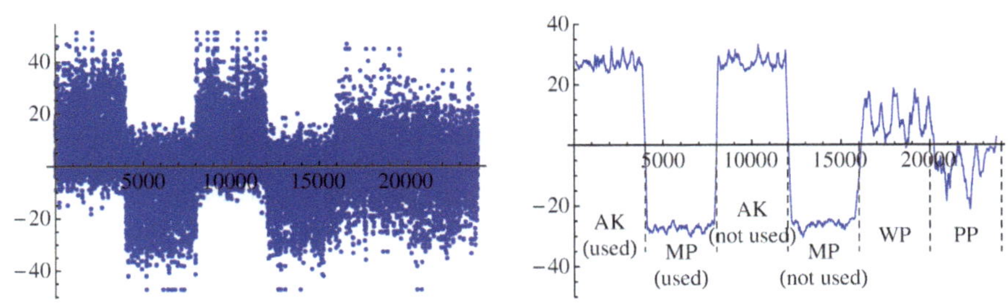

Fig. 7 Tolstoy versus Austen: graphs of $\tilde{R}_n$ and running average of $\tilde{R}_n$

3. 4000 letter excerpt from "Anna Karenina" not used in training,
4. 4000 letter excerpt from "Mansfield Park" not used in training
5. 4000 letter excerpt from "War and Peace" by Leo Tolstoy,
6. 4000 letter excerpt from "Pride and Prejudice" by Jane Austen.

In conclusion, in the present paper we discussed various approaches to modelling random processes with variable memory. We presented a method for estimating Gibbs potentials, which also provides a method to evaluate two-sided conditional probabilities. Following the idea of [8], we evaluated performance of our algorithm in binary classification of unknown linguistic samples. Performance is on par and in some cases even better than the standard method based on scoring the one-sided conditional probabilities.

References

1. Bejerano, G., Yona, G.: Variations on probabilistic suffix trees: statistical modeling and prediction of protein families. Bioinformatics **17**(1), 23 (2001)
2. Bramson, M., Kalikow, S.: Nonuniqueness in g-functions. Isr. J. Math. **84**(1), 153–160 (1993)
3. Fernández, F., Viola, A., Weinberger, M.J.: Efficient algorithms for constructing optimal bi-directional context sets. In: IEEE Data Compression Conference, pp. 179–188 (2010)
4. Fernández, R.: Gibbsianness and non-Gibbsianness in lattice random fields. Math. Stat. Phys. Session **LXXXIII**, 731–799 (2006) (Elsevier)
5. Fernández, R., Maillard, G.: Chains with complete connections: general theory, uniqueness, loss of memory and mixing properties. J. Stat. Phys. **118**(3), 555–588 (2005)
6. Galves, A., Galves, C., Garcia, N.L., Leonardi, F.: Context tree selection and linguistic rhythm retrieval from written texts. Ann. Appl. Stat. **6**(1), 186–209 (2012)
7. Georgii, H.O.: Gibbs measures and phase transitions. Walter de Gruyter, Berlin (2011)
8. Haixiao, C., Kulkarni, S.R., Verdu, S.: Universal divergence estimation for finite-alphabet sources. IEEE Trans. Inf. Theory **52**(8), 3456–3475 (2006)
9. Kozlov, O.K.: Gibbs description of a system of random variables. Probl. Peredachi Infor. **10**(3), 94–103 (1974)
10. Ordentlich, E., Weinberger, M.J., Weissman, T.: Multi-directional context sets with applications to universal denoising and compression, In: Proceedings of International Symposium on Information Theory, 2005. ISIT 2005, pp. 1270–1274 (2005)
11. Rissanen, J.: A universal prior for integers and estimation by minimum description length. Ann. Stat. **11**(2), 416–431 (1983)
12. Verbitskiy, E.: Thermodynamics of hidden Markov processes. In: Entropy of Hidden Markov Processes and Connections to Dynamical Systems. London Mathematical Society, Lecture note series, vol. 385, pp. 258–272. Cambridge University Press, Cambridge (2011)
13. Weissman, T., Ordentlich, E., Seroussi, G., Verdú, S., Weinberger, M.J.: Universal discrete denoising: known channel. IEEE Trans. Infor. Theory **51**(1), 5–28 (2005)
14. Willems, F.M.J., Shtarkov, Y.M., Tjalkens, T.J.: The context-tree weighting method: basic properties. IEEE Trans. Infor. Theory **41**(3), 653–664 (1995)
15. Yu, J., Verdú, S.: Schemes for bidirectional modeling of discrete stationary sources. IEEE Trans. Infor. Theory **52**(11), 4789–4807 (2006)

Need for Mathematics Researchers in Industry: From Standpoint of an Industrial Researcher

Shinichiro Nakamura

Abstract This note discusses the need for mathematics researchers in industry from the standpoint of an industrial researcher, based on my experience in the areas of molecular science simulations in the chemical industry. I will explain why mathematics researchers are vital to industry, as well as prove why working in industry is appealing to mathematics researchers. Mathematicians are the key of success in fundamental and applied researches in industry.

Keywords Molecular science · Research volume · Industrial problem

1 Introduction

This short note discusses the need for mathematics researchers in industry from the standpoint of an industrial researcher, based on my experience in the areas of computational science, quantum chemistry, and molecular science simulations in the chemical industry for more than 26 years. I hope that my experiences will explain why mathematics researchers are vital to industry, as well as prove why working in industry is appealing to mathematics researchers, which would no doubt contribute to the success of fundamental and applied researches in industry by mathematicians. The success of Japanese industries (heavy industry, car and mechanical industries, electronics and computer industries) after the chaos of post-war Japan is attributed to the enhancement of applied mathematics at the researcher level. In those days, the mathematical skills sought were mostly based on classical mechanics. However, from the early 1980s, major chemical companies, especially those in Europe and US, began to hire researchers of quantum chemistry as well as molecular simulation.

S. Nakamura (✉)
Nakamura Laboratory, RIKEN Research Cluster for Innovation, 2-1 Hirosawa, Wako, Saitama 351-0198, Japan
e-mail: snakamura@riken.jp

M. Wakayama et al. (eds.), *The Impact of Applications on Mathematics*, Mathematics for Industry 1, DOI: 10.1007/978-4-431-54907-9_29,

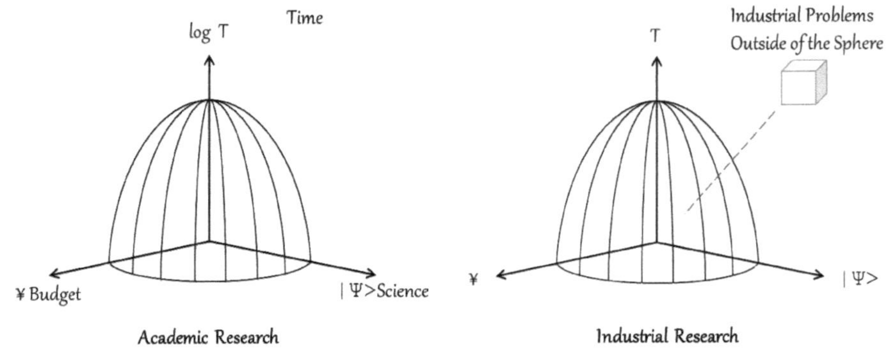

Fig. 1 Scheme of research volume

Around the same time, use of the supercomputer started to spread all over the world. The same trend was also seen immediately in Japan. My experience, as the first generation of quantum chemistry researcher in the Japanese industry, also began to accumulate during this time. Two examples of achievements by my colleagues and myself are cited as references [1, 2].

2 Research Volume and Mathematics

Figure 1 shows the most important argument in discussing the reasons for why industry is in need for mathematics researchers. I believe that every research area except mathematics can be expressed by this figure. The two spheres represent the research volume carried out in academia (left) and industry (right), respectively. The volumes are defined by three axes; time, science, and budget. Note that time in academia is in log-scale, whereas that in industry is real.

The mission of academic research is to increase the size of this volume in three conditions (axes). On the contrary, if the subject is outside this sphere, the papers cannot be judged by reliable referees. This results in the delay in acceptance for publication, until supportive experimental evidences become available. It is necessary for papers by academic researchers to have certainty of continuous. On the other hand, if industrial problems (future commercial products) were inside the sphere, there will be no potential appeal to consumers. In fact, if an outside problem exists, it will provide the chance for new products to emerge.

Industrial research would have more chance for success if these outside problems could be resolved by some approach. In other words, although scientific solutions are available for problems inside or on the surface of the sphere, for every scientific hypothesis except mathematics, the final criterion depends on experimental evidences. If the problem is located outside the sphere, namely out of the scope of available experimental methods, it means that the problem cannot be resolved in a sound physical, chemical, or biological way. This raises the question of what is the ideal means of resolving outside problems. The answer is only mathematics. In most research areas, only mathematics can be performed without the need for experimental evidence.

3 Concluding Remark

Finally, I would like to demonstrate a novel example of molecular music [3] which was accidentally obtained in molecular dynamics simulations. This is a kind of byproduct which appeared by chance. We, therefore, have no established method for pursuing this subject, and believe that only mathematics can unveil the reason for the existence of this kind of music in molecular motions.

References

1. Nakamura, S., Kobayashi, T., Takata, A., Uchida, K., Asano, Y., Murakami, A., Goldberg, A., Guillaumont, D., Yokojima, S., Kobatake, S., Irie, M.: Quantum yields and potential energy surfaces: a theoretical study. J. Phys. Org. Chem. **20**, 821–829 (2007)
2. Uchida, K., Nishikawa, N., Izumi, N., Yamazoe, S., Mayama, H., Kojima, Y., Yokojima, S., Nakamura, S., Tsujii, K., Irie, M.: Phototunable diarylethene microcrystalline surfaces: lotus and petal effects upon wetting. Angew. Chem. Int. Ed. **49**(34), 5942–5944 (2010)
3. Arai, A., Nakamura, S., Ide, H.: Sonification of Molecules (2013) (Mainichi Shinbun)

Index

M. Wakayama et al. (eds.), *The Impact of Applications on Mathematics*,
Mathematics for Industry 1, DOI: 10.1007/978-4-431-54907-9,